MATH *for* MACHINISTS

JOHN TEMPLETON | MARK W. HUTH

Second Edition

Publisher
The Goodheart-Willcox Company, Inc.
Tinley Park, IL
www.g-w.com

Preface

In his original preface, Mr. Mark W. Huth stated "all machinists must be able to do math." I would add that skill in math will help aspiring machinists achieve career success. Math is used to determine dimensions when reading prints and inspecting parts, to calculate positions for CNC toolpaths, to make adjustments in machining processes, and to determine the speeds and feeds of tools, to name but a few examples. Math is an everyday component in a machinist's workday.

In my many years of teaching and tutoring, I have found that mathematics can cause stress and anxiety for students, some of whom have grown to fear the subject. However, when ideas are broken down into smaller segments and knowledge is built from a foundation of concepts, math can be a subject students not only understand but also enjoy. *Math for Machinists* is written with this in mind, assuming very little prior math experience. The aim is to help students master the content of every unit and develop their mathematical ability, from simple concepts to more complex manufacturing applications. Like the acquisition of any other skill, math requires practice and perseverance. Don't be afraid of making mistakes—it's part of the learning process. However, be sure to learn how or why a mistake was made. Most importantly: Ask questions when you don't understand something.

Math for Machinists was written for students who are learning the machining trades in high schools, technical colleges, or apprentice programs in industry. It can be used in a math class to supplement a core mathematics textbook, as the basis for an applied math class for machining students, or as a math supplement to a core machine shop textbook. Each section of *Math for Machinists* is a grouping of short units focusing on a single topic. Each unit is meant to be studied as a single lesson. New key terms introduced in the section are listed on the section opener page. Each unit begins with learning objectives to be covered, and concepts and operations are explained simply and concisely. Examples illustrate each new topic. Units end with a fill-in-the-blanks review and a selection of practice and application math and machine shop word problems.

This updated and improved second edition has been strengthened with dozens of new math problems and significant new, expanded, and revised content based on instructor feedback and reviews of the first edition. New features and concepts, such as "shop talk," practical shop problems, and the Cartesian coordinate system have been added. A revised trigonometry section with step-by-step calculator examples walks students through the process of solving every problem. Sections covering thread formulas and speeds and feeds have been revised and expanded. An abundance of practice and application problems have been added to every unit. This allows instructors to customize content for classwork, quizzes, and homework assignments. For students, the extra questions allow more practice and application of the new concepts learned, with the aim of attaining a thorough understanding of each unit.

I believe *Math for Machinists* will help students understand the importance of math in the shop, give students confidence in their mathematical abilities, and help develop the future skilled machinists the industry desperately needs.

John Templeton

About the Authors

John Templeton brings over 25 years of machining experience to this text. Beginning with a four-year apprenticeship in the United Kingdom, Mr. Templeton graduated from Darlington College of Technology before progressing through a variety of fields in the machining industry. Positions of note include work as a Prototype/R&D Machinist, working with advanced aerospace metals and materials; Tool & Die Maker; CNC Programmer; Lead Machinist; and Machining Consultant. Mr. Templeton has almost 15 years teaching experience as an instructor at NTMA Training Centers of Southern California, attaining Career Technical Education (CTE) teaching credentials through the University of San Diego. Aside from the mathematics, blueprint reading, and CNC programming classes in the machinist program, Mr. Templeton has also taught apprenticeship classes (LAUSD) in math and blueprint reading.

First edition author Mark W. Huth brought a great deal of relevant experience from multiple perspectives. He served seven years in the US Navy as an aviation machinist's mate, where he was recognized for proficiency in his craft. Following the Navy, he received a baccalaureate degree from the State University of New York at Oswego. Mr. Huth taught high school for several years before beginning a career in educational publishing. His publishing career provided him the opportunity to work with some of the most successful teachers across North America. Mr. Huth authored many career and technical education titles, as well as successful textbooks on math topics.

Reviewers

The author and publisher wish to thank the following industry and teaching professionals for their valuable input into the development of *Math for Machinists*.

Brendan Anderson
Alfred State College
Alfred, NY

Rick Calverley
Lincoln College of
Technology
Grand Prairie, TX

Jason S. Carpenter
Tennessee College of
Applied Technology
Morristown
Morristown, TN

E.J. Daigle
Dunwoody College of
Technology
Minneapolis, MN

Jim Gilliam
Northwest Mississippi
Community College
Senatobia, MS

Gene Harr
Tennessee College of
Applied Technology at
Chattanooga State
Chattanooga, TN

Henry J. Hatem
Renton Technical College
Renton, WA

Noah McCoy
Texas State Technical College
Harlingen, TX

Daniel Morales
South Texas College
McAllen, TX

Justin Owen
Danville Community
College
Danville, VA

Len Walsh
Goodwin University
East Hartford, CT

Josh Worthley
Danville Community
College
Danville, VA

New to This Edition

A great deal of work has gone into making the second edition of *Math for Machinists* even more useful for instructors, students, and aspiring machinists. New to this edition is a focus on "shop talk," providing students practice using math-related terms they will hear and experience in a professional machine shop. A new Shop Math section includes five new units that provide practical, machining-related math skills and practice, so students can apply their new trigonometry skills. These five new units are: Sine Bars and Sine Plates, Drill Point Angles, Center-to-Center Distances, Dovetails, and Tapers.

Another new section, Coordinate Systems, includes three new units: CNC Milling, CNC Turning, and Bolt Circles. These units teach and reinforce the concept of a coordinate system and how it applies to machining practices, including the movements of CNC machines and determining ordered-pair coordinates of holes on bolt circles.

The units that cover thread formulas, speeds and feeds, and right triangle trigonometry have been revamped, providing more detail and information. Revised appendixes allow students to quickly find formulas and conversions needed in the shop. Formulas in Appendix B include the unit in which the information is found, should students want to revisit the concept.

Aside from the new content focused on machining-related math, this new edition has been packed full with more problems at the end of every unit, to ensure students and instructors have enough to work with toward the attainment of machine math knowledge. A fill-in-the-blanks review has been added at the end of each unit to reinforce the unit's key concepts.

Here is a summary of what is new in this edition:

- Eight new units focusing on math machinists encounter in the shop, including five new units focusing on practical, machining-related uses of trigonometry, and three new units focusing on coordinate systems as they apply to machining and CNC use.
- Revised unit for thread formulas and calculating measurements over threads.
- Revised unit for speeds and feeds, adding drills, reamers, and taps.
- Revised unit for right triangle trigonometry, providing improved explanations and questions.
- New emphasis on "shop talk," preparing students for work in a machine shop.
- New fill-in-the-blank unit reviews and dozens of new review questions in every unit.
- Updated appendixes with formulas and conversions that can be quickly referenced when needed.

Features of the Textbook

The instructional design of this textbook includes student-focused learning tools to help you succeed. This visual guide highlights these features.

Section and Unit Opening Materials

Each section opener contains a **Section Outline** of the units in sequence and a list of **Key Terms** appearing in that section.

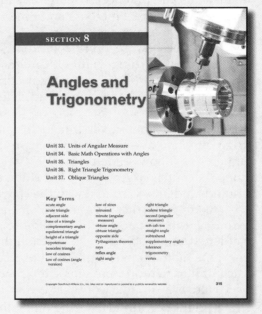

Each unit opener includes **Objectives** that clearly identify the knowledge and skills to be obtained upon completing the unit.

Additional Features

Examples in each unit demonstrate presented concepts, showing the mathematical work required to solve a problem.

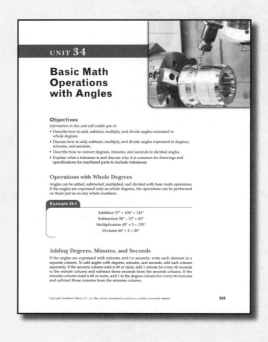

End-of-Unit Content

End-of-unit material provides an opportunity for review and application of concepts. Fill-in-the-blanks **Review Questions** reinforce the key concepts from the unit.

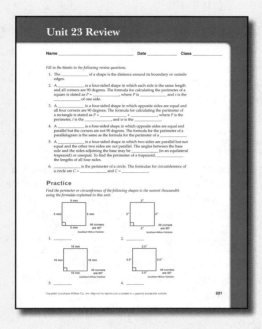

Practice questions allow you to demonstrate knowledge, identification, and comprehension of unit material.

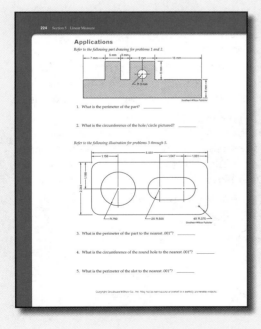

Applications extend your learning and help you analyze and apply knowledge in practical, machining-related math problems.

Illustrations have been designed to clearly and simply communicate the specific topic and tie mathematical concepts to practical machining situations. Many new machining-related illustrations have been added for this edition.

TOOLS FOR STUDENT AND INSTRUCTOR SUCCESS

Student Tools

Student Text

Math for Machinists assumes almost no math experience and takes students from basic math skills to the mathematical knowledge and skills they will need in the machine shop. Topics covered range from the simple addition of whole numbers to the knowledge of formulas necessary to determine speeds and feeds and drill point angles, to the working knowledge of trigonometry necessary to solve problems in the machine shop relating to work with sine bars, dovetails, and bolt circles. Coverage of coordinate systems allows students a better understanding of how CNC milling and turning (lathe) machines operate and how point coordinates are described based on quadrants.

G-W Digital Companion

E-flash cards and vocabulary exercises allow interaction with content to create opportunities to increase achievement.

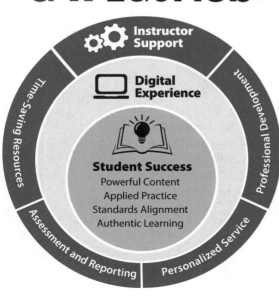

Instructor Tools

LMS Integration

Integrate Goodheart-Willcox content within your Learning Management System for a seamless user experience for both you and your students. EduHub LMS-ready content in Common Cartridge® format facilitates single sign-on integration and gives you control of student enrollment and data. With a Common Cartridge integration, you can access the LMS features and tools you are accustomed to using and G-W course resources in one convenient location—your LMS.

G-W Common Cartridge provides a complete learning package for you and your students. The included digital resources help your students remain engaged and learn effectively:

- **eBook**
- **Drill and Practice** vocabulary activities

When you incorporate G-W content into your courses via Common Cartridge, you have the flexibility to customize and structure the content to meet the educational needs of your students. You may also choose to add your own content to the course.

For instructors, the Common Cartridge includes the Online Instructor Resources. QTI® question banks are available within the Online Instructor Resources for import into your LMS. These prebuilt assessments help you measure student knowledge and track results in your LMS gradebook. Questions and tests can be customized to meet your assessment needs.

Online Instructor Resources (OIR)

- The **Instructor Resources** provide instructors with time-saving preparation tools such as answer keys, editable lesson plans, and other teaching aids.
- **Instructor's Presentations for PowerPoint®** are fully customizable, richly illustrated slides that help you teach and visually reinforce the key concepts from each unit.
- Administer and manage assessments to meet your classroom needs using **Assessment Software with Question Banks**, which include hundreds of matching, completion, multiple choice, and short answer questions to assess student knowledge of the content in each unit.

See **www.g-w.com/math-for-machinists-2024** for a list of all available resources.

Professional Development

- Expert content specialists
- Research-based pedagogy and instructional practices
- Options for virtual and in-person Professional Development

Brief Contents

Contents

x

SECTION 6
Formulas

SECTION 7
Powers and Roots

SECTION 8
Angles and Trigonometry

Whole Numbers

Key Terms

abstract number	dividend	positive
Arabic number system	divisor	product
concrete number	multiplicand	quotient
decimal number system	multiplication table	remainder
denominate number	multiplier	sum
difference	negative	zero

UNIT 1

Number Systems

Objectives

Information in this unit will enable you to:

- Recognize and understand the difference between positive and negative numbers.
- Explain what the Arabic number system is and how it works.
- Describe how abstract numbers differ from concrete numbers, and how denominate numbers work.

Positive and Negative Numbers

The numerals we write on a page are symbols to represent some value. The numeral 0 has no value. If we see the numeral 3, we know that it is a symbol for the value 3 of whatever units we might be working with. It could represent 3 inches, 3 pounds, or 3 cows. Let's consider pounds of force. If a balloon has helium gas applying 3 pounds of force to raise the balloon into the air, we can consider that a **positive** value. If a weight is tied to the balloon, that weight is applying a **negative** force. If the negative force is 2 pounds, there is still 1 more positive pound of force than the negative force. The end result is a value of 1 pound of positive force (+1).

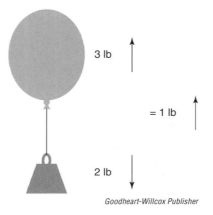

3 lb

= 1 lb

2 lb

Goodheart-Willcox Publisher

The principles of positive, zero, and negative numbers are often shown on a number line. On the number line, **zero** is in the center, negative numbers are to the left, and positive numbers are to the right.

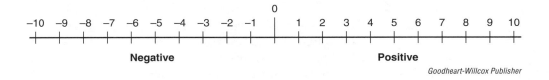

Goodheart-Willcox Publisher

As you move toward the right on the number line, the numbers become more positive. For example, 7 is more positive than 3. In fact, it is four numbers more positive. The same is true on the negative side of the number line. Negative 3 (–3) is four numbers more positive (less negative) than negative 7 (–7).

What happens when you cross the 0 point? Let's move from 2 to –5. We have moved seven numbers to the left (toward the negative). Negative 5 is seven numbers more negative than positive 2. If we start with a value of 3 and move five numbers toward the negative side (to the left), we arrive at –2.

These principles will apply to all numbers as you learn to do mathematical operations, like adding and subtracting.

Arabic Number System

The number system we use is based on ten digits: 0, 1, 2, 3, 4, 5, 6, 7, 8, 9. It is called the **Arabic number system**, because it was developed centuries ago in Arabia. It is also called a **decimal number system**. The word "decimal" is derived from Latin and means "based on ten."

Numbers of 0 through 9 are indicated simply by the digit representing the number. To show a number greater than 9, a digit is placed to the left of the first digit. The digit in the first position (farthest right) is in the ones position. The next position to the left is the tens position. Digits in the tens position represent multiples of 10. If the digit in the tens position is 1, the number is 10 plus whatever is in the ones position. If the digit in the tens position is 3, the number is 30 (3 times 10) plus whatever is in the ones position.

Example 1-1

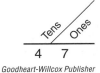

Goodheart-Willcox Publisher

The next position to the left is the hundreds, then thousands, and so on.

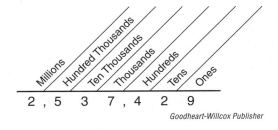

Goodheart-Willcox Publisher

(Continued)

To make large numbers easier to read, it is common to insert a comma after every three positions to the left. Using this convention, the number in the previous figure is written 2,537,429. The number would be pronounced two million five hundred thirty-seven thousand four hundred twenty-nine. A thin space is used in the SI system (metric) instead of a comma. With spaces, the number is written 2 537 429.

Abstract, Concrete, and Denominate Numbers

As was discussed at the beginning of this unit, numbers are symbols to represent values. The number alone does not convey any real meaning. Numbers with nothing attached to them are called **abstract numbers**. When the number designates a quantity of some specific thing like dollars, inches, or pieces, it conveys some very specific information. A number designating the quantity of something is called a **concrete number**.

Example 1-2

7 does not tell us much. It is just a number.
7 dollars is a very specific amount of money.

When a concrete number is associated with a unit of measurement, it is called a **denominate number**. Most of the numbers a machinist works with are denominate numbers. When working with concrete numbers, two rules will help eliminate problems:

- Always write or say the units every time you write or say the number (4 mm or 4 millimeters, not just 4).
- Do not try to mix units. (Remember the old adage "Don't mix apples and oranges.") You cannot add millimeters and ounces. If the units can be converted to be the same, do that. If the units cannot be converted to be the same, you are trying to do impossible math.

Example 1-3

6 inches and 1 foot can be converted to either 18 inches or 1 1/2 feet.
6 inches and 1 pound cannot be used in the same mathematical operation.

Unit 1 Review

Name _____ Date _____ Class _____

Fill in the blanks in the following review questions.

1. On the number line, _____ is in the center, _____ numbers are to the left, and _____ numbers are to the right.

2. The number system we use that is based on 10 digits is called the _____ number system. It is also called the _____ number system.

3. A(n) _____ number is a number with nothing attached to it.

4. A _____ number is a number designating the quantity of something.

5. Machinists mostly work with _____ numbers, which are concrete numbers associated with units of measurement.

Practice

Greater Positive Value
Which number has a greater positive value in the following pairs?

1. 1 or 3   2. 4 or 6   3. 9 or 7   4. 6 or 2

   _____   _____   _____   _____

Greater Negative Value
Which number has a greater negative value in the following pairs?

5. −4 or −2   6. −1 or −5   7. −8 or −9   8. −7 or −6

   _____   _____   _____   _____

Value of a Move
What is the value of the move on the number line for the following numbers?

9. From 1 to 7   10. From 3 to 6   11. From 2 to 5

   _____   _____   _____

12. From −2 to −4   13. From −6 to −9   14. From −1 to −7

   _____   _____   _____

15. From 9 to 1   16. From 7 to 2   17. From 5 to 3

   _____   _____   _____

18. From −8 to −4   19. From −4 to −1   20. From −6 to −3

   _____   _____   _____

Position of Each Digit

Label the position of each digit in the following numbers according to their place values.

21. 5 2

22. 4 7 1

23. 8 6 3

24. 2 5 7
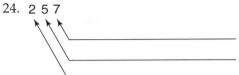

25. 1,2 8 4
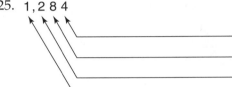

26. 3,4 6 9
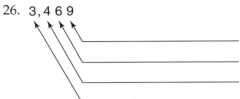

27. 1 0,2 1 8
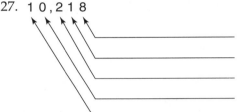

28. 3 4,6 2 1
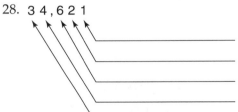

29. 4 1 0,5 2 8
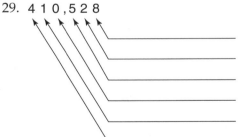

30. 7 3 4,1 9 6
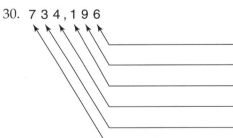

31. 3,4 8 2,9 5 1
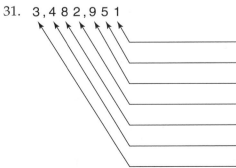

32. 7,1 5 3,4 6 2
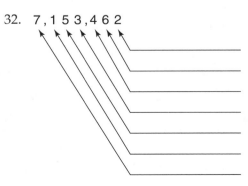

Name _____   **Date** _____   **Class** _____

Abstract, Concrete, or Denominate?

Identify the following as abstract, concrete, or denominate numbers.

33. 490

34. 2 miles

35. 0.5 gallon

36. 3.337

37. 8 men

38. 4 groups

39. 16,197

40. 1 inch

41. 15 mm

42. 81 km

43. 782

44. 14 liters

45. 43

46. 26 people

47. 18 inches

48. 102,346

49. 19 apples

50. 3

51. 128 oz.

52. 57 meters

53. 5 cats

54. 27 grams

55. 12 cm

56. 6 buildings

Conversions

Write True if the following pairs can be converted and False if they cannot.

57. inches/feet

58. milliliters/kilometers

59. gallons/cups

60. dollars/cents

61. inches/mm

62. cm/meters

63. inches/oz.

64. pounds/cups

65. gallons/mm

66. miles/km

67. Fahrenheit/Celsius

68. Fahrenheit/inches

69. meters/yards

Work Space/Notes

UNIT 2

Adding Whole Numbers

Objectives

Information in this unit will enable you to:

- Describe how to add numbers less than 10.
- Explain how to add numbers greater than 10.

Adding Numbers of Less Than 10

To add numbers of any value, you will need to memorize the results of adding numbers of less than 10. For example, the only way to add 4 plus 3 is to know that the result of adding those two numbers is 7. The result of addition is called the **sum**.

To add numbers of less than 10, write the numbers in a column, add them, and write the sum below. If you are adding denominate numbers, remember to always include the units when you write the number.

Example 2-1

Add 5 plus 3.

$$
\begin{array}{r}
5 \text{ hours} \\
+\ 3 \text{ hours} \\
\hline
8 \text{ hours (sum)}
\end{array}
$$

When any number in the problem is 10 or more, line up the ones column. Digits to the left of the ones column represent tens, hundreds, or greater, depending on the column they are in. When the sum of a column is greater than 10 or more, write the digit farthest to the right in that place and add the digit on the left to the next column.

Example 2-2

Add 9 plus 4.

$$\begin{array}{r} 9 \text{ mm} \\ + 4 \text{ mm} \\ \hline 13 \text{ mm} \end{array}$$

Nine plus 4 is 13, so write the 3 in the ones place, at the bottom of the column being added, then add the 1 to the tens place.

Example 2-3

Add 2, 4, and 8.

$$\begin{array}{r} 2 \\ 4 \\ + 8 \\ \hline 14 \end{array}$$

Two plus 4 is 6 and 6 plus 8 is 14. Align the ones column and write the leftover 1 in the tens column.

Math problems are often stated as word problems, such as "A pin projects 2 inches above the surface of a 5-inch cylinder. What is the total height of the assembly?" This one is pretty simple. You just need to add 2 inches and 5 inches. Some problems are more complex and require some thought to decipher just what the relevant facts are and what math needs to be done. The statement of the problem might include a lot of information that is not needed to solve the problem. In such a case, the first step might be to write a sentence that simply states the math problem. The next step would be to write the problem in a mathematical form to be solved.

Example 2-4

In the figure below, what is overall length of the part?

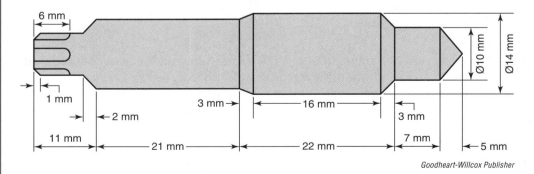

Goodheart-Willcox Publisher

The first step is to sort out the relevant information. The sizes of the chamfers, the diameters, and the 6 mm grooves at the left end are not needed to solve the problem.

(Continued)

Write the problem in a form to be added and do the math.

$$
\begin{array}{r}
11 \text{ mm} \\
21 \text{ mm} \\
22 \text{ mm} \\
7 \text{ mm} \\
+\ 5 \text{ mm} \\
\hline
66 \text{ mm}
\end{array}
$$

As the problems become more complex, it will become important to remember to sort out the problem and write it as a form to be solved.

Adding Numbers Greater Than 10

When numbers in the problem are 10 or greater, align the ones column and the digits to the left will fall into their proper places. Add the ones column first. Write the digit farthest to the right in that sum in the ones column and add anything to its left to the tens column. Repeat this for as many places as necessary.

Example 2-5

Add 212, 96, and 5.

$$
\begin{array}{r}
{}^{1\ 1} \\
212 \\
96 \\
+\ \ \ 5 \\
\hline
313
\end{array}
$$

Two plus 6 plus 5 is 13. Write the 3 in the ones place and add the leftover 1 to the tens column. One plus 9 plus the leftover 1 is 11. Write 1 in the tens place and add the leftover 1 to the hundreds column. Two plus the leftover 1 is 3.

Example 2-6

Add 384, 176, and 67.

$$
\begin{array}{r}
{}^{2\ 1} \\
384 \\
176 \\
+\ \ 67 \\
\hline
627
\end{array}
$$

The numbers 4 plus 6 plus 7 add to 17. Write the 7 in the ones place and add the leftover 1 to the tens column. The numbers 8 plus 7 plus 6 plus the leftover 1 add to 22. Write the 2 in the tens place and add the leftover 2 to the hundreds column. Finally, 3 plus 1 plus the leftover 2 add to 6, bringing the final result: 627.

Work Space/Notes

Unit 2 Review

Name _____ **Date** _____ **Class** _____

Fill in the blanks in the following review questions.

1. The result of addition is called the _____.

2. To add numbers of any value, you need to _____ the results of adding numbers of less than 10.

3. When any number in an addition problem is 10 or more, line up the _____ column.

4. With word problems and more complex problems, it is important to sort out the problem and write it as a form to be _____.

Practice

Add the numbers.

1. $\begin{array}{r} 5 \\ +\ 2 \\ \hline \end{array}$
2. $\begin{array}{r} 6 \\ +\ 3 \\ \hline \end{array}$
3. $\begin{array}{r} 8 \\ +\ 7 \\ \hline \end{array}$
4. $\begin{array}{r} 17 \\ +\ 9 \\ \hline \end{array}$

5. $\begin{array}{r} 4 \\ +\ 9 \\ \hline \end{array}$
6. $\begin{array}{r} 1 \\ +\ 4 \\ \hline \end{array}$
7. $\begin{array}{r} 2 \\ +\ 6 \\ \hline \end{array}$
8. $\begin{array}{r} 3 \\ +\ 7 \\ \hline \end{array}$

9. $\begin{array}{r} 23 \\ +18 \\ \hline \end{array}$
10. $\begin{array}{r} 31 \\ +\ 11 \\ \hline \end{array}$
11. $\begin{array}{r} 44 \\ +\ 35 \\ \hline \end{array}$
12. $\begin{array}{r} 278 \\ +\ 136 \\ \hline \end{array}$

13. $\begin{array}{r} 416 \\ +\ 171 \\ \hline \end{array}$
14. $\begin{array}{r} 592 \\ +\ 386 \\ \hline \end{array}$
15. $\begin{array}{r} 839 \\ +\ 527 \\ \hline \end{array}$
16. $\begin{array}{r} 1,042 \\ +\ 287 \\ \hline \end{array}$

17. $\begin{array}{r} 3,198 \\ +\ 472 \\ \hline \end{array}$
18. $\begin{array}{r} 6,456 \\ +\ 2,189 \\ \hline \end{array}$
19. $\begin{array}{r} 9,012 \\ +\ 3,471 \\ \hline \end{array}$
20. $\begin{array}{r} 10,264 \\ +\ 1,475 \\ \hline \end{array}$

21. $\begin{array}{r} 12,971 \\ +\ 3,245 \\ \hline \end{array}$
22. $\begin{array}{r} 36,472 \\ +\ 9,159 \\ \hline \end{array}$
23. $\begin{array}{r} 318 \\ 427 \\ +\ 114 \\ \hline \end{array}$
24. $\begin{array}{r} 721 \\ 1,445 \\ 4,451 \\ +\ 22 \\ \hline \end{array}$

25. 21 and 12

_____

26. 36 and 19

_____

27. 49 and 32

_____

28. 67 and 55

_____

29. 85 and 16

_____

30. 17 and 122

_____

31. 39 and 209

_____

32. 163 and 302

_____

33. 241 and 357

_____

Applications

Solve the problems.

1. If 3 inches, 11 inches, 9 inches, and 6 inches are cut from a 36-inch bar of tool steel, how many total inches have been cut off? (Do not allow for saw kerfs.)

_____

2. If a 48″ bar of Ø1″ aluminum has pieces cut off that measure 4″, 12″, 7″, and 6″ in length, how much material in total has been cut off? (Do not allow for saw kerfs.)

_____

3. If five pieces of 1018 cold-rolled steel are cut to lengths of 10″, 5″, 8″, 13″, and 7″, how much material in total is cut? (Do not allow for saw kerfs.)

_____

Refer to the figure below for the following problems.

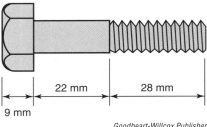

22 mm   28 mm

9 mm

Goodheart-Willcox Publisher

4. What is the overall length of the bolt?

_____

5. What is the length of the bolt body (threaded and unthreaded) without the head?

_____

6. What is the length from the top of the head to where the threads start?

_____

Name _____ **Date** _____ **Class** _____

Refer to the figure below for the following problems.

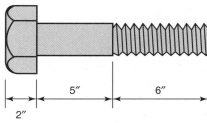

Goodheart-Willcox Publisher

7. What is the overall length of the bolt?

   _____

8. What is the length of the bolt body (threaded and unthreaded) without the head?

   _____

9. What is the length from the top of the head to where the threads start?

   _____

Refer to the figure below for the following problems.

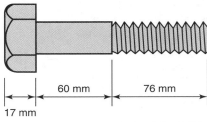

Goodheart-Willcox Publisher

10. What is the overall length of the bolt?

    _____

11. What is the length of the bolt body (threaded and unthreaded) without the head?

    _____

12. What is the length from the top of the head to where the threads start?

    _____

13. A tool chest contains 4 drift punches, 2 calipers, 4 sine bars, 2 steel rules, 5 screwdrivers, 9 Allen wrenches, and 2 hammers. How many tools are in the chest?

_____

14. A tool store has 3 hacksaws, 10 hacksaw blades, 5 center punches, 9 steel rules, and 1 bottle of layout fluid in stock available to purchase. How many items are for sale?

_____

15. If Customer X requires 10 parts to be manufactured, Customer Y requires 7 parts, and Customer Z requires 16 parts, how many parts need to be manufactured in total?

_____

Refer to the figure below to calculate the dimensions in the following problems.

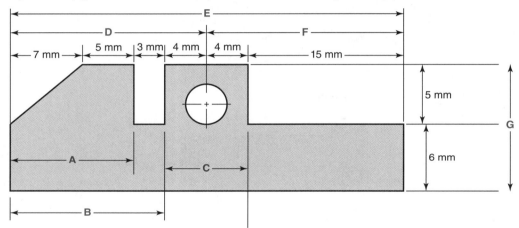

Goodheart-Willcox Publisher

16. Dimension A

_____

17. Dimension B

_____

18. Dimension C

_____

19. Dimension D

_____

20. Dimension E

_____

21. Dimension F

_____

22. Dimension G

_____

Name _____   **Date** _____   **Class** _____

Refer to the drawing below to answer the following problems.

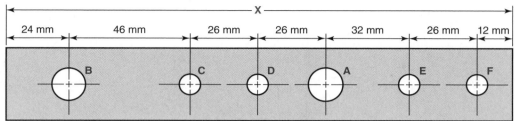

Goodheart-Willcox Publisher

23. What is the total length of the part represented by dimension X?

_____

Calculate the distance between the centers of the following holes.

24. Holes A and B

_____

25. Holes A and C

_____

26. Holes B and D

_____

27. Holes A and F

_____

28. Holes B and E

_____

29. Holes B and F

_____

30. Holes C and E

_____

31. Holes C and F

_____

32. Holes D and E

_____

33. Holes D and F

_____

Work Space/Notes

UNIT 3

Subtracting Whole Numbers

Objectives

Information in this unit will enable you to:

- Discuss how to subtract numbers of less than 10.
- Explain how to use borrowing in subtraction when the bottom numeral is greater than the upper numeral.
- Describe how to subtract more than one amount from a number.

Subtracting Numbers of Less Than 10

Subtraction of whole numbers is much like addition of whole numbers. The numbers are written one above the other, with the larger number on top and the number being subtracted from it on the bottom. The result of subtracting one number from another is called the **difference**. Unlike addition, subtraction is limited to two numbers, with one being subtracted from the other.

Example 3-1

$$
\begin{array}{r}
8 \\
-\,5 \\
\hline
3 \text{ (difference)}
\end{array}
$$

Borrowing When the Bottom Numeral Is Greater Than the Upper Numeral

When the value of the digit being subtracted (the one on the bottom) is greater than the value of the digit from which it is being subtracted (the one on the top), add 10 to the ones place and subtract 1 from the tens place in the top number. It is easiest to keep track of this "borrowing" if you write small numerals above the digit being adjusted.

Example 3-2

$$\overset{3}{\cancel{4}}\overset{1}{2}$$
$$-19$$
$$\overline{23}$$

Nine cannot be subtracted from 2, so 1 is subtracted from the 4 in 42 to make it 3, and because we took 1 from the tens place, 10 is added to the ones place, increasing the 2 to 12. Nine from 12 is 3. The 4 on the top line was reduced to 3, so the subtraction in that place is 1 from 3 is 2.

Subtracting More Than One Amount

It is often necessary to subtract more than one amount from a number. There are two ways such a problem can be solved. One way is that the amounts to be subtracted can be added together, and then that sum can be subtracted.

Example 3-3

If 3 inches, 8 inches, and 5 inches are cut from a 24-inch piece, how many inches remain?

Step 1: Add 3 inches, 8 inches, and 5 inches.

$$\begin{array}{r} 3 \text{ inches} \\ 8 \text{ inches} \\ + \ 5 \text{ inches} \\ \hline 16 \text{ inches} \end{array}$$

Step 2: Subtract that amount from the original 24-inch piece.

$$\begin{array}{r} 24 \text{ inches} \\ - \ 16 \text{ inches} \\ \hline 8 \text{ inches} \end{array}$$

The other way to solve this problem is to subtract each cut piece separately from whatever is left after the previous piece was cut. (This method is usually only practical when no more than two or three values have to be subtracted.)

Example 3-4

Step 1: Subtract 3 inches from 24 inches.

$$\begin{array}{r} 24 \text{ inches} \\ -\ 3 \text{ inches} \\ \hline 21 \text{ inches} \end{array}$$

Step 2: Subtract 8 inches from 21 inches.

$$\begin{array}{r} 21 \text{ inches} \\ -\ 8 \text{ inches} \\ \hline 13 \text{ inches} \end{array}$$

Step 3: Subtract the final amount, 5 inches, from 13 inches to find the amount that remains after all three cuts.

$$\begin{array}{r} 13 \text{ inches} \\ -\ 5 \text{ inches} \\ \hline 8 \text{ inches} \end{array}$$

Checking Subtraction

To check the result of subtraction, add the subtracted number to the answer. The result should be the number from which the original subtraction was done.

$$\begin{array}{r} 5 \\ -4 \\ \hline 1 \\ +4 \\ \hline 5 \end{array} \qquad \begin{array}{r} 62 \\ -37 \\ \hline 25 \\ +37 \\ \hline 62 \end{array} \qquad \begin{array}{r} 103 \\ -87 \\ \hline 16 \\ +87 \\ \hline 103 \end{array}$$

Checking Addition

Subtraction can also be used to check addition. When two numbers have been added, subtract one of the added numbers from the other. The result should be the other added number.

$$\begin{array}{r} 2 \\ +8 \\ \hline 10 \\ -8 \\ \hline 2 \end{array} \qquad \begin{array}{r} 5 \\ +3 \\ \hline 8 \\ -3 \\ \hline 5 \end{array} \qquad \begin{array}{r} 73 \\ +48 \\ \hline 121 \\ -48 \\ \hline 73 \end{array}$$

When three or more numbers are added, the result can be checked by adding from bottom to top. The result should be the same in both directions.

To add / To check

$$\begin{array}{r} 927 \\ 318 \\ 426 \\ 183 \\ \hline 927 \end{array}$$

Work Space/Notes

Unit 3 Review

Name _____ **Date** _____ **Class** _____

Fill in the blanks in the following review questions.

1. The result of subtracting one number from another is called the _____.

2. If it is necessary to subtract more than one amount from a number, there are _____ ways such a problem can be solved. One way is to _____ the amounts to be subtracted and then _____ that resulting sum.

3. To check the result of subtraction, add the subtracted number to the _____. The _____ should be the number from which the original subtraction was done.

Practice

Subtract the numbers and label all denominate numbers.

1. 7
   − 5

2. 6
   − 3

3. 8 man-hours
   − 3 man-hours

4. 7 pieces
   − 1 piece

5. 5 gallons
   − 2 gallons

6. 57
   − 13

7. 99
   − 4

8. 114
   − 108

9. 56
   − 25

10. 22
    − 15

11. 6 feet
    − 2 feet

12. 18 parts
    − 6 parts

13. 27
    − 9

14. 41 mm
    − 9 mm

15. 66 inches
    − 13 inches

16. 79
    − 24

17. 81
    − 16

18. 133
    − 38

19. 184 meters
    − 76 meters

20. 231
    − 49

21. 317
    − 138

22. 442
    − 352

23. 611 miles
    − 246 miles

24. 914
    − 785

25. 1243 yards
    − 456 yards

26. 3897
    − 1249

27. 5276
    − 3485

28. 7862
    − 4193

29. Subtract 7 from 8.

_____

30. Subtract 3 from 7.

_____

31. Subtract 5 from 19.

_____

32. Subtract 14 from 31.

_____

33. Subtract 2 from 9.

_____

34. Subtract 1 from 6.

_____

35. Subtract 2 from 5.

_____

36. Subtract 11 from 25.

_____

37. Subtract 18 from 42.

_____

38. Subtract 23 from 29.

_____

39. Subtract 34 from 57.

_____

40. Subtract 101 from 276.

_____

41. Subtract 385 from 842.

_____

42. Subtract 1038 from 2741.

_____

43. Subtract 2367 from 5958.

_____

44. Subtract 927 from 9065.

_____

45. Subtract 2 and 9 from 23.

_____

46. Subtract 13 and 24 from 41.

_____

47. Subtract 3, 29, and 67 from 150.

_____

48. Subtract 34, 78, and 92 from 275.

_____

49. Subtract 100 mm, 150 mm, and 275 mm from 750 mm.

_____

50. Subtract 4 inches and 7 inches from 15 inches.

_____

51. Subtract 2 pieces, 5 pieces, and 3 pieces from 20 pieces.

_____

52. Subtract 14 bolts, 20 bolts, 12 bolts, and 9 bolts from 100 bolts.

_____

Name _____ **Date** _____ **Class** _____

Applications

Solve the problems.

1. Sixteen ounces of cutting oil is taken from a 128-ounce container. How much cutting oil is left?

    _____

2. If a 72″ bar of Ø1″ aluminum has pieces cut off that measure 6″, 8″, 5″, and 14″ in length, how much Ø1″ aluminum material is left? (Do not allow for saw kerfs.)

    _____

3. From a full 55-gallon drum of mineral-based oil, 2 gallons are used in a vertical mill, 1 gallon is used in a conventional lathe, and 5 gallons are used in the CNC machining center. How much mineral-based oil is left in the 55-gallon drum?

    _____

4. If 125 parts are manufactured in a shift but the Quality Control department determines 37 parts need reworking and 6 are scrap, how many acceptable parts were machined in total?

    _____

5. After reworking 37 parts, the Quality Control inspector finds that 8 parts still do not meet the required dimensions and 3 parts are scrap. How many of the reworked parts are now deemed acceptable and can be shipped to customers?

    _____

6. What is the inside diameter of the sleeve bushing?

    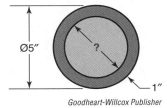

    Ø5″   ?   1″

    Goodheart-Willcox Publisher

    _____

Refer to the figure below to answer the questions that follow.

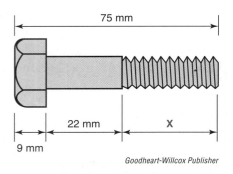

Goodheart-Willcox Publisher

7. What is the threaded length of the bolt, represented by dimension X?

_____

8. What is the length of the bolt body (threaded and unthreaded) without the head?

_____

Refer to the figure below to answer the questions that follow.

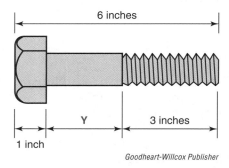

Goodheart-Willcox Publisher

9. What is the unthreaded length of the bolt, represented by dimension Y?

_____

10. What is the length of the bolt body (threaded and unthreaded) without the head?

_____

Name _____   **Date** _____   **Class** _____

Refer to the figure below to calculate the dimensions in the following problems.

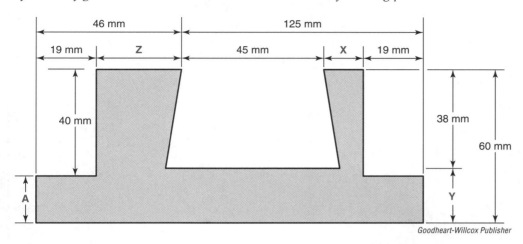

Goodheart-Willcox Publisher

11. Dimension X   12. Dimension Y   13. Dimension Z

_____   _____   _____

14. Dimension A   15. Difference between dimensions Y and A

_____   _____

Refer to the figure below to calculate the answers for the following problems.

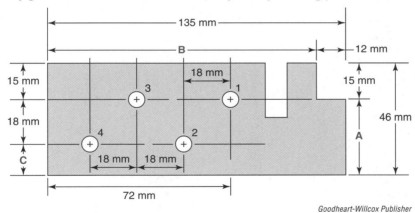

Goodheart-Willcox Publisher

16. Dimension A   17. Dimension B   18. Dimension C

_____   _____   _____

19. Distance of hole 2 from left edge   20. Distance of hole 3 from left edge
     of part   of part

_____   _____

21. Distance of hole 4 from left edge of part

_____

Work Space/Notes

UNIT 4

Multiplying Whole Numbers

Objectives

Information in this unit will enable you to:

- Identify the symbols used to indicate multiplication.
- Explain what a multiplication table is and how it works.
- Name the elements of multiplication and describe how to multiply by numbers less than 10.
- Discuss how to multiply by 10 or more.

Multiplication Symbols

Multiplication can be shown in several different ways. All of them have the same meaning. The list below shows multiplying 4 times 5.

$$4 \times 5 = 20$$
$$4 \cdot 5 = 20$$
$$4(5) = 20$$

Multiplication Table

It is necessary to memorize basic multiplication facts before you can efficiently multiply even the simplest numbers. The table shown on the following page is often called a **multiplication table**. If you follow the line across from one number along the left side to the column beneath a number on the top edge, the cell where the two lines intersect shows the result of multiplying those two numbers.

0	1	2	3	4	5	6	7	8	9	10	11	12
1	1	2	3	4	5	6	7	8	9	10	11	12
2	2	4	6	8	10	12	14	16	18	20	22	24
3	3	6	9	12	15	18	21	24	27	30	33	36
4	4	8	12	16	20	24	28	32	36	40	44	48
5	5	10	15	20	25	30	35	40	45	50	55	60
6	6	12	18	24	30	36	42	48	54	60	66	72
7	7	14	21	28	35	42	49	56	63	70	77	84
8	8	16	24	32	40	48	56	64	72	80	88	96
9	9	18	27	36	45	54	63	72	81	90	99	108
10	10	20	30	40	50	60	70	80	90	100	110	120
11	11	22	33	44	55	66	77	88	99	110	121	132
12	12	24	36	48	60	72	84	96	108	120	132	144

Goodheart-Willcox Publisher

Example 4-1

7 times 4 is 28.

The multiplication table can be expanded in either direction. To multiply a number on the left edge times 13, you can simply add that number to the value shown in the 12 column.

Example 4-2

Nine times 13 is 108 plus one more 9 or 117. Nine times 14 would be one more 9 or 126.

Not knowing the multiplication table is one of the principal reasons that students struggle with basic math. Refer to the table above if you have trouble. Through practice, you will memorize the multiplication facts. It may help you study to make flash cards with a multiplication problem on the front and the answer on the back. See the example in the following figure.

Multiplication Flash Cards

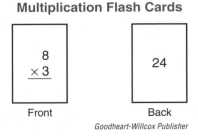

Front   Back

Goodheart-Willcox Publisher

Multiplication by Less Than 10

In multiplication, the number being multiplied is called the **multiplicand**. The number by which it is multiplied is the **multiplier**. The result of multiplication is called the **product**. The problem is set up with the multiplicand on top and the multiplier on the second line, with the ones place of each aligned vertically. The product should also be written with the ones place aligned to the right.

$$
\begin{array}{r}
2 \\
\times\,3 \\
\hline
6
\end{array}
$$

2 ← Multiplicand
×3 ← Multiplier
6 ← Product

The multiplication starts on the right, multiplying the far right (ones place) numeral by the numeral on the far right of the multiplier. If the product of multiplying the ones place is more than 9, the numerals to the left of the ones place are added to the tens place in the product when the tens place is multiplied. This adding to the next place can be repeated for as many places as necessary.

Example 4-3

Step 1: 8 times 7 is 56. Write the 6 in the ones place and the 5 in the tens place.

$$
\begin{array}{r}
27 \\
\times\,8 \\
\hline
56
\end{array}
$$

Step 2: 8 times 2 is 16. Write 16 in the tens place beneath the 5 from step 1. Add the digits in their respective columns.

$$
\begin{array}{r}
27 \\
\times\,8 \\
\hline
56 \\
+\,16 \\
\hline
216
\end{array}
$$

Example 4-4

Step 1: 6 times 8 is 48. Write the 8 in the ones place and the 4 in the tens place.

$$
\begin{array}{r}
88 \\
\times\,6 \\
\hline
48
\end{array}
$$

Step 2: 6 times 8 is 48. Write 48 in the tens place beneath the 4 from step 1. Add the digits in their respective columns.

$$
\begin{array}{r}
88 \\
\times\,6 \\
\hline
48 \\
+\,48 \\
\hline
528
\end{array}
$$

It is not generally necessary to write each step separately. What is shown in the example as step 2 is usually the only step that needs to be written out as you do the multiplication.

Multiplication by 10 or More

When the multiplier is 10 or more, the problem is set up in the same manner, but there are additional steps to find the product. Begin with the multiplier numeral in the ones place, as above. When multiplication by the ones place is complete, do the same process to multiply by the tens place, but write the farthest right digit of the product in the tens place. Place a 0 in the ones place of the second product, just to help keep everything in its proper place. Being particularly careful to keep all digits in their proper places, add the product of the ones multiplication and the tens multiplication. Add a comma between the hundreds and thousands to make the product easier to read.

Example 4-5

$$
\begin{array}{r}
436 \\
\times\ 24 \\
\hline
1{,}744 \\
+\ 8{,}720 \\
\hline
10{,}464
\end{array}
$$

This can be carried out for as many places as necessary.

Example 4-6

$$
\begin{array}{r}
3{,}243 \\
\times\,7{,}524 \\
\hline
12{,}972 \\
64{,}860 \\
1{,}621{,}500 \\
+\ 22{,}701{,}000 \\
\hline
24{,}400{,}332
\end{array}
$$

It does not matter which number is the multiplicand and which is the multiplier. The product will be the same either way. If one is greater than the other, it is usually easiest to use the one with the fewest places as the multiplier. That reduces the number of steps necessary to do the multiplication.

Unit 4 Review

Name _____ **Date** _____ **Class** _____

Fill in the blanks in the following review questions.

1. In multiplication, the number being multiplied is called the _____.

2. The number by which a number is multiplied is the _____.

3. The result of multiplication is called the _____.

Practice

Perform the indicated multiplication. Show your work.

1. 7
    $\times\,3$

2. 5
    $\times\,4$

3. 8
    $\times\,6$

4. 9
    $\times\,5$

5. 3
    $\times\,6$

6. 4(5)

7. 2(8)

8. 4(3)

9. 11(3)

10. 12(6)

11. 1(9)

12. 12(11)

13. $3 \cdot 3$

14. $6 \cdot 2$

15. $5 \cdot 7$

16. $9 \cdot 3$

17. $5 \cdot 10$

18. $13 \cdot 4$

19. $17 \cdot 6$

20.   6
    $\times\,9$

21. 4
    $\times\,7$

22. 2
    $\times\,9$

23. 11
    $\times\,7$

24. 19
    $\times\,9$

25.   24 inches
　　× 6

26.   51
　　× 4

27.   43
　　× 5

28.   35
　　× 3

29.   47 dollars
　　× 4

30.   66
　　× 7

31.   94
　　× 5

32.   24
　　× 8

33.   41 man-hours
　　× 23

34.   65
　　× 42

35.   47
　　× 23

36.   86 pieces
　　× 17

37.   72
　　× 56

38.   125 mm
　　× 50

39.   12 feet
　　× 81

40.   105
　　× 91

41.   142
　　× 73

42.   224
　　× 335

43.   658
　　× 572

44.   18
　　× 2,564

45.   21
　　× 752

46.   516 rivets
　　× 99

47.   342
　　× 157

48.   66 parts
　　× 1,382

49.   116
　　× 2,896

50.   225
　　× 4,167

51.   437
　　× 5,842

52.   509
　　× 6,214

53.   1,014 bolts
　　× 3,197

54.   1,862
　　× 5,391

55.   3,141
　　× 7,263

56.   2,976
　　× 9,031

Name _____   **Date** _____   **Class** _____

Applications

Solve the following problems. Show your work.

1. Fifteen 3-inch-long pins are to be turned from bar stock. What is the total length of bar stock needed? No allowance is to be made for kerfs or chucking.

   _____

2. Forty-three bolts, weighing 3 ounces each, are packed for shipping. What is the total weight of the bolts?

   _____

3. A threaded rod has 11 threads per inch. How many threads are there in 16 inches of the rod?

   _____

4. If a part requires 3 hours on the lathe, 4 hours on the milling machine, and 2 hours on a grinder, how many hours are needed on each machine tool to make 165 parts?

   Lathe

   _____

   Milling machine

   _____

   Grinder

   _____

5. If 26 4-inch-long bolts are to be turned from ⌀1" cold-rolled steel, how much bar stock is needed to manufacture the bolts? (Do not allow for kerfs or chucking.)

    _____

6. If it takes 18 seconds to cut each of the bolts from problem 5 on the band saw, how many seconds will it take in total to cut the parts?

    _____

7. If it takes an apprentice 48 seconds to turn each of the 4-inch-long bolts on a conventional lathe, how long will it take the apprentice to complete the 26 bolts?

    _____

8. If 1 box of set screws holds 50 pieces, and there are 12 boxes in stock, how many set screws in total are in stock?

    _____

9. If each inch of length of a ⌀5" bar of 6061 aluminum weighs 2 pounds, how much does a 144-inch-long bar weigh?

    _____

10. If a ⌀5" bar of 6061 aluminum is cut on a horizontal band saw to create 17 pieces and each cut takes 58 seconds to complete, how long will it take in total to cut all the pieces?

    _____

11. If a socket head cap screw has 2 inches of threaded length with 28 threads per inch, how many threads are on the screw?

    _____

Name _____   **Date** _____   **Class** _____

Refer to the drawing below for problems 12 through 17.

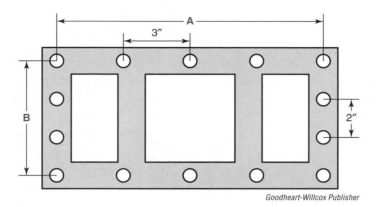

Goodheart-Willcox Publisher

12. A shop made 368 mounting plates. Each plate has 14 holes 5/16" in diameter.
    How many holes are there in the total order?

    _____

13. If it takes 14 seconds to drill one hole and move into position to drill the next
    hole, how many seconds does it take to drill all of the holes in one plate?

    _____

14. Referring to question 13, how many seconds does it take to drill all of the
    holes in the total order?

    _____

15. If the holes are equally spaced horizontally, what is dimension A?

    _____

16. If the holes are equally spaced vertically, what is dimension B?

    _____

17. It takes the drill 3 pecks to drill each hole. How many pecks total does the drill need to make to complete all 14 holes for each plate?

_____

18. You work 37 hours per week, starting at 8:00 a.m. each day, except Fridays, when you start at 7:30 a.m. You work a 5-day workweek. How many hours have you worked in a 6-week period?

_____

19. How many inches of brass rod are needed to make 50 of the parts shown here?

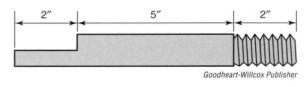

Goodheart-Willcox Publisher

_____

A widget company operates 24 hours a day, 5 days per week. Production is split into three 8-hour shifts per day, and the production line can make 65 parts every hour.

20. How many parts are made per shift?

_____

21. How many parts are made per day, providing the production line runs nonstop?

_____

22. How many parts can the production line potentially produce each workweek?

_____

UNIT 5

Dividing Whole Numbers

Objectives

Information in this unit will enable you to:

- Recognize and properly use terminology associated with division.
- Discuss how to divide whole numbers.
- Explain what a remainder is and what to do when you have one.
- Describe how to use multiplication to check your division.

The Language of Division

Division is the opposite of multiplication. It is the process of dividing a number into equal parts. For example, a pile of 10 pins can be divided into two piles of 5 pins each. The multiplication table that was explained in Unit 4 is useful for division. There are several ways to state a division problem.

- 6 goes into 30 five times
- 30 divided by 6 equals five
- There are 6 fives in 30

There are also several ways to write the problem.

- $6\overline{)30}$ with quotient 5

- $\dfrac{30}{6} = 5$

- $30 \div 6 = 5$

A few key terms are necessary in order to explain how to do division. The **dividend** is the number being divided. The **divisor** is the number by which the dividend is divided. The result of division is the **quotient**.

$$\text{Divisor} \longrightarrow 6\overline{)30} \begin{smallmatrix} \longleftarrow \text{Quotient (5)} \\ \longleftarrow \text{Dividend} \end{smallmatrix}$$

Division of Whole Numbers

To set the problem up, write the dividend inside the division symbol and the divisor to the left of the symbol. Now follow these steps:

1. Cover all but the left-most digit of the dividend with your finger. Ask yourself, how many times will the divisor go into the number showing? Write that number above the digit of the dividend. If it will not fit into the number showing, leave the space blank and go to step 2.

$$4\overline{)3120}$$

2. Slide your finger one place to the right, so there are two digits of the dividend showing. How many times will the divisor go into that two-digit number? Write the number of times above the second digit of the dividend. It is the first digit of the quotient.

$$\begin{array}{r} 7 \\ 4\overline{)3120} \end{array}$$

3. Multiply the divisor times the number showing in the quotient (the number you found in step 2) and write it below the portion of the dividend that has been divided so far. Subtract the number you just wrote from the portion of the divided portion of the dividend. If the difference is equal to or greater than the divisor, increase the number you just wrote in the quotient.

$$\begin{array}{r} 7 \\ 4\overline{)3120} \\ -28 \\ \hline 3 \end{array}$$

4. Draw an arrow down from the next digit of the dividend and write that number as the next place to the right of the difference you found in step 3.

$$\begin{array}{r} 7 \\ 4\overline{)3120} \\ -28 \\ \hline 32 \end{array}$$

5. How many times will the divisor go into the number at the bottom (the one created through subtraction and placing the numeral from the dividend at the right)? Write the number of times as the next digit to the right in the quotient. If the divisor will not fit into the number, write a 0 in the quotient. That 0 is necessary to keep the places.

$$\begin{array}{r} 78 \\ 4\overline{)3120} \\ -28 \\ \hline 32 \end{array}$$

6. Repeat the processes of multiplying the divisor by the last numeral written in the quotient and subtracting, as in step 3. In this example, the subtraction shows a difference of 0. Draw an arrow from the next digit to the right in the dividend (the last digit to the right in this case) and write that numeral as the far-right digit at the bottom of your work. In this example, the subtraction

yielded 0 and the number written at the bottom of the arrow is 0. The divisor will not go into 0 at all, so a 0 is written as the far-right digit of the quotient. These steps can be repeated as far as necessary to divide the entire dividend.

$$
\begin{array}{r}
780 \\
4\overline{)3120} \\
-28 \\
\hline
32 \\
-32 \\
\hline
00
\end{array}
$$

Division with a Remainder

In the example explained above, the divisor went into the dividend exactly 780 times. Frequently, there is something left over after all the places in the dividend have been divided. That amount is called a **remainder**. When the last subtraction is done at the bottom of the problem and the last numeral in the quotient is written above the ones place of the dividend, the difference remaining at the bottom is the remainder. Write the remainder with the rest of the quotient, but use an "r" to indicate that it is a remainder.

Example 5-1

$$
\begin{array}{r}
438 \text{ r}1 \\
4\overline{)1753} \\
-16 \\
\hline
15 \\
-12 \\
\hline
33 \\
-32 \\
\hline
1
\end{array}
$$

Checking Division with Multiplication

As was stated at the beginning of this unit, division is the opposite of multiplication. You can check to see if your division is correct by multiplying the divisor by the quotient, then adding any remainder.

Example 5-2

Check the division in the above example.

$$
\begin{array}{r}
438 \\
\times\ \ 4 \\
\hline
1,752 \\
+\ \ 1 \\
\hline
1,753
\end{array}
$$

Work Space/Notes

Unit 5 Review

Name _____ **Date** _____ **Class** _____

Fill in the blanks in the following review questions.

1. The _____ is the number being divided.

2. The _____ is the number by which the _____ is divided.

3. The result of division is the _____.

4. Frequently, there is something _____ after all the places in the _____ have been divided. That amount is called a _____.

Practice

Perform the indicated division. Check your answers by multiplication. Show your work.

1. $4\overline{)28}$

2. $3\overline{)18}$

3. $6\overline{)72}$

_____   _____   _____

4. $7\overline{)56}$

5. $9\overline{)63}$

6. $12\overline{)204}$

_____   _____   _____

7. $17\overline{)459}$

8. $\dfrac{60}{5}$

9. $\dfrac{66}{11}$

_____   _____   _____

10. $\dfrac{84}{4}$

11. $\dfrac{120}{15}$

12. $\dfrac{195}{13}$

_____   _____   _____

13. $\dfrac{324}{18}$

14. $\dfrac{459}{17}$

15. $\dfrac{36}{12}$

_____   _____   _____

16. $\dfrac{27}{8}$

17. $15 \div 3$

18. $27 \div 9$

_____   _____   _____

19. 40 ÷ 5

_____

20. 48 ÷ 6

_____

21. 72 ÷ 8

_____

22. 132 ÷ 12

_____

23. 143 ÷ 11

_____

24. 378 ÷ 14

_____

25. 176 divided by 8

_____

26. 286 divided by 11

_____

27. 506 divided by 23

_____

28. 648 divided by 18

_____

29. 357 divided by 17

_____

30. 654 divided by 12

_____

31. 421 ÷ 10

_____

32. 17,626 ÷ 14

_____

33. 22,607 ÷ 37

_____

34. 5,508 ÷ 18

_____

35. 36,360 ÷ 202

_____

36. 42,750 ÷ 190

_____

37. 51,851 ÷ 19

_____

38. 65,790 ÷ 102

_____

39. $\dfrac{1,116}{31}$

_____

40. $\dfrac{2,226}{53}$

_____

41. $\dfrac{2,937}{89}$

_____

42. $\dfrac{4,224}{96}$

_____

43. $\dfrac{6,832}{112}$

_____

44. $\dfrac{5,786}{526}$

_____

45. $\dfrac{24,990}{490}$

_____

Name _____   **Date** _____   **Class** _____

Applications

Solve the following problems. Check your answers by multiplication. Show your work.

1. A piece of bar stock 84 inches long is cut into 7 pieces of equal size. What is the length of the cut pieces?

   _____

2. If you receive $168 for working an 8-hour day, how much did you receive for 1 hour?

   _____

3. Five studs are pressed into a plate 54 millimeters long. The holes are equally spaced and the distance from the end of the plate to the center of the first stud is the same as the center-to-center distance between the studs. What is the center-to-center distance between the studs (dimension x)?

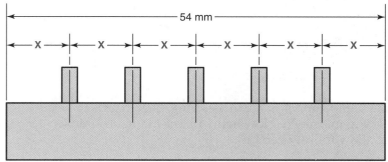

Goodheart-Willcox Publisher

   _____

4. If 80 pieces are finished on the grinder in 16 hours, how many are finished in 1 hour?

   _____

5. How many minutes does it take to finish grinding one of the parts mentioned in problem 4? (There are 60 minutes in an hour.)

   _____

6. A box of 50 cap screws costs 12 dollars. What does 1 cap screw cost? (Hint: 12 dollars is 1,200 cents.)

   _____

7. How many of the studs shown in the figure below can be made from a piece of stock 4 meters long? No studs can be made of any remainder less than the length of a full stud. (A meter is 1,000 millimeters.)

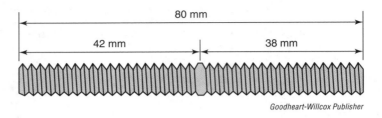

80 mm

42 mm

38 mm

Goodheart-Willcox Publisher

_____

8. How many 4-inch-long tools can be cut from a piece of tool steel 50 inches long? No tools can be made from a remainder of less than 4 inches.

_____

9. If four machinists working 7 hours can produce 112 parts, how many parts does one machinist make in 7 hours?

_____

10. Referring to problem 9, how many parts does one machinist make in one hour?

_____

11. A square bar of copper is 120 inches long. Each part is to be cut 3 inches long before being machined. How many pieces in total could be cut from the bar? (Do not allow for saw kerfs.)

_____

12. If you receive $1,440 biweekly (every 2 weeks) for working 40 hours per week, 5 days per week (10 days total), how much do you receive for working one week?

_____

13. Referring to problem 12, how much do you earn for each day worked (each 8-hour shift)?

_____

14. Referring to problem 12, what is the hourly wage you are paid?

_____

15. If a box of 100 hex nuts costs 26 dollars, how much does it cost per hex nut?

_____

Name _____ **Date** _____ **Class** _____

16. The company wants to determine whether manufacturing hex nuts in-house would be cheaper than buying them. If an estimated cost of 30 dollars is calculated to produce 120 hex nuts, how much would it cost to produce each hex nut in-house? Is the result cheaper than buying the hex nuts (the answer to problem 15)?

    _____

17. How many pins can be cut from a 120-inch-long brass bar if each part is

    10 inches long?   8 inches long?   6 inches long?

    _____   _____   _____

    4 inches long?   3 inches long?

    _____   _____

18. If six machinists working 8 hours per day can produce 576 parts, how many parts does each machinist make per day?

    _____

19. Referring to problem 18, how many parts does each machinist manufacture per hour?

    _____

20. Each part assembly requires 8 springs. If 156 springs are in stock, how many parts could be assembled? How many springs would remain, if any?

    _____

    _____

21. If a machine shop makes 1,248 parts per 8-hour shift, and a quality control inspector checks 26 parts per hour, how long will it take the inspector to check every part from a single shift?

    _____

22. Referring to problem 21, after checking all 1,248 parts, the inspector finds that one in every 48 parts is out of tolerance. How many parts in total are out of tolerance?

    _____

23. Referring to problem 21, the company decides it is not practical or economical to check every part. Instead, the inspector will check 1 out of every 13 parts. With the new policy, how many of the 1,248 parts would be inspected?

    _____

Work Space/Notes

UNIT 6

Combined Operations

Objectives

Information in this unit will enable you to:

- Recognize when you need to use more than one type of operation to solve a math problem.
- Describe the order in which the operations should be performed based on the specifics of the problem.

Problems Involving More Than One Type of Operation

Many of the math problems you will encounter in the machine shop will require you to use a combination of addition, subtraction, multiplication, and division. For example, you might have to add several lengths of bar stock, subtract another length, then multiply the results to find the weight.

Example 6-1

You draw the following lengths of bar stock from the supply room: two pieces at 48 inches, three pieces at 60 inches, and one piece that was 72 inches before someone cut 24 inches off from it. If the bar stock weighs 3 pounds (48 ounces) per foot, and it costs 3 dollars per pound, what is the cost of the bar stock you will use?

Step 1: Add all lengths

$$\begin{array}{r} 48'' \\ 48'' \\ 60'' \\ 60'' \\ 60'' \\ +\,72'' \\ \hline 348'' \end{array}$$

(Continued)

Step 2: Subtract the amount taken by another job

$$
\begin{array}{r}
348'' \\
-\ \ 24'' \\
\hline
324''
\end{array}
$$

Step 3: Divide the weight per foot by the number of inches in a foot (12) to find the weight per inch in ounces.

$$
\begin{array}{r}
4 \\
12\overline{)48} \\
-48 \\
\hline
00
\end{array}
$$

Step 4: Multiply the weight per inch by the number of inches of bar stock.

$$
\begin{array}{r}
324 \\
\times\ \ \ \ 4 \\
\hline
1{,}296 \text{ ounces}
\end{array}
$$

Step 5: Divide the total number of ounces (1,296) by the number of ounces in a pound (16) to find the number of pounds.

$$
\begin{array}{r}
81 \\
16\overline{)1296} \\
-128\downarrow \\
\hline
16 \\
-16 \\
\hline
00
\end{array}
$$

Step 6: Multiply the number of pounds of bar stock (81) by the cost per pound ($3) to find the total cost.

$$
\begin{array}{r}
81 \\
\times\ \ 3 \\
\hline
243 \text{ dollars}
\end{array}
$$

The order in which you do the math operations can be very important. It is not possible to specify one order of operations to cover all math problems. Every problem must be considered on its own. It is generally helpful to write down all of the relevant facts in the problem before you do anything else. Then write the problem with the steps in the order they need to be taken. The acronym PEMDAS is helpful in determining the order in which operations should be performed. The letters in PEMDAS represent the following:

P   Parentheses (Do the operations in parentheses first.)

E   Exponents (Powers and roots of numbers. These are explained in a later unit.)

MD   Multiplication and Division (Do multiplication and division from left to right.)

AS   Addition and Subtraction (Do addition and subtraction from left to right.)

Example 6-2

Solve the following.

$$7 + (4 \times 3) - 3^2$$

Do the operations in parentheses first.

$$7 + 12 - 3^2$$

Reduce the exponent next (fully explained in units 31 and 32). The exponent is the small numeral written near the upper-right corner of the base number, such as 2 in the case of 3^2, or 3 to the second power. The power of a number is the product of multiplying that number by itself (3×3).

$$7 + 12 - 9$$

Do addition and subtraction from left to right.

$$7 + 12 - 9 = 10$$

Once the problem is solved, ask yourself, "Does this answer seem reasonable?"

Example 6-3

Solve the following.

$$(4 \times 9) + 7 \times 3 - (8 \times 3)$$

Do the operations in parentheses first.

$$36 + 7 \times 3 - 24$$

Reduce the exponents next. There are no exponents in this example, so you can skip to the next operation.

Do multiplication and division from left to right.

$$36 + 21 - 24$$

Do addition and subtraction from left to right.

$$36 + 21 - 24 = 33$$

Work Space/Notes

Unit 6 Review

Name _____ Date _____ Class _____

Fill in the blanks in the following review questions.

1. Many of the math problems you will encounter in the machine shop will require you to use a _____ of addition, subtraction, multiplication, and division.

2. The acronym PEMDAS is helpful in determining the order in which operations should be performed. In PEMDAS, the P stands for _____, the E stands for _____, M and D stand for _____ and _____, and A and S stand for _____ and _____.

Practice

Solve the following problems. Show your work.

1. $(12 - 3) \div 3 + 8$

   _____

2. $3 \times 6 + 2 \times 4$

   _____

3. $10 \times 5 - 8 \div 2$

   _____

4. $56 - 23 - (7 \times 3)$

   _____

5. $12 \times (8 - 7 + 3) \div 6$

   _____

6. $9 \div (7 - 5 + 1) + (9 + 4) \times 6$

   _____

7. $(8 \times 7 + 5)$

   _____

8. $3 + (4^2 + 5) - 13$

   _____

9. $5^2 \times (3 + 6) \div (2 + 1)$

   _____

10. $6^2 + (2^2 + 7) - 3^2$

    _____

11. $8 \times (4^2 \div 2 \times 8)$

    _____

12. $1 + 5 \times (7 + 8^2 \div 4)$

    _____

13. $9 - 5 + (7 \times 3^2 - 6)$

    _____

14. $(4^2 + 9) \times (1 + 7 - 3)$

    _____

15. $(8 \div 2) \times (3^2 + 1) \div 4$

    _____

16. $5 \times (8^2 \div 4 \times 3)$

    _____

17. $(3^2 + 5^2) - (4^2 \div 2)$

    _____

18. $(2^2 \times 3^2 + 6^2) - (4^2 - 5)$

    _____

Applications

Solve the following problems. Show your work.

1. You earn $18 per hour, work 7 hours per day for 5 days, and $4 per hour is withheld for taxes. What is your take-home pay for one week?

   _____

2. Referring to problem 1, what is your take-home pay for one year? (There are 52 weeks in one year.)

   _____

3. Referring to problem 1, how much is withheld for taxes in one year?

   _____

4. The supply room has five pieces of 1/4-inch diameter drill rod in 36-inch lengths only. How much waste is left if you cut fifteen 11-inch pieces?

   _____

5. If the supply room has seven bars of Ø2″ aluminum that are each 10′ long, and each part to be cut is 9″ long, how many pieces can be cut per bar? (Hint: There are 12 inches in 1 foot.)

   _____

6. Referring to problem 5, how many pieces total can be cut from the seven bars?

   _____

Name _____   **Date** _____   **Class** _____

7. A customer has placed a job order for 96 parts. Each part measures 64 mm
    in length after being machined on a conventional lathe. Before machining,
    the part is cut on the saw with 3 mm added for the machining operation.
    Each saw cut requires 2 mm. How much total material in mm is required
    for the job?

    _____

8. Referring to problem 7, if bars of material are available in 3-meter lengths,
    how many bars are needed to complete the job? (Hint: There are
    1,000 millimeters in 1 meter.)

    _____

Use this figure for the following problems.

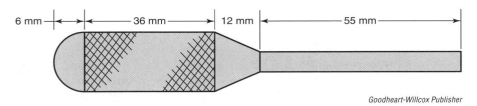

9. How much stock is needed to produce 24 of the parts shown in the figure?
    Allow 2 millimeters for saw kerfs to make each part.

    _____

10. How many of the parts shown in the figure can be produced from a 1-meter rod?
    Allow 2 millimeters for saw kerfs to make each part.

    _____

11. How much scrap will there be from making the 24 parts? Allow 2 millimeters
    for saw kerfs to make each part.

    _____

12. There are 28 pounds of bolts in a bin. If each bolt weights 4 ounces and each box used to ship the bolts can hold 25 bolts, how many full boxes of bolts are available to be shipped? (Hint: There are 16 ounces in 1 pound.)

_____

13. A customer requires 4,800 parts. If each part takes 12 seconds to assemble, how many hours does it take to fulfill the job order?

_____

14. A part manufactured on a lathe takes 10 seconds to face to length, 23 seconds to turn the diameter to size, 36 seconds to thread, and 17 seconds to knurl before it is parted off, which adds a further 7 seconds. How many minutes does it take the operator to complete an order for 120 of these parts?

_____

15. A Ø3-mm high-speed steel drill bit drills holes 21 mm deep using 3 pecks per hole. Each part has 8 such holes on a bolt circle of Ø100 mm. The drill bit costs $4 and can drill 32 holes before it needs to be replaced. How much would be spent on tooling to produce 352 parts?

_____

16. Referring to problem 15, it takes the operator 9 seconds to position the drill and drill each hole, 18 seconds to take the part out and set up the next part in the correct position, and 120 seconds to replace or resharpen the drill bit. How many minutes does it take to complete the order for 352 parts?

_____

17. You are paid $18 per hour. From that, $4 per hour is withheld for taxes. You work 38 hours per week, 52 weeks per year. How much do you have left per year, after paying $210 per month on a loan and $650 per month for your share of rent?

_____

Name _____ **Date** _____ **Class** _____

18. Bandsaw blade stock is available in 100-foot rolls to be cut and welded in the shop. How much scrap will there be from making as many 108-inch blades as possible?

_____

19. Castings are finished on a belt grinder. Twenty-five castings are finished per hour, and the grinder belt lasts 18 hours. If a higher-grade belt will last 30 hours, how many more castings can be finished on one grinder belt, using the higher-grade belt?

_____

20. If the lower-grade belt in problem 8 costs $7 and the higher grade belt costs $11, which costs less per casting finished?

_____

21. How many of the parts shown in the figure below can be made from a piece of stock 850 millimeters long?

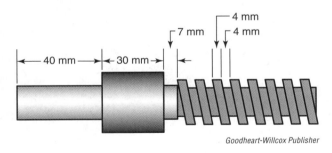

Goodheart-Willcox Publisher

_____

Work Space/Notes

Common Fractions

Key Terms

common denominator
denominator
equivalent fractions
fraction bar
improper fraction

invert
least common
  denominator
lowest terms
mixed number

numerator
proper fraction
reducing
whole

UNIT 7

Parts of a Fraction

Objectives

Information in this unit will enable you to:

- Identify and describe the parts of a fraction.
- Discuss the characteristics of equivalent fractions.

Parts of a Fraction

A fraction is a part of something larger. The larger object is called the **whole**. If a brass rod is cut into 4 equal parts, each part is 1/4 of the whole rod.

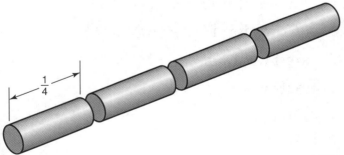

Goodheart-Willcox Publisher

Notice there are three parts to the fraction: a top number, a bar, and a bottom number. The top number is called the **numerator**. The numerator represents the number of parts; in the example above, it is 1. The bottom number is called the **denominator**. The denominator is the number of parts there are in the whole; in the example above, it is 4. The **fraction bar** separates the numerator from the denominator.

Fraction bar ⟶ $\dfrac{1}{4}$ ⟵ Numerator, Denominator

If the numerator in the example were 3, the fraction would be 3/4, representing 3 parts of the whole.

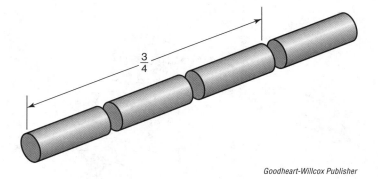

Goodheart-Willcox Publisher

If the denominator were 2 and the numerator 1, that would indicate that the rod was divided into 2 parts and the fraction would represent 1 of the 2 parts, or 1/2.

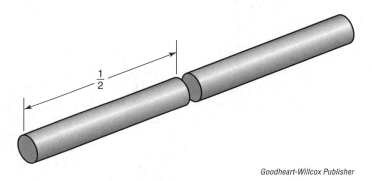

Goodheart-Willcox Publisher

Equivalent Fractions

If two fractions are written using different numbers, but they represent the same value, they are **equivalent fractions**.

Example 7-1

$$\frac{1}{2} = \frac{2}{4} = \frac{4}{8} = \frac{8}{16} = \frac{16}{32} = \frac{32}{64} = \frac{64}{128}.$$

All represent one-half of a whole.

If our brass rod is divided into 4 parts, 2 of those parts would be 2/4 of the whole bar. The whole contains two sets of 2/4, so 2/4 is 1/2 of the whole bar. 2/4 and 1/2 are equivalent fractions.

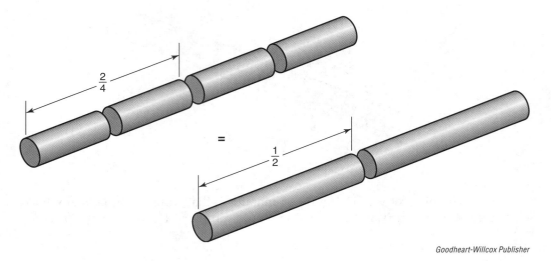

If both the denominator and the numerator of a fraction are multiplied or divided by the same number, the result is an equivalent fraction. If, for example, the denominator and numerator of 2/3 are multiplied by 4, the result would be a fraction of 8/12. 8/12 is an equivalent fraction to 2/3. They both have the same value. If the denominator and numerator of 8/12 are divided by the same amount, the result will also be an equivalent fraction. For example, divide 8 and 12 each by 2. The resulting fraction is 4/6, which is an equivalent fraction to 2/3 and 8/12.

Many fractions can be made easier to work with by **reducing** them to their **lowest terms**. This means that a fraction can be written as an equivalent fraction with a smaller denominator and numerator. This is done by dividing the numerator and the denominator by the same value. Of course the numbers must divide evenly (that is, with no remainder). For example, 12/16 can be reduced to 3/4 as in the following examples.

$$\frac{12 \text{ divided by } 4}{16 \text{ divided by } 4} = \frac{3}{4}$$

$$\frac{145 \text{ divided by } 5}{200 \text{ divided by } 5} = \frac{29}{40}$$

Divisibility Rules

Following are a few rules to remember to make reducing fractions easier. Ask your instructor for other tests of divisibility.

If the last digit is even (0, 2, 4, 6, 8), you can divide by 2.

If the sum of the digits is evenly divisible by 3, you can divide by 3.

$$213: 2 + 1 + 3 = 6$$
$$6 \div 3 = 2$$

So 213 can be divided by 3.

$$323: 3 + 2 + 3 = 8$$
$$8 \div 3 = \frac{8}{3} = 2\frac{2}{3}$$

So 323 cannot be divided by 3.

If the last two digits are evenly divisible by 4, you can divide by 4.

$$908: 08 \div 4 = 2$$

So 908 can be divided by 4.

$$625: 25 \div 4 = 6\frac{1}{4}$$

So 625 cannot be divided by 4.

If the last digit is a 0 or 5, you can divide by 5.

Work Space/Notes

Unit 7 Review

Name _____ Date _____ Class _____

Fill in the blanks in the following review questions.

1. A fraction is a part of something larger. The larger object is called the
   _____.

2. The top number in a fraction is called the _____.

3. The bottom number in a fraction is called the _____.

4. The _____ separates the numerator from the denominator.

5. If two fractions are written using different numbers but represent the same
   value, they are called _____.

6. Many fractions can be made easier to work with by _____ them to
   their lowest terms.

Practice

In the following statements, what is the numerator and what is the denominator?

1. Three quarts of oil are taken from an 8-quart container.

   numerator _____, denominator _____

2. Twelve pounds of acetylene are used from a 40-pound cylinder.

   numerator _____, denominator _____

3. One hundred twenty-one feet of wire remain on a 500-foot spool.

   numerator _____, denominator _____

4. A block is 7/16-inch thick.

   numerator _____, denominator _____

5. Twenty-one inches of bar stock is cut from a 96-inch bar.

   numerator _____, denominator _____

Equivalent fractions

Express the following fractions as sixty-fourths:

6. $\dfrac{1}{4} = \dfrac{}{64}$ 7. $\dfrac{3}{8} = \dfrac{}{64}$ 8. $\dfrac{5}{16} = \dfrac{}{64}$ 9. $\dfrac{14}{32} = \dfrac{}{64}$

10. $\dfrac{1}{2} = \dfrac{}{64}$ 11. $\dfrac{36}{128} = \dfrac{}{64}$ 12. $\dfrac{9}{16} = \dfrac{}{64}$ 13. $\dfrac{7}{8} = \dfrac{}{64}$

Express the following fractions as thirty-seconds:

14. $\dfrac{1}{2} = \dfrac{}{32}$

15. $\dfrac{1}{8} = \dfrac{}{32}$

16. $\dfrac{1}{16} = \dfrac{}{32}$

17. $\dfrac{12}{64} = \dfrac{}{32}$

18. $\dfrac{3}{16} = \dfrac{}{32}$

19. $\dfrac{32}{128} = \dfrac{}{32}$

20. $\dfrac{7}{8} = \dfrac{}{32}$

21. $\dfrac{12}{16} = \dfrac{}{32}$

Express the following fractions as sixteenths:

22. $\dfrac{3}{4} = \dfrac{}{16}$

23. $\dfrac{4}{8} = \dfrac{}{16}$

24. $\dfrac{2}{32} = \dfrac{}{16}$

25. $\dfrac{5}{8} = \dfrac{}{16}$

26. $\dfrac{1}{2} = \dfrac{}{16}$

27. $\dfrac{12}{32} = \dfrac{}{16}$

28. $\dfrac{1}{4} = \dfrac{}{16}$

29. $\dfrac{24}{128} = \dfrac{}{16}$

Express the following fractions as eighths:

30. $\dfrac{2}{4} = \dfrac{}{8}$

31. $\dfrac{1}{2} = \dfrac{}{8}$

32. $\dfrac{4}{16} = \dfrac{}{8}$

33. $\dfrac{8}{32} = \dfrac{}{8}$

34. $\dfrac{40}{64} = \dfrac{}{8}$

35. $\dfrac{12}{32} = \dfrac{}{8}$

36. $\dfrac{2}{16} = \dfrac{}{8}$

37. $\dfrac{112}{128} = \dfrac{}{8}$

Write the missing term for the equivalent fractions.

38. $\dfrac{1}{4} = \dfrac{}{8}$

39. $\dfrac{3}{9} = \dfrac{}{18}$

40. $\dfrac{5}{8} = \dfrac{}{32}$

41. $\dfrac{}{10} = \dfrac{30}{100}$

42. $\dfrac{}{32} = \dfrac{14}{64}$

43. $\dfrac{1}{2} = \dfrac{3}{}$

44. $\dfrac{7}{16} = \dfrac{21}{}$

45. $\dfrac{3}{25} = \dfrac{12}{}$

46. $\dfrac{2}{5} = \dfrac{8}{}$

47. $\dfrac{1}{6} = \dfrac{5}{}$

48. $\dfrac{2}{3} = \dfrac{}{15}$

49. $\dfrac{4}{5} = \dfrac{}{20}$

50. $\dfrac{3}{4} = \dfrac{}{16}$

51. $\dfrac{3}{4} = \dfrac{}{12}$

52. $\dfrac{1}{8} = \dfrac{}{16}$

53. $\dfrac{}{4} = \dfrac{24}{32}$

Name _____ **Date** _____ **Class** _____

54. $\frac{1}{5} = \frac{}{25}$

55. $\frac{}{7} = \frac{10}{35}$

56. $\frac{}{32} = \frac{6}{64}$

57. $\frac{7}{} = \frac{14}{16}$

58. $\frac{}{16} = \frac{52}{64}$

59. $\frac{}{8} = \frac{42}{56}$

60. $\frac{5}{} = \frac{15}{18}$

61. $\frac{2}{} = \frac{48}{72}$

Reduce the fractions to their lowest terms.

62. $\frac{8}{12} =$ _____

63. $\frac{9}{30} =$ _____

64. $\frac{6}{20} =$ _____

65. $\frac{8}{40} =$ _____

66. $\frac{30}{50} =$ _____

67. $\frac{4}{100} =$ _____

68. $\frac{3}{12} =$ _____

69. $\frac{16}{18} =$ _____

70. $\frac{48}{54} =$ _____

71. $\frac{4}{20} =$ _____

72. $\frac{9}{27} =$ _____

73. $\frac{3}{33} =$ _____

74. $\frac{85}{135} =$ _____

75. $\frac{6}{346} =$ _____

76. $\frac{54}{63} =$ _____

77. $\frac{24}{600} =$ _____

78. $\frac{40}{720} =$ _____

79. $\frac{64}{100} =$ _____

80. $\frac{6}{180} =$ _____

81. $\frac{21}{45} =$ _____

82. $\frac{75}{90} =$ _____

83. $\frac{6}{8} =$ _____

84. $\frac{24}{28} =$ _____

85. $\frac{52}{100} =$ _____

86. $\frac{18}{32} =$ _____

87. $\frac{125}{1,000} =$ _____

88. $\frac{228}{440} =$ _____

Applications

What fraction of the following bars is shaded?

1. = _____

Goodheart-Willcox Publisher

2. = _____

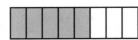

Goodheart-Willcox Publisher

3. = _____

Goodheart-Willcox Publisher

4. = _____

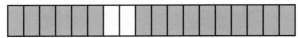

Goodheart-Willcox Publisher

Shade in the proper number of sections in the following bars to match the fractions.

5. $\frac{1}{2}$

Goodheart-Willcox Publisher

6. $\frac{7}{8}$

Goodheart-Willcox Publisher

7. $\frac{4}{10}$

Goodheart-Willcox Publisher

8. $\frac{5}{12}$

Goodheart-Willcox Publisher

Name _____ **Date** _____ **Class** _____

What fraction of the following circles is shaded?

9. = _____

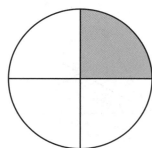

Goodheart-Willcox Publisher

10. = _____

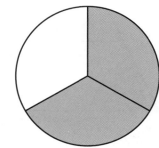

Goodheart-Willcox Publisher

11. = _____

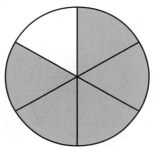

Goodheart-Willcox Publisher

12. = _____

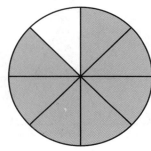

Goodheart-Willcox Publisher

Shade in the proper number of sections in the following circles to match the fractions.

13. $\frac{1}{5}$

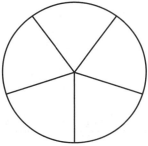

Goodheart-Willcox Publisher

14. $\frac{1}{3}$

Goodheart-Willcox Publisher

15. $\frac{3}{8}$

Goodheart-Willcox Publisher

16. $\frac{7}{12}$

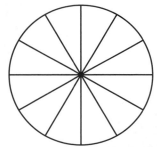

Goodheart-Willcox Publisher

What fraction of the following triangles is shaded?

17. = _____

Goodheart-Willcox Publisher

18. = _____

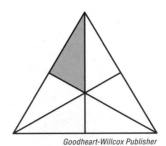

Goodheart-Willcox Publisher

19. = _____

Goodheart-Willcox Publisher

20. = _____

Goodheart-Willcox Publisher

Shade in the proper number of sections in the following triangles to match the fractions.

21. $\frac{1}{2}$

Goodheart-Willcox Publisher

22. $\frac{3}{5}$

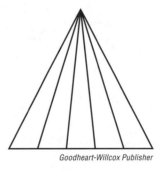

Goodheart-Willcox Publisher

23. $\frac{2}{9}$

Goodheart-Willcox Publisher

24. $\frac{9}{16}$

Goodheart-Willcox Publisher

Name _____   **Date** _____   **Class** _____

What fraction of the following squares is shaded?

25. = _____

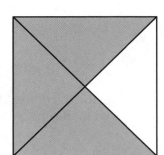
Goodheart-Willcox Publisher

26. = _____

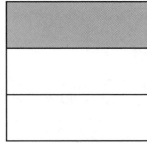
Goodheart-Willcox Publisher

27. = _____

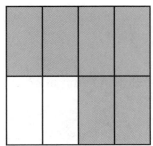
Goodheart-Willcox Publisher

28. = _____

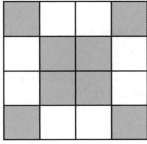
Goodheart-Willcox Publisher

Shade in the proper number of sections in the following squares to match the fractions.

29. $\frac{5}{9}$

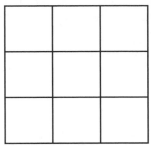
Goodheart-Willcox Publisher

30. $\frac{16}{25}$

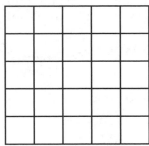
Goodheart-Willcox Publisher

31. $\frac{11}{16}$

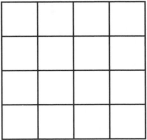
Goodheart-Willcox Publisher

32. $\frac{6}{10}$

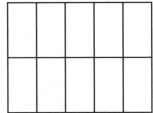
Goodheart-Willcox Publisher

Write the fraction for each of the following.

33. If 18 gallons of soluble oil (coolant) is taken from a 55-gallon drum, what fraction of coolant is taken?

      _____

What fraction remains in the drum?

      _____

34. A round aluminum bar is cut into 16 equal pieces. Write the fraction of the bar represented by the following numbers of pieces, reduced to the lowest terms.

3 pieces: _____   4 pieces: _____   6 pieces: _____

8 pieces: _____   10 pieces: _____   12 pieces: _____

14 pieces: _____

35. What fraction of a dollar is represented by the following amounts? (Reduce to the lowest terms.)

25 cents: _____   35 cents: _____   50 cents: _____

66 cents: _____   72 cents: _____   85 cents: _____

36. There are 128 US fluid ounces in 1 US liquid gallon. What fraction of a gallon is represented by the following amounts? (Reduce to the lowest terms.)

16 fluid ounces: _____   22 fluid ounces: _____

44 fluid ounces: _____   56 fluid ounces: _____

64 fluid ounces: _____   88 fluid ounces: _____

112 fluid ounces: _____

37. Three pounds of bolts are taken from a 10-pound box. What fraction of the bolts is taken?

_____

What fraction remains in the box?

_____

Name _____   **Date** _____   **Class** _____

38. A flat steel bar is cut into seven pieces. Write the fraction representing two of
     those pieces.

     _____

39. There are 8 pints in a gallon. What fraction of a gallon is 3 pints?

     _____

Refer to this figure for problems 40 and 41.

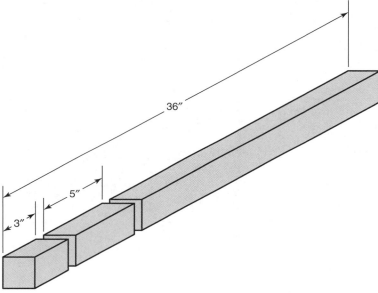

Goodheart-Willcox Publisher

40. What fraction of the bar is cut off for the two small pieces?

     _____

41. What fraction of the bar remains after the two pieces are cut off?

     _____

Work Space/Notes

UNIT 8

Proper Fractions, Improper Fractions, and Mixed Numbers

Objectives

Information in this unit will enable you to:

- Discuss the difference between proper and improper fractions.
- Explain how to reduce an improper fraction and what a mixed number is.
- Describe how to change a mixed number into an improper fraction.

Proper and Improper Fractions

A **proper fraction** is a fraction in which the numerator is smaller than the denominator. The fractions 1/2, 3/4, 2/3, and 25/64 are examples of proper fractions. An **improper fraction** is when the numerator is greater than or equal to the denominator. If you have oil in 16-ounce cans, but you need 20 ounces, you will need 20/16 of a can of oil. The fractions 4/3, 15/10, and 150/45 are other examples of improper fractions.

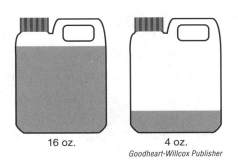

16 oz.   4 oz.
Goodheart-Willcox Publisher

Mixed Numbers

Improper fractions are reduced by dividing the numerator by the denominator. This division will yield a whole number and sometimes a remainder. If there is a remainder, it becomes the new numerator of the fraction. See the examples below.

$$\frac{25}{8} \qquad\qquad \frac{18}{3} \qquad\qquad \frac{115}{20}$$

$$8\overline{)25} \qquad\qquad 3\overline{)18} \qquad\qquad 20\overline{)115}$$

$$\begin{array}{r} 3\ r1 \\ 8\overline{)25} \\ -24 \\ \hline 1 \end{array} \qquad\qquad \begin{array}{r} 6 \\ 3\overline{)18} \\ -18 \\ \hline 0 \end{array} \qquad\qquad \begin{array}{r} 5\ r15 \\ 20\overline{)115} \\ -100 \\ \hline 15 \end{array}$$

$$3\frac{1}{8} \qquad\qquad 6 \qquad\qquad 5\frac{15}{20} = 5\frac{3}{4}$$

In the example on the far right, the division yielded a whole number of 5 and a fraction of 15/20. The numerator and the denominator could each be divided by 5, reducing the fraction to 3/4.

Numbers that include a whole number and a fraction, such as those produced by reducing an improper fraction, are called **mixed numbers**. They mix whole numbers and fractions. Examples for mixed numbers are:

$$2\frac{7}{8},\ 4\frac{1}{2},\ \text{and}\ 112\frac{13}{16}$$

Changing Mixed Numbers to Improper Fractions

Often when working with mixed numbers, it is easiest to first change them to improper fractions. The following example explains how to make this change.

Example 8-1

Change 5 3/4 to an improper fraction.

   Step 1: Multiply the whole number by the denominator.

$$5 \times 4 = 20\ \text{(There are 20 fourths in 5 wholes.)}$$

   Step 2: Add the numerator to the product of multiplying the whole number by the denominator.

$$20 + 3 = 23\ \text{(There are 20 + 3 fourths in the mixed}$$
$$\text{number: 20 in the whole and 3 in the fraction.)}$$

   Step 3: Use the sum from step 2 as the numerator and the original denominator for the improper fraction.

$$5\frac{3}{4} = \frac{23}{4}$$

Unit 8 Review

Fill in the blanks in the following review questions.

1. A(n) _____ fraction is a fraction in which the numerator is smaller than the denominator.

2. A(n) _____ fraction is a fraction in which the numerator is greater than or equal to the denominator.

3. Numbers that include a _____ number and a fraction are called mixed numbers.

4. When working with mixed numbers, it is often easiest to first change them to _____ fractions.

Practice

Label the following as proper fraction, improper fraction, or mixed number.

1. $\frac{2}{3}$

   _____

2. $\frac{27}{32}$

   _____

3. $\frac{21}{15}$

   _____

4. $3\frac{1}{3}$

   _____

5. $\frac{62}{97}$

   _____

6. $\frac{5}{3}$

   _____

7. $17\frac{22}{32}$

   _____

8. $\frac{59}{64}$

   _____

9. $\frac{255}{4}$

   _____

10. $\frac{8}{4}$

    _____

11. $\frac{3}{11}$

    _____

12. $\frac{7}{6}$

    _____

13. $\frac{9}{3}$

    _____

14. $\frac{8}{5}$

    _____

15. $1\frac{1}{4}$

    _____

16. $\frac{17}{16}$

    _____

17. $2\frac{3}{15}$

    _____

18. $\frac{7}{2}$

    _____

19. $\frac{8}{7}$

_____

20. $\frac{2}{9}$

_____

21. $8\frac{6}{16}$

_____

22. $\frac{6}{64}$

_____

23. $\frac{18}{16}$

_____

24. $\frac{3}{27}$

_____

25. $4\frac{7}{12}$

_____

26. $3\frac{1}{3}$

_____

27. $\frac{8}{25}$

_____

28. $\frac{1}{64}$

_____

29. $\frac{6}{4}$

_____

30. $\frac{2}{7}$

_____

31. $\frac{5}{8}$

_____

32. $\frac{6}{5}$

_____

33. $\frac{8}{3}$

_____

34. $1\frac{9}{12}$

_____

35. $\frac{1}{16}$

_____

36. $\frac{64}{32}$

_____

37. $2\frac{5}{19}$

_____

38. $\frac{30}{8}$

_____

39. $\frac{1}{32}$

_____

Reduce the improper fractions. Show your work.

40. $\frac{9}{3} =$ _____

41. $\frac{45}{20} =$ _____

42. $\frac{141}{8} =$ _____

43. $\frac{75}{16} =$ _____

44. $\frac{28}{18} =$ _____

45. $\frac{222}{35} =$ _____

Name _____   **Date** _____   **Class** _____

46. $\frac{4}{1} =$ _____

47. $\frac{3}{2} =$ _____

48. $\frac{1,431}{63} =$ _____

49. $\frac{222}{22} =$ _____

50. $\frac{88}{46} =$ _____

51. $\frac{18}{8} =$ _____

52. $\frac{66}{12} =$ _____

53. $\frac{111}{102} =$ _____

54. $\frac{70}{32} =$ _____

55. $\frac{102}{50} =$ _____

56. $\frac{62}{52} =$ _____

57. $\frac{82}{64} =$ _____

58. $\frac{98}{16} =$ _____

59. $\frac{56}{30} =$ _____

60. $\frac{52}{4} =$ _____

61. $\frac{94}{32} =$ _____

62. $\frac{46}{32} =$ _____

63. $\frac{101}{16} =$ _____

64. $\frac{38}{32} =$ _____

65. $\frac{12}{5} =$ _____

66. $\frac{29}{7} =$ _____

67. $\frac{42}{5} =$ _____

68. $\frac{38}{7} =$ _____

69. $\frac{39}{6} =$ _____

70. $\frac{12}{1} =$ _____

71. $\frac{32}{11} =$ _____

72. $\frac{44}{12} =$ _____

Change the mixed numbers to improper fractions. Show your work.

73. $3\frac{1}{3}$ = _____

74. $1\frac{1}{4}$ = _____

75. $12\frac{31}{32}$ = _____

76. $8\frac{8}{16}$ = _____

77. $17\frac{11}{32}$ = _____

78. $9\frac{15}{16}$ = _____

79. $1\frac{27}{100}$ = _____

80. $14\frac{9}{14}$ = _____

81. $210\frac{2}{3}$ = _____

82. $10\frac{8}{10}$ = _____

83. $2\frac{1}{6}$ = _____

84. $4\frac{2}{7}$ = _____

85. $5\frac{2}{5}$ = _____

86. $9\frac{1}{8}$ = _____

87. $3\frac{6}{7}$ = _____

88. $3\frac{4}{16}$ = _____

89. $5\frac{1}{6}$ = _____

90. $10\frac{8}{32}$ = _____

91. $6\frac{1}{16}$ = _____

92. $7\frac{1}{64}$ = _____

93. $20\frac{1}{5}$ = _____

94. $41\frac{1}{2}$ = _____

95. $110\frac{1}{4}$ = _____

96. $63\frac{2}{5}$ = _____

97. $55\frac{1}{3}$ = _____

98. $36\frac{3}{10}$ = _____

99. $2\frac{10}{11}$ = _____

Least Common Denominator

Objectives

Information in this unit will enable you to:

- Explain what a common denominator is and how to obtain it.
- Describe how to reduce denominators to obtain the least common denominator and discuss why it is beneficial to do so.

Least Common Denominators

Some math operations with two or more fractions require that all of the fractions have the same denominator. When the denominators are the same, the fractions are said to have a **common denominator**. For example, 1/6, 3/6, and 9/6 have 6 as their common denominator. When the denominators are not the same, a common denominator can be found by multiplying the numerator and denominator of each fraction by the denominator of the other fraction.

Example 9-1

Find a common denominator of 1/4 and 3/5.

$$\frac{1 \times 5}{4 \times 5} \qquad \frac{2 \times 4}{5 \times 4}$$

$$\frac{5}{20} \qquad \frac{8}{20}$$

The common denominator is 20.

Example 9-2

Find a common denominator of 2/3 and 4/7.

$$\frac{2 \times 7}{3 \times 7} \qquad \frac{4 \times 3}{7 \times 3}$$

$$\frac{14}{21} \qquad \frac{12}{21}$$

The common denominator is 21.

If there are more than two fractions, a common denominator can be found by multiplying the numerator and denominator of each fraction by the denominators of all the other fractions.

Example 9-3

Find a common denominator of 2/4, 3/5, and 4/9.

$$\frac{2 \times 5 \times 9}{4 \times 5 \times 9} \qquad \frac{3 \times 4 \times 9}{5 \times 4 \times 9} \qquad \frac{4 \times 4 \times 5}{9 \times 4 \times 5}$$

$$\frac{90}{180} \qquad\qquad \frac{108}{180} \qquad\qquad \frac{80}{180}$$

To simplify further work, the fractions with common denominators can sometimes be reduced to fractions with lower common denominators. The reduced denominators are called the lowest common denominator or **least common denominator**. The least common denominator is the smallest of all the possible common denominators. To find the least common denominator, divide each of the numerators and each of the denominators by 2, 3, or 5.

Example 9-4

Find a common denominator of 1/6, 3/5, and 5/8.

$$\frac{1 \times 5 \times 8}{6 \times 5 \times 8} \qquad \frac{3 \times 6 \times 8}{5 \times 6 \times 8} \qquad \frac{5 \times 6 \times 5}{8 \times 6 \times 5}$$

$$\frac{40}{240} \qquad\qquad \frac{144}{240} \qquad\qquad \frac{150}{240}$$

$$\frac{40 \div 2}{240 \div 2} \qquad \frac{144 \div 2}{240 \div 2} \qquad \frac{80 \div 2}{240 \div 2}$$

The least common denominator is

$$\frac{20}{120} \qquad\qquad \frac{72}{120} \qquad\qquad \frac{75}{120}$$

Unit 9 Review

Name _____ **Date** _____ **Class** _____

Fill in the blanks in the following review questions.

1. Some math operations with two or more fractions require that all the fractions have the same _____.

2. Fractions are said to have a common _____ when their _____ are the same.

3. A _____ denominator can be found by _____ the numerator and denominator of each fraction by the _____ of the other fraction.

4. To simplify work with fractions with common denominators, the denominators can sometimes be reduced to the _____ common denominators, or _____ common denominators.

Find the common denominators of the pairs of fractions. Show your work.

1. $\frac{1}{2}$ and $\frac{1}{3}$ _____

2. $\frac{3}{4}$ and $\frac{1}{10}$ _____

3. $\frac{3}{5}$ and $\frac{1}{16}$ _____

4. $\frac{5}{12}$ and $\frac{19}{23}$ _____

5. $\frac{1}{4}$ and $\frac{1}{9}$ _____

6. $\frac{3}{8}$ and $\frac{11}{15}$ _____

7. $\frac{7}{40}$ and $\frac{19}{32}$ _____

8. $\frac{4}{5}$ and $\frac{5}{7}$ _____

9. $\frac{5}{4}$ and $\frac{7}{5}$ _____

10. $\frac{11}{8}$ and $\frac{12}{10}$ _____

11. $\frac{1}{2}$ and $\frac{3}{5}$ _____

12. $\frac{1}{4}$ and $\frac{1}{5}$ _____

13. $\frac{1}{8}$ and $\frac{2}{5}$ _____

14. $\frac{15}{16}$ and $\frac{1}{3}$ _____

15. $\frac{1}{3}$ and $\frac{9}{20}$ _____

16. $\frac{1}{20}$ and $\frac{1}{10}$ _____

17. $\frac{5}{18}$ and $\frac{1}{3}$ _____

18. $\frac{4}{15}$ and $\frac{3}{5}$ _____

19. $\frac{7}{9}$ and $\frac{2}{15}$ _____

20. $\frac{1}{3}$ and $\frac{11}{17}$ _____

21. $\frac{4}{11}$ and $\frac{1}{4}$ _____

22. $\frac{7}{13}$ and $\frac{3}{4}$ _____

23. $\frac{6}{11}$ and $\frac{4}{7}$ _____

24. $\frac{7}{13}$ and $\frac{2}{5}$ _____

25. $\frac{9}{19}$ and $\frac{3}{5}$ _____

26. $\frac{7}{16}$ and $\frac{3}{10}$ _____

27. $\frac{13}{5}$ and $\frac{9}{2}$ _____

28. $\frac{17}{9}$ and $\frac{3}{2}$ _____

29. $\frac{15}{7}$ and $\frac{12}{3}$ _____

30. $\frac{31}{16}$ and $\frac{8}{3}$ _____

Name _____ **Date** _____ **Class** _____

Find the least common denominators of the sets. Show your work.

31. $\frac{1}{3}$ $\frac{2}{5}$ $\frac{3}{8}$ _____

32. $\frac{8}{12}$ $\frac{14}{15}$ $\frac{20}{30}$ _____

33. $\frac{6}{9}$ $\frac{6}{12}$ $\frac{6}{18}$ _____

34. $\frac{3}{5}$ $\frac{10}{15}$ $\frac{4}{9}$ $\frac{7}{8}$ _____

35. $\frac{7}{8}$ $\frac{11}{5}$ $\frac{4}{5}$ $\frac{2}{3}$ _____

36. $\frac{3}{4}$ $2\frac{1}{5}$ $\frac{1}{8}$ _____

37. $\frac{5}{9}$ $\frac{5}{12}$ $3\frac{2}{3}$ _____

38. $1\frac{1}{2}$ $3\frac{1}{8}$ $5\frac{4}{5}$ _____

39. $1\frac{3}{7}$ $1\frac{14}{100}$ $6\frac{1}{3}$ _____

40. $\frac{3}{4}$ $\frac{1}{3}$ $10\frac{1}{2}$ _____

41. $\frac{1}{2}$ $\frac{3}{5}$ $\frac{4}{7}$ _____

42. $\frac{1}{2}$ $\frac{2}{3}$ $\frac{1}{5}$ _____

43. $\frac{2}{3}$ $\frac{2}{5}$ $\frac{3}{7}$ _____

44. $\frac{1}{3}$ $\frac{7}{10}$ $\frac{5}{6}$ _____

45. $\frac{4}{5}$ $\frac{9}{10}$ $\frac{13}{16}$ _____

46. $\frac{1}{4}$ $\frac{1}{10}$ $\frac{15}{16}$ _____

47. $\frac{1}{2}$ $\frac{2}{3}$ $\frac{1}{4}$ $\frac{1}{3}$ _____

48. $\frac{1}{3}$ $\frac{3}{4}$ $\frac{2}{5}$ $\frac{5}{6}$ _____

49. $\frac{3}{2}$ $\frac{9}{16}$ $\frac{7}{5}$ $\frac{1}{6}$ _____

50. $\frac{4}{3}$ $\frac{2}{5}$ $\frac{9}{8}$ $\frac{1}{32}$ _____

51. $1\frac{1}{2}$ $\frac{2}{3}$ $\frac{1}{16}$ _____

52. $\frac{3}{5}$ $4\frac{1}{16}$ $1\frac{1}{4}$ _____

53. $2\frac{1}{8}$ $\frac{3}{16}$ $\frac{2}{5}$ _____

54. $\frac{3}{4}$ $4\frac{4}{5}$ $\frac{1}{9}$ _____

55. $2\frac{1}{4}$ $1\frac{1}{3}$ $4\frac{1}{6}$ _____

56. $4\frac{1}{7}$ $3\frac{1}{9}$ $6\frac{1}{8}$ _____

57. $5\frac{3}{8}$ $3\frac{1}{50}$ $9\frac{3}{16}$ _____

58. $4\frac{1}{2}$ $5\frac{2}{5}$ $7\frac{1}{7}$ _____

59. $\frac{3}{16}$ $\frac{1}{32}$ $12\frac{1}{2}$ _____

60. $\frac{1}{8}$ $2\frac{3}{5}$ $5\frac{1}{9}$ _____

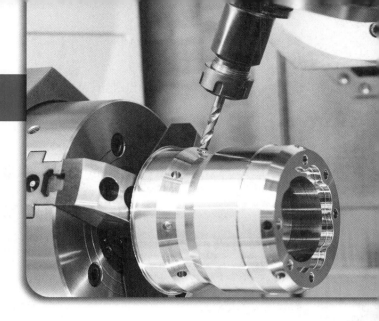

UNIT 10

Adding Fractions

Objective

Information in this unit will enable you to:

- Discuss the steps involved in adding multiple fractions with different denominators.

Learning the Principles

Machinists often need to perform basic mathematical operations with fractions, such as adding fractions. Fractions must have a common denominator to add them. In some cases, the job might only involve fractions that all have the same denominator, but more often they will have different denominators. The first step will be to use what you learned in Unit 9 to write all of the fractions with a common denominator. Then the numerators are added using the same common denominator. See the following example.

Example 10-1

Three bars with thicknesses of 1/8″, 1/4″, and 1/2″ are stacked. What is the height of the stack?

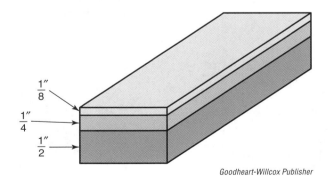

Goodheart-Willcox Publisher

$$\frac{1''}{8} + \frac{1''}{4} + \frac{1''}{2}$$

$$\frac{1''}{8} + \frac{2''}{8} + \frac{4''}{8} = \frac{7''}{8}$$

If the sum of adding the fractions yields an improper fraction (1 or more), it can be written as a mixed number.

Example 10-2

Three bars with thicknesses of 1/4″, 1/2″, and 3/8″ are stacked. What is the height of the stack?

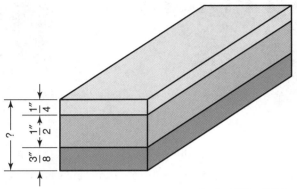

Goodheart-Willcox Publisher

$$\frac{1''}{4} + \frac{1''}{2} + \frac{3''}{8}$$

$$\frac{2''}{8} + \frac{4''}{8} + \frac{3''}{8} = \frac{9''}{8}$$

$$\frac{9''}{8} = 1\frac{1''}{8}$$

If the problem includes one or more mixed numbers, there are two options for the solution of the problem. In the first option, all of the mixed numbers are converted to improper fractions, and then the common denominator is found and the numerators are added.

Example 10-3

Three bars with thicknesses of 3/4″, 1 1/2″, and 1 1/8″ are stacked. What is the height of the stack?

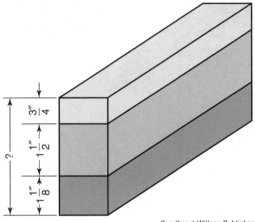

Goodheart-Willcox Publisher

Option 1

$$\frac{3''}{4} + 1\frac{1''}{2} + 1\frac{1''}{8}$$

$$\frac{3''}{4} + \frac{3''}{2} + \frac{9''}{8}$$

$$\frac{6''}{8} + \frac{12''}{8} + \frac{9''}{8} = \frac{27''}{8}$$

$$\frac{27''}{8} = 3\frac{3''}{8}$$

In the second option, the whole numbers are added, the remaining fractions are added as proper fractions, and then the sum of the proper fractions is added to the sum of the whole numbers.

Option 2

$$\frac{3''}{4} + 1\frac{1''}{2} + 1\frac{1''}{8}$$

$$1'' + 1'' + \frac{3''}{4} + \frac{1''}{2} + \frac{1''}{8}$$

$$2'' + \frac{6''}{8} + \frac{4''}{8} + \frac{1''}{8} = 2\frac{11''}{8}$$

$$2\frac{11''}{8} = 3\frac{3''}{8}$$

Work Space/Notes

Name _____ Date _____ Class _____

Fill in the blanks in the following review questions.

1. To add fractions, they must have a common _____. Then, the _____ are added.

2. If the sum of added fractions yields a(n) _____ fraction (1 or more), it can be written as a _____ number.

3. If a problem includes mixed numbers, there are two options for the solution. In the first option, all mixed numbers are converted to _____ fractions, the common _____ is found, and the numerators are added.

4. In the second option to solve problems including mixed numbers, the whole numbers are _____, the remaining fractions are added as _____ fractions, and then the sum of the _____ fractions is added to the sum of the _____ numbers.

Practice

Add the fractions. Write all results as proper fractions or mixed numbers.

1. $\dfrac{1}{3} + \dfrac{3}{5} =$ _____

2. $\dfrac{3}{4} + \dfrac{3}{8} =$ _____

3. $\dfrac{1}{4} + \dfrac{5}{8} =$ _____

4. $\dfrac{7}{12} + \dfrac{3}{5} =$ _____

5. $\dfrac{1}{15} + \dfrac{3}{21} =$ _____

6. $\dfrac{1}{2} + \dfrac{1}{3} =$ _____

7. $\dfrac{5}{6} + \dfrac{1}{8} =$ _____

8. $\dfrac{1}{3} + \dfrac{1}{8} =$ _____

9. $\dfrac{1}{6} + \dfrac{2}{7} =$ _____

10. $\dfrac{1}{5} + \dfrac{2}{3} =$ _____

11. $\dfrac{1}{4} + \dfrac{1}{3} =$ _____

12. $\dfrac{1}{7} + \dfrac{3}{8} =$ _____

13. $\dfrac{1}{9} + \dfrac{3}{10} =$ _____

14. $\dfrac{3}{7} + \dfrac{5}{9} =$ _____

15. $\frac{11}{12} + \frac{2}{3} =$ _____

16. $\frac{3}{32} + \frac{1}{64} =$ _____

17. $\frac{5}{32} + \frac{1}{8} =$ _____

18. $\frac{7}{16} + \frac{3}{64} =$ _____

19. $\frac{1}{16} + \frac{2}{3} =$ _____

20. $\frac{5}{32} + \frac{1}{10} =$ _____

21. $\frac{5}{8} + \frac{2}{3} + \frac{1}{6} =$ _____

22. $\frac{3}{16} + \frac{5}{8} + \frac{5}{12} =$ _____

23. $1\frac{1}{8} + \frac{5}{16} =$ _____

24. $\frac{9}{16} + 15\frac{1}{4} =$ _____

25. $2\frac{2}{3} + 3\frac{4}{6} + 3\frac{1}{12} =$ _____

26. $\frac{1}{2} + \frac{1}{3} + \frac{1}{4} =$ _____

27. $\frac{2}{3} + \frac{3}{4} + \frac{2}{5} =$ _____

28. $\frac{1}{4} + \frac{3}{5} + \frac{5}{6} =$ _____

29. $\frac{4}{5} + \frac{3}{8} + \frac{1}{10} =$ _____

30. $\frac{1}{8} + \frac{2}{10} + \frac{4}{9} =$ _____

31. $\frac{7}{8} + \frac{3}{16} + \frac{5}{32} =$ _____

32. $2\frac{1}{4} + \frac{7}{32} =$ _____

33. $\frac{9}{16} + 4\frac{1}{2} =$ _____

34. $5\frac{1}{3} + \frac{5}{64} =$ _____

35. $12\frac{1}{2} + \frac{15}{32} =$ _____

36. $22\frac{1}{4} + 1\frac{1}{3} =$ _____

Name _____   **Date** _____   **Class** _____

37. $33\frac{1}{8} + 4\frac{3}{5} =$ _____

38. $1\frac{1}{4} + 4\frac{1}{3} + 2\frac{1}{10} =$ _____

39. $3\frac{9}{10} + 7\frac{2}{3} + 4\frac{1}{8} =$ _____

40. $5\frac{1}{2} + 8\frac{8}{9} + 6\frac{5}{12} =$ _____

41. $\frac{5}{8} + 2\frac{1}{2} + 1\frac{1}{3} =$ _____

42. $\frac{19}{56} + 2\frac{1}{4} + 3\frac{21}{24} =$ _____

43. $\frac{57}{125} + 2\frac{19}{48} =$ _____

44. $15\frac{1}{8} + 27\frac{2}{3} =$ _____

45. $\frac{1}{1000} + \frac{5}{100} =$ _____

46. $\frac{3}{5} + 3\frac{1}{8} + 2\frac{9}{10} =$ _____

47. $5\frac{1}{5} + 8\frac{11}{15} + 1\frac{4}{9} =$ _____

48. $3\frac{12}{15} + 1\frac{4}{9} + 4\frac{3}{8} =$ _____

49. $2\frac{9}{16} + 7\frac{1}{6} + 5\frac{1}{2} =$ _____

50. $4\frac{1}{7} + 4\frac{3}{32} + 8\frac{1}{16} =$ _____

51. $6\frac{1}{6} + 9\frac{1}{25} + 2\frac{3}{11} =$ _____

52. $4\frac{43}{100} + 3\frac{7}{150} =$ _____

53. $9\frac{1}{250} + 12\frac{3}{500} =$ _____

54. $19\frac{15}{16} + 22\frac{35}{64} =$ _____

55. $24\frac{1}{4} + 33\frac{1}{16} =$ _____

56. $44\frac{7}{8} + 67\frac{1}{3} =$ _____

57. $100\frac{1}{2} + 27\frac{57}{64} =$ _____

58. $\frac{3}{1000} + \frac{1}{100} + \frac{9}{10} =$ _____

59. $\frac{3}{250} + \frac{9}{500} + \frac{873}{1000} =$ _____

60. $10\frac{1}{100} + 15\frac{189}{250} + 42\frac{1}{3} =$ _____

Applications

Solve the problems.

1. Use the chart to answer the following questions about the time Emil worked on job 277A.

Emil Rodriguez: 277A	Week 1 Hours	Week 2 Hours	Week 3 Hours	Week 4 Hours
Monday	$7\frac{1}{2}$	$5\frac{3}{4}$	8	$6\frac{5}{6}$
Tuesday	$6\frac{7}{8}$	$7\frac{1}{3}$	$6\frac{11}{12}$	$5\frac{1}{4}$
Wednesday	$8\frac{1}{6}$	$6\frac{2}{3}$	$7\frac{1}{3}$	$4\frac{1}{2}$
Thursday	$7\frac{5}{6}$	$8\frac{1}{4}$	$6\frac{1}{4}$	$7\frac{2}{3}$
Friday	$4\frac{3}{4}$	$5\frac{1}{6}$	$4\frac{5}{12}$	$3\frac{1}{12}$

Goodheart-Willcox Publisher

How much time was spent on the job during week 1? _____

How much time was spent on the job during week 2? _____

How much time was spent on the job during week 3? _____

How much time was spent on the job during week 4? _____

How much time was spent on the job in total (all four weeks)?

_____

2. The following lengths are cut from a 36-inch brass rod: 1 1/2 inches, 5/8 inch, 3 2/3 inches. How much was cut from the rod?

_____

3. A 2 1/2-inch round piece is necked down 1/4 inch, then 3/8 inch more, and then another 7/16 inch. How much was the final step necked down from the original piece?

_____

Name _____   **Date** _____   **Class** _____

4. A low-carbon steel flat bar is cut into six pieces. The pieces measure 1 3/4 inches, 1 1/8 inches, 2 1/3 inches, 1/2 inch, 4 1/12 inches, and 6 inches in length. How much total material was cut? (Ignore kerfs.)

   _____

5. Five parts are made based on the shape in the following figure. Use the figure and table to calculate the total length of each part.

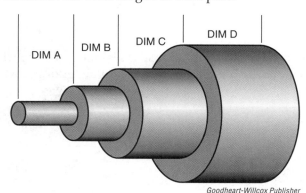

Goodheart-Willcox Publisher

Part Number	Dimension A	Dimension B	Dimension C	Dimension D
One	$1\frac{5}{9}$	$1\frac{13}{16}$	$3\frac{1}{8}$	$3\frac{1}{2}$
Two	$2\frac{1}{9}$	$3\frac{3}{16}$	$1\frac{1}{4}$	$4\frac{1}{3}$
Three	$1\frac{7}{8}$	$3\frac{5}{8}$	$4\frac{1}{16}$	$2\frac{1}{8}$
Four	$3\frac{1}{16}$	$1\frac{13}{32}$	$2\frac{1}{3}$	$5\frac{1}{64}$
Five	$2\frac{3}{32}$	$4\frac{31}{64}$	$3\frac{1}{8}$	$3\frac{1}{4}$

Goodheart-Willcox Publisher

What is the total length of part one? _____

What is the total length of part two? _____

What is the total length of part three? _____

What is the total length of part four? _____

What is the total length of part five? _____

6. A machined steel tube has an inside diameter of 13/32 inch and a wall thickness of 7/64 inch. What is the outside diameter of the tube?

_____

7. An assembly is made up of four pieces having lengths of 3/4", 7/16", 2 5/8", and 5/16". What is the total length of the assembly?

_____

8. A 316 stainless steel washer has an inside diameter of 1/2 inch and a wall thickness of 5/16 inch. What is the outside diameter of the washer?

_____

Refer to this drawing for problems 9 through 17.

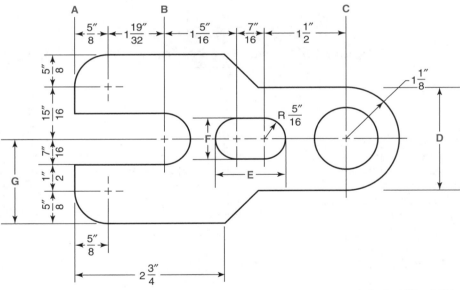

Goodheart-Willcox Publisher

9. What is the distance from B to C in the drawing?

_____

Name _____   **Date** _____   **Class** _____

10. What is the total length of the part in the drawing?

_____

11. What is the total width of the part in the drawing?

_____

12. What is the distance from A to B in the drawing?

_____

13. What is the distance from A to C in the drawing?

_____

14. What is the distance represented by D in the drawing?

_____

15. What is the distance represented by E in the drawing?

_____

16. What is the distance represented by F in the drawing?

_____

17. What is the distance represented by G in the drawing?

_____

18. A stainless steel rod is cut into five pieces of the following lengths: 2 1/8",
    4 9/16", 4 5/32", 3 17/24", and 7 5/9". What was the original length of the
    rod, allowing 1/8" waste for each cut?

    _____

19. The weights of several aluminum castings are 112 1/3 pounds, 370 1/4 pounds,
    725 pounds, 87 1/2 pounds, and 56 2/3 pounds. What is the total weight of the
    castings?

    _____

20. Four studs require machining. If the studs measure 2 3/4 inches, 3 5/8 inches,
    1 3/16 inches, and 4 1/8 inches in length, and we also allow 1 3/8 inches
    (total) for kerfs and facing operations, how long a piece of steel is required to
    produce the studs?

    _____

21. A 3-inch-diameter shaft is turned down in six different cuts on a lathe.
    The first cut takes off 5/32 inch from the diameter, the second cut removes
    1/8 inch, the third removes 3/32 inch, the fourth removes 1/16 inch, the
    fifth removes 1/32 inch, and the finish pass removes 1/64 inch. How much
    material in total was taken from the diameter?

    _____

UNIT 11

Subtracting Fractions

Objective

Information in this unit will enable you to:

- Explain how to subtract fractions with different denominators.

Learning the Principles

To subtract fractions, they must have a common denominator. Once the fractions have a common denominator, one numerator can be subtracted from the other. The denominator of the difference will be the common denominator of the problem.

Example 11-1

Shims of assorted thickness are mounted on a shaft to make a total of 3/16″. What remains on the shaft if 1/10″ is removed?

$$\frac{3}{16}'' - \frac{1}{10}''$$

$$\frac{30}{160}'' - \frac{16}{160}'' = \frac{14}{160}''$$

$$\frac{14}{160}'' = \frac{7}{80}''$$

Example 11-2

Shims of assorted thickness are mounted under a cutting tool to make a total of 5/64″. What remains under the tool if 1/20″ is removed?

$$\frac{5}{64}'' - \frac{1}{20}''$$

$$\frac{100}{1280}'' - \frac{64}{1280}'' = \frac{36}{1280}''$$

$$\frac{36}{1280}'' = \frac{9}{320}''$$

If the problem involves a mixed number, it is changed to an improper fraction, as in the following example.

Example 11-3

If 3/4″ is cut off a 2 1/2″ pin, how much of the pin remains?

Step 1: Change 2 1/2″ to an improper fraction.

$$2\frac{1}{2}'' = \frac{5''}{2}$$

Step 2: Find a common denominator of 3/4″ and 5/2″.

$$\frac{3''}{4} \text{ and } \frac{10''}{4}$$

Step 3: Subtract the numerators.

$$\frac{10''}{4} - \frac{3''}{4} = \frac{7''}{4}$$

Step 4: Express 7/4″ as a mixed number.

$$\frac{7''}{4} = 1\frac{3}{4}''$$

Example 11-4

If 5/8″ is cut off a 3 7/16″ bolt, how much of the bolt remains?

Step 1: Change 3 7/16″ to an improper fraction.

$$3\frac{7}{16}'' = \frac{55''}{16}$$

Step 2: Find a common denominator of 5/8″ and 55/16″.

$$\frac{10''}{16} \text{ and } \frac{55''}{16}$$

Step 3: Subtract the numerators.

$$\frac{55''}{16} - \frac{10''}{16} = \frac{45''}{16}$$

Step 4: Express 45/16″ as a mixed number.

$$\frac{45''}{16} = 2\frac{13}{16}''$$

Unit 11 Review

Name _____ Date _____ Class _____

Fill in the blanks in the following review questions.

1. To subtract fractions, they must have a common _____.

2. Once the fractions have a common _____, one _____ can be subtracted from the other.

3. The _____ of the resulting difference will be the common _____ of the problem.

4. If the problem involves a mixed number, that number is changed to a(n) _____ fraction.

Practice

Subtract the fractions. Write all results as proper fractions or mixed numbers reduced to their lowest terms.

1. $\dfrac{7}{8}$
   $-\dfrac{3}{8}$

2. $\dfrac{13}{16}$
   $-\dfrac{5}{16}$

3. $\dfrac{27}{50}$
   $-\dfrac{22}{50}$

4. $2\dfrac{1}{8}$
   $-\dfrac{3}{4}$

5. $3\dfrac{51}{60} - 1\dfrac{1}{4} = $ _____

6. $\dfrac{9}{16} - \dfrac{5}{16} = $ _____

7. $\dfrac{15}{32} - \dfrac{5}{32} = $ _____

8. $\dfrac{39}{64} - \dfrac{7}{64} = $ _____

9. $\dfrac{19}{52} - \dfrac{7}{52} = $ _____

10. $\dfrac{19}{24} - \dfrac{5}{24} = $ _____

11. $\dfrac{47}{48} - \dfrac{23}{48} = $ _____

12. $\dfrac{87}{100} - \dfrac{33}{100} = $ _____

13. $\dfrac{63}{64} - \dfrac{41}{64} = $ _____

14. $\dfrac{437}{500} - \dfrac{73}{100} = $ _____

15. $4\frac{1}{4} - \frac{2}{3} =$ _____

16. $7\frac{1}{3} - \frac{7}{8} =$ _____

17. $5\frac{1}{8} - \frac{3}{5} =$ _____

18. $2\frac{17}{32} - 1\frac{1}{6} =$ _____

19. $6\frac{35}{64} - 3\frac{1}{9} =$ _____

20. $8\frac{1}{100} - 7\frac{1}{32} =$ _____

21. $24\frac{1}{2} - 2\frac{5}{8} =$ _____

22. $3\frac{3}{32} - 1\frac{1}{8} =$ _____

23. $6\frac{5}{7} - 2\frac{9}{20} =$ _____

24. $11\frac{1}{3} - 3\frac{11}{12} =$ _____

25. $2\frac{3}{4} - \frac{14}{31} =$ _____

26. $12\frac{1}{4} - 5\frac{1}{3} =$ _____

27. $35\frac{1}{8} - 10\frac{1}{6} =$ _____

28. $26\frac{1}{5} - 20\frac{1}{4} =$ _____

29. $8\frac{3}{64} - 5\frac{1}{10} =$ _____

30. $11\frac{1}{4} - 2\frac{11}{16} =$ _____

31. $41\frac{1}{32} - 18\frac{1}{8} =$ _____

32. $14\frac{1}{6} - 1\frac{1}{25} =$ _____

33. $7\frac{1}{8} - 1\frac{93}{100} =$ _____

34. $56\frac{1}{4} - 39\frac{1}{250} =$ _____

35. $100\frac{1}{2} - 22\frac{17}{20} =$ _____

36. $9\frac{1}{100} - 2\frac{1}{1000} =$ _____

37. $81\frac{237}{500} - 19\frac{127}{250} =$ _____

38. $3\frac{2}{3} - \frac{19}{43} =$ _____

39. $5\frac{1}{17} - 1\frac{1}{52} =$ _____

40. $92\frac{1}{18} - 61\frac{1}{19} =$ _____

Name _____   **Date** _____   **Class** _____

Applications

Refer to the figure and use the table to calculate dimension A (the length of the bolt under the head) and dimension B (the length of thread) for each bolt size.

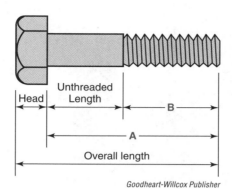

Goodheart-Willcox Publisher

Bolt Size	Thickness of Head	Unthreaded Length	Overall Length
#10 – 24	$\frac{1}{8}''$	1″	$2\frac{5}{8}''$
Ø1/4″ – 28	$\frac{11}{64}''$	$3\frac{3}{4}''$	$4\frac{43}{64}''$
Ø5/16″ – 18	$\frac{13}{64}''$	$3\frac{5}{8}''$	$4\frac{45}{64}''$
Ø3/8″ – 24	$\frac{1}{4}''$	$3\frac{15}{16}''$	$5\frac{1}{2}''$
Ø7/16″ – 14	$\frac{9}{32}''$	$4\frac{7}{8}''$	$6\frac{9}{32}''$
Ø7/8″ – 9	$\frac{35}{64}''$	$6\frac{3}{4}''$	$9\frac{35}{64}''$
Ø1$\frac{3}{4}''$ – 5	$1\frac{3}{32}''$	$5\frac{45}{64}''$	$10\frac{27}{32}''$

Goodheart-Willcox Publisher

1. #10 – 24

    Dim A: _____

    Dim B: _____

2. Ø1/4″ – 28

    Dim A: _____

    Dim B: _____

3. Ø5/16″ – 18

    Dim A: _____

    Dim B: _____

4. Ø3/8″ – 24

    Dim A: _____

    Dim B: _____

5. Ø7/16″ – 14

    Dim A: _____

    Dim B: _____

6. Ø7/8″ – 9

    Dim A: _____

    Dim B: _____

7. Ø1$\frac{3}{4}''$ – 5

    Dim A: _____

    Dim B: _____

Refer to this figure for problems 8 through 11.

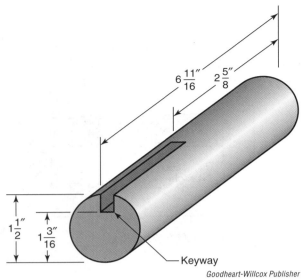

Goodheart-Willcox Publisher

8. What is the length of the keyway?

_____

9. What is the depth of the keyway?

_____

10. Referring to question 8, if the part is revised and the 2 5/8″ dimension is changed to 3 33/64″, what would be the new length of the keyway?

_____

11. Referring to question 9, if the part is revised and the diameter dimension of 1 1/2″ is changed to 1 5/8″, what would be the new depth of the keyway?

_____

12. You normally work 37 1/2 hours per week, but last week you took 4 3/4 hours off for personal business. How many hours did you work last week?

_____

Name _____   **Date** _____   **Class** _____

Refer to this figure for problems 13 and 14.

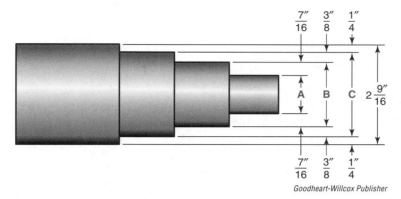

Goodheart-Willcox Publisher

13. A Ø2 9/16" shaft is turned down in three steps as shown in the figure. What is the diameter (thickness) of each of the following parts?

    A. _____

    B. _____

    C. _____

14. Referring to question 13, if the engineer redesigns the part so the shaft now measures Ø2 19/32", the 1/4" measurement is now 3/16", the 3/8" measurement is now 13/32", and the 7/16" measurement is now 29/64", what would the new diameters be for A, B, and C?

    A: _____

    B: _____

    C: _____

15. A 48"-long brass round has the following lengths removed: 1 1/8", 14 3/4", 12 2/3", and 4 1/16". How much remains of the original round? (Ignore kerfs.)

_____

16. A 72"-long copper bar has the following lengths removed: 2 3/16", 10 3/8", 31 1/4", 17 1/2", and 8 3/4". How much material remains after the parts are cut off. (Ignore kerfs.)

_____

17. A titanium washer has an outside diameter of 5/8 inch and a wall thickness of 7/32 inch. What is the inside diameter of the washer?

_____

18. As an apprentice, you are assigned the task of topping off coolant levels for machines around the shop. Two CNC mills require filling, one with 3 1/3 gallons and the other with 4 1/2 gallons. Two CNC lathes also require coolant, one needing 2 1/4 gallons and the other 1 3/8 gallons. If the drum had 29 gallons of coolant when you started, how much remains?

_____

19. A blueprint calls for the diameter on a part to be turned down from Ø6 7/16" to Ø4 3/32". How much material was removed on the diameter?

_____

20. The diameter of 4 3/32" from question 19 has a tolerance of ± 1/64". What is the smallest diameter dimension the part can be machined to?

_____

Name _____   **Date** _____   **Class** _____

Refer to this figure for problem 21.

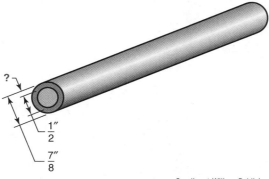

1" / 2

7" / 8

Goodheart-Willcox Publisher

21. A 7/8" steel round has a 1/2" hole bored its entire length, creating a tube.
    What is the wall thickness of the tube?

    _____

22. The following amounts are taken from a 5-gallon container of cutting oil:
    1 1/4 quarts, 7/8 quart, and 1 1/2 quarts. How much remains? (A gallon is
    4 quarts.)

    _____

23. If you earn $21.25 (twenty-one and one-fourth dollars) per hour and $4.50
    (four and one-half dollars) is withheld, how much is left for your take-home pay?

    _____

Work Space/Notes

UNIT 12

Multiplying Fractions

Objectives

Information in this unit will enable you to:

- Explain how to multiply proper fractions.
- Describe the steps involved in multiplying mixed numbers.

Multiplying Proper Fractions

To multiply two fractions, multiply the denominators by each other and multiply the numerators by each other. Then reduce the product to its lowest terms.

Example 12-1

$$\frac{3}{8} \times \frac{1}{3} = \frac{3}{24}$$

$$\frac{3}{24} = \frac{1}{8}$$

If there are more than two fractions, they are multiplied the same way. Multiply all of the denominators and multiply all of the numerators.

Example 12-2

$$\frac{1}{4} \times \frac{2}{5} \times \frac{5}{9} = \frac{1 \times 2 \times 5}{4 \times 5 \times 9}$$

$$\frac{1 \times 2 \times 5}{4 \times 5 \times 9} = \frac{10}{180}$$

$$\frac{10}{180} = \frac{1}{18}$$

It is often easier to reduce the fractions before beginning multiplication. If all of the fractions making up the problem are combined into one fraction, it is easier to see opportunities to reduce the fractions. In the following example, once the fractions are combined, it is easier to see that the 12 in the denominator can be evenly divided by the 3 in the numerator, reducing those two terms to an equivalent of 1/4.

Example 12-3

$$\frac{3}{4} \times \frac{5}{12} \times \frac{1}{2} = \frac{\overset{1}{3} \times 5 \times 1}{4 \times \underset{4}{12} \times 2}$$

$$\frac{1 \times 5 \times 1}{4 \times 4 \times 2} = \frac{5}{32}$$

Multiplying Mixed Numbers

As with addition and subtraction, mixed numbers must be changed to improper fractions before being multiplied.

Example 12-4

$$4\frac{3}{8} \times 2\frac{3}{7} = \frac{\overset{5}{35}}{8} \times \frac{17}{\underset{1}{7}}$$

$$\frac{5}{8} \times \frac{17}{1} = \frac{85}{8}$$

$$\frac{85}{8} = 10\frac{5}{8}$$

Multiplying a whole number by a mixed number is done in the same way. Write the whole number as an improper fraction by using the whole number as the numerator and 1 as the denominator. Change the mixed number to an improper fraction, then multiply the denominators and the numerators, as explained above.

Example 12-5

$$6 \times 2\frac{1}{2} = \frac{6}{1} \times \frac{5}{2}$$

$$\frac{6}{1} \times \frac{5}{2} = \frac{30}{2}$$

Reduce to lowest terms.

$$\frac{30}{2} = \frac{15}{1} \text{ or } 15$$

Unit 12 Review

Fill in the blanks in the following review questions.

1. To multiply fractions, multiply the _____ by each other and multiply the _____ by each other.

2. After multiplying, reduce the _____ to its lowest terms.

3. As with addition and subtraction, mixed numbers and whole numbers must be changed to _____ fractions before being multiplied.

4. Write the whole number as a(n) _____ fraction by using the whole number as the _____ and 1 as the _____.

Practice

Multiply the fractions. Reduce your answers to lowest terms. Show your work.

1. $\frac{1}{4} \times \frac{1}{2} =$ _____

2. $\frac{3}{7} \times \frac{3}{9} =$ _____

3. $\frac{7}{8} \times \frac{24}{25} =$ _____

4. $\frac{2}{5} \times \frac{111}{60} =$ _____

5. $\frac{7}{9} \times \frac{45}{50} =$ _____

6. $\frac{1}{3} \times \frac{1}{4} =$ _____

7. $\frac{1}{2} \times \frac{2}{3} =$ _____

8. $\frac{3}{4} \times \frac{1}{5} =$ _____

9. $\frac{2}{5} \times \frac{1}{7} =$ _____

10. $\frac{2}{7} \times \frac{5}{6} =$ _____

11. $\frac{1}{3} \times \frac{6}{7} =$ _____

12. $\frac{9}{10} \times \frac{3}{8} =$ _____

13. $\frac{4}{5} \times \frac{8}{9} =$ _____

14. $\frac{3}{16} \times \frac{2}{3} =$ _____

15. $\frac{2}{9} \times \frac{7}{3} =$ _____

16. $\frac{5}{6} \times \frac{11}{8} =$ _____

17. $\frac{4}{3} \times \frac{19}{16} =$ _____

18. $\frac{3}{25} \times \frac{43}{32} =$ _____

19. $\frac{15}{32} \times \frac{9}{16} =$ _____

20. $\frac{7}{64} \times \frac{5}{9} =$ _____

21. $\frac{1}{8} \times \frac{2}{3} \times \frac{3}{4} =$ _____

22. $\frac{2}{3} \times \frac{1}{15} \times \frac{21}{32} \times \frac{1}{4} =$ _____

23. $\frac{13}{16} \times \frac{3}{5} \times \frac{3}{8} \times \frac{2}{3} =$ _____

24. $\frac{27}{20} \times \frac{4}{3} \times \frac{1}{2} =$ _____

25. $\frac{120}{10} \times \frac{13}{16} \times \frac{2}{3} \times \frac{4}{10} =$ _____

26. $\frac{1}{5} \times \frac{1}{2} \times \frac{5}{6} =$ _____

27. $\frac{2}{3} \times \frac{3}{7} \times \frac{3}{8} =$ _____

28. $\frac{3}{5} \times \frac{1}{9} \times \frac{2}{3} =$ _____

29. $\frac{1}{4} \times \frac{3}{8} \times \frac{2}{5} \times \frac{1}{3} =$ _____

30. $\frac{1}{8} \times \frac{3}{10} \times \frac{5}{16} \times \frac{3}{4} =$ _____

31. $\frac{7}{8} \times \frac{1}{5} \times \frac{3}{4} \times \frac{5}{6} =$ _____

32. $\frac{15}{64} \times \frac{3}{4} \times \frac{7}{8} \times \frac{1}{16} =$ _____

33. $\frac{9}{8} \times \frac{13}{10} \times \frac{1}{3} =$ _____

34. $\frac{12}{9} \times \frac{7}{6} \times \frac{4}{3} =$ _____

35. $\frac{25}{16} \times \frac{5}{8} \times \frac{1}{4} \times \frac{2}{3} =$ _____

36. $\frac{10}{3} \times \frac{31}{32} \times \frac{4}{5} \times \frac{4}{50} =$ _____

Name _____ **Date** _____ **Class** _____

37. $2\frac{1}{3} \times 1\frac{1}{4} =$ _____

38. $5\frac{1}{2} \times 4\frac{5}{8} =$ _____

39. $31\frac{1}{4} \times 2\frac{5}{12} =$ _____

40. $4\frac{7}{8} \times 2\frac{1}{3} \times 4\frac{16}{21} =$ _____

41. $11\frac{1}{3} \times 3\frac{3}{8} \times 5\frac{5}{17} =$ _____

42. $3\frac{1}{2} \times 2\frac{3}{4} =$ _____

43. $1\frac{3}{4} \times 3\frac{2}{3} =$ _____

44. $4\frac{1}{5} \times 2\frac{1}{6} =$ _____

45. $3\frac{2}{3} \times 4\frac{5}{6} =$ _____

46. $5\frac{1}{5} \times 2\frac{1}{8} =$ _____

47. $8\frac{1}{12} \times 5\frac{1}{10} =$ _____

48. $7\frac{1}{7} \times 5\frac{8}{9} =$ _____

49. $11\frac{15}{16} \times 22\frac{3}{5} =$ _____

50. $37\frac{1}{32} \times 8\frac{3}{4} =$ _____

51. $3\frac{1}{8} \times 4\frac{1}{6} \times 2\frac{1}{5} =$ _____

52. $5\frac{5}{6} \times 4\frac{1}{3} \times 3\frac{3}{8} =$ _____

53. $2\frac{1}{6} \times 5\frac{4}{9} \times 6\frac{3}{16} =$ _____

54. $12\frac{1}{4} \times 2\frac{2}{7} \times 3\frac{1}{64} =$ _____

Applications

Solve the following problems. Reduce your answers to lowest terms. Show your work.

1. Eighteen pins must be cut from a rod. Each pin is to be 13/16" long, including the saw kerf. How long must the rod be?

_____

2. You are to make 14 nuts. Each nut is to be 3/8" thick and each nut requires an extra 1/16" for cutting. How long must a 5/8" hex bar be to make these nuts?

_____

3. How much stock is needed to make seven chisels that are 6 3/4" long? Allow a 1/16" saw kerf for each chisel.

_____

4. A power hacksaw blade has 12 teeth per inch. How many teeth are there in 6 1/4 inches of the blade?

_____

5. If washers come in boxes of 100 and 3 3/4 boxes are delivered to each of six workstations, what is the total number of boxes of washers delivered?

_____

6. A job order requests 24 threaded bolts be made, each 1 1/32" long. Ignoring the amount of material needed for saw kerf, how much material in total length will be cut?

_____

7. You are making 18 heavy-duty washers. Each washer is 15/64" thick and requires an extra 5/32" for cutting and facing to size. How long must the bar be to make the washers?

_____

Name _____   **Date** _____   **Class** _____

8. How much stock is required to make 14 pin punches that are 5 1/4" long? Allow 1/8" saw kerf for each punch.

    _____

9. What is the overall length of the part shown in the figure below if the center-to-center distance between each hole is 15/16" and the end holes are each 5/8" from the ends of the part?

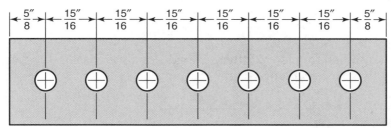

    Goodheart-Willcox Publisher

    _____

10. In question 9, if the center-to-center distance between holes is changed to 3/4", what is the new overall length of the part?

    _____

11. Three 17/32" blocks are stacked on top of one another. What is 1/2 the height of the stack?

    _____

12. If you normally work 8 hours per day and 1/10 of a day is spent in training and safety meetings each week, how many days are spent in these meetings in a 50-week year?

    _____

13. In a certain gear train, for every revolution of the input shaft the output shaft turns 2/3 revolution. How many revolutions will the output shaft turn in a minute if the input shaft turns 750 revolutions per minute?

_____

14. There are 28 castings weighing 2 1/3 pounds each on a pallet. What is the total weight on the pallet?

_____

15. A flat bar of titanium is 1/2″ thick, 1″ wide, and weighs 63/64 pound per foot. What will a bar 8 feet long weigh?

_____

16. A manufacturer is making 38 polypropylene unthreaded spacers with a diameter of 1/2″ and length of 2 1/4″. If 1/8″ is allowed for saw kerfs and facing operations, how much material is required total?

_____

17. If you accumulate 1/5 day of vacation time with each completed week of work, how many vacation days do you accrue after 50 weeks of work?

_____

18. If each cut creates saw kerf of 1/16 inch, how much material is wasted after making 453 cuts?

_____

19. A multispindle screw machine can make 180 parts per hour. If production works a shift of 7 1/3 hours, how many parts can be made per shift?

_____

UNIT 13

Dividing Fractions

Objectives

Information in this unit will enable you to:

- Explain how to divide proper fractions.
- Discuss the steps involved in dividing mixed numbers.

Dividing Proper Fractions

To divide fractions, the divisor (the number doing the dividing) is inverted. To **invert** something is to turn it upside down. To invert the divisor when dividing fractions, the numerator and denominator are swapped: 3/4 inverted is 4/3. With the divisor inverted, the terms (the two parts of the problem) are multiplied, the same way fractions were multiplied in Unit 12.

Example 13-1

Divide 5/8 by 2/3. 2/3 is the divisor, so invert it, and write the problem as a multiplication problem.

$$\frac{5}{8} \div \frac{2}{3}$$

$$\frac{5}{8} \times \frac{3}{2} = \frac{15}{16}$$

Example 13-2

Divide 3/4 by 2/5. The divisor is 2/5, so invert it, and write the problem as a multiplication problem.

$$\frac{3}{4} \div \frac{2}{5}$$

$$\frac{3}{4} \times \frac{5}{2} = \frac{15}{8}$$

$$\frac{15}{8} = 1\frac{7}{8}$$

Dividing Mixed Numbers

As with multiplication, mixed numbers must be converted to improper fractions before division can be done. With the problem written with proper and improper fractions, invert the divisor and change the division sign to a multiplication sign. If the result is an improper fraction, it should be converted to a mixed number. Express the answer in its lowest terms.

Example 13-3

$$3\frac{1}{4} \div \frac{2}{3} = \frac{13}{4} \div \frac{2}{3}$$

$$\frac{13}{4} \times \frac{3}{2} = \frac{39}{8}$$

$$\frac{39}{8} = 4\frac{7}{8}$$

Example 13-4

$$7\frac{1}{3} \div \frac{3}{4} = \frac{22}{3} \div \frac{3}{4}$$

$$\frac{22}{3} \times \frac{4}{3} = \frac{88}{9}$$

$$\frac{88}{9} = 9\frac{7}{9}$$

If either of the terms is a whole number, remember that a whole number can be written as a fraction with a denominator of 1.

Example 13-5

$$5 \div \frac{3}{5} = \frac{5}{1} \div \frac{3}{5}$$

$$\frac{5}{1} \times \frac{5}{3} = \frac{25}{3}$$

$$\frac{25}{3} = 8\frac{1}{3}$$

Unit 13 Review

Name _____ Date _____ Class _____

Fill in the blanks in the following review questions.

1. To divide fractions, the divisor (the number doing the dividing) is _____, which means it is turned upside down.

2. To invert the divisor when dividing fractions, the _____ and _____ are swapped.

3. With the _____ inverted, the terms (the two parts of the problem) are multiplied.

4. As with multiplication, mixed numbers must be converted to _____ fractions before division can be done.

5. If the result is a(n) _____ fraction, it should be converted to a mixed number. Express the answer in its _____ terms.

6. If either of the terms is a whole number, remember that a whole number can be written as a fraction with a denominator of _____.

Practice

Perform the indicated division. Show your work.

1. $\frac{3}{4} \div \frac{3}{8} =$ _____

2. $\frac{2}{3} \div \frac{1}{6} =$ _____

3. $\frac{5}{9} \div \frac{7}{8} =$ _____

4. $\frac{1}{10} \div \frac{3}{10} =$ _____

5. $\frac{21}{64} \div \frac{3}{4} =$ _____

6. $\frac{431}{600} \div \frac{1}{2} =$ _____

7. $\frac{1}{4} \div \frac{1}{2} =$ _____

8. $\frac{2}{3} \div \frac{1}{2} =$ _____

9. $\frac{1}{2} \div \frac{3}{4} =$ _____

10. $\frac{1}{2} \div \frac{1}{5} =$ _____

11. $\frac{1}{3} \div \frac{2}{5} =$ _____

12. $\frac{5}{6} \div \frac{3}{4} =$ _____

13. $\frac{3}{7} \div \frac{2}{3} =$ _____

14. $\frac{4}{5} \div \frac{5}{6} =$ _____

15. $\frac{3}{8} \div \frac{5}{7} =$ _____

16. $\frac{1}{9} \div \frac{5}{8} =$ _____

17. $\frac{6}{7} \div \frac{7}{10} =$ _____

18. $\frac{11}{12} \div \frac{2}{9} =$ _____

19. $\frac{5}{16} \div \frac{3}{8} =$ _____

20. $\frac{3}{32} \div \frac{1}{16} =$ _____

21. $\frac{15}{32} \div \frac{7}{8} =$ _____

22. $\frac{31}{32} \div \frac{7}{8} =$ _____

23. $\frac{3}{100} \div \frac{1}{3} =$ _____

24. $\frac{29}{64} \div \frac{9}{10} =$ _____

25. $3\frac{1}{2} \div \frac{2}{3} =$ _____

26. $1\frac{3}{4} \div \frac{5}{8} =$ _____

27. $3\frac{5}{16} \div 1\frac{1}{2} =$ _____

28. $9\frac{1}{8} \div 2\frac{1}{8} =$ _____

29. $7\frac{2}{5} \div 3 =$ _____

30. $4\frac{5}{8} \div 8 =$ _____

31. $2\frac{1}{4} \div \frac{5}{6} =$ _____

32. $1\frac{1}{3} \div \frac{3}{4} =$ _____

33. $5\frac{1}{6} \div \frac{3}{5} =$ _____

34. $4\frac{1}{3} \div \frac{2}{7} =$ _____

35. $3\frac{3}{4} \div \frac{2}{3} =$ _____

36. $5\frac{4}{5} \div \frac{7}{8} =$ _____

37. $4\frac{3}{8} \div 2\frac{3}{5} =$ _____

38. $1\frac{5}{8} \div 2\frac{5}{6} =$ _____

39. $6\frac{3}{32} \div 1\frac{9}{16} =$ _____

40. $7\frac{1}{12} \div 1\frac{1}{6} =$ _____

41. $10\frac{1}{16} \div 4\frac{3}{8} =$ _____

42. $3\frac{1}{9} \div 4\frac{1}{10} =$ _____

43. $6 \div 1\frac{11}{16} =$ _____

44. $8\frac{1}{32} \div 4 =$ _____

45. $11\frac{1}{8} \div 2\frac{1}{16} =$ _____

46. $13 \div 7\frac{1}{16} =$ _____

47. $9\frac{1}{8} \div 3 =$ _____

48. $3\frac{1}{16} \div 15 =$ _____

Name _____ **Date** _____ **Class** _____

Applications

Solve the following problems. Show your work.

1. A steel bar 9 1/3 feet long weighs 55 1/2 pounds. What does it weigh per foot?

    _____

2. How many 2 3/4″ dowels can be cut from a 24″ rod? (Hint: Nothing can be cut from any remainder of less than 2 3/4″.)

    _____

3. How many 2 1/4″ long pieces of ∅1/8″ drill rod can be cut from a piece 32 3/8″ long if 1/16″ is allowed for the kerf as each piece is cut?

    _____

4. A hexagonal titanium bar measuring 1 1/2 inches across flats and 6 1/3 feet long weighs 24 3/4 pounds. What does it weigh per foot?

    _____

5. How many 3 3/32″ pins can be cut from a 36 1/2″ rod? (Hint: Nothing can be cut from any remainder of less than 3 3/32″.)

    _____

Consider the illustration below to answer the following questions.

6. The pitch of a screw thread is the distance between corresponding points on adjacent threads. In the example below, there are 10 threads over a length of 1″. Therefore, the pitch is equal to 1/10″.

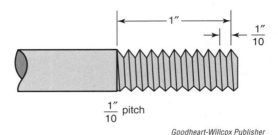

Goodheart-Willcox Publisher

A. If a threaded shaft has 36 threads in 3 inches, what is its pitch?

_____

B. A #10 socket head screw has 30 threads over a threaded length of 1 1/4″. What is the pitch of the threads?

_____

C. A Ø1/4″ alloy steel set screw has 35 threads over its threaded length of 1 3/4″. What is the pitch of the threads?

_____

D. A Ø7/8″ threaded rod 6″ in length has 54 threads. What is the pitch of the threads?

_____

E. A steel coupling nut has 20 threads tapped through its length of 5/8″. What is the pitch of the threads?

_____

F. If a threaded shaft 8 1/2″ long has 51 threads, what is its pitch?

_____

G. If the threaded portion of a spindle is 2 3/4″ long and it has 77 threads, what is its pitch?

_____

Name _____ **Date** _____ **Class** _____

7. How many revolutions will it take for a lathe tool to advance 8 5/8″ if the tool advances 1/50″ with each revolution?

   _____

8. The feed on a lathe is set for 1/32″ per revolution. How many revolutions will it take for the lathe tool to advance 8 5/8″?

   _____

9. A grinder polishes 22 1/2 square inches per minute. How long will it take to polish 125 square inches?

   _____

10. A box of reamers weighs 50 2/3 pounds. If each reamer weighs 6 1/3 pounds, how many reamers does the box contain?

   _____

11. A pallet of aluminum castings weighs 264 1/3 pounds. If each casting weighs 6 7/9 pounds, how many castings are on the pallet?

   _____

12. How many 1 1/3″ strips can be cut from a piece of shim stock 12″ wide?

   _____

13. How many parts can be manufactured from a 120″ length of bar stock if each part is 3 1/13″ long. Ignore saw kerfs.

   _____

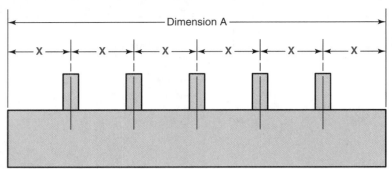

14. Five studs are pressed into a plate, as shown above. The holes are equally spaced, and the distance from the end of the plate to the center of the first stud is the same as the center-to-center distance between the studs (dimension X).

    A. What is the center-to-center distance between the studs (dimension X) if dimension A = 7 1/2"?

    _____

    B. If the engineer revises the part and changes dimension A to 12 2/3", what is the center-to-center distance between the studs (dimension X)?

    _____

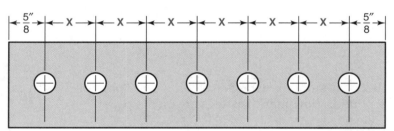

15. The overall length of the part shown above is 6 1/2". The center-to-center distance (dimension X) between each hole is equal.

    A. Calculate the center-to-center distance between each hole (dimension X).

    _____

    B. The part is redesigned and only six equally spaced holes are drilled instead of seven. What would the new center-to-center distance between holes be if the overall length of 6 1/2" and the 5/8" dimension from the edge to the end hole remain the same?

    _____

Decimal Fractions

Key Terms

decimal

decimal fraction

decimal point

nearer fractional equivalent

UNIT 14

Decimal System

Objectives

Information in this unit will enable you to:

- Explain what decimals are and how to use a decimal point.
- Describe how zeros are used in decimal numbers.
- Discuss why machinists use decimals.

The Decimal Point

The word **decimal** means based on 10. As was explained in Unit 1, our number system is based on tens. Ten is 10 times greater than 1 and 100 is 10 times greater than 10. The same is true going in the opposite direction. A **decimal point** is a period used to separate whole numbers from numbers less than 1. Previous units have discussed fractions, numbers less than 1. Numbers less than 1 can be **decimal fractions**, like 1/10, 1/100, and 1/1,000.

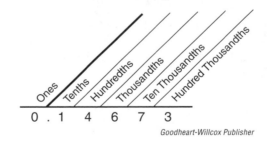

Goodheart-Willcox Publisher

Decimal fractions are most often referred to simply as "decimals." The number 0.245 is an example of a decimal. It is two hundred forty-five thousandths. Numbers that include digits on both sides of the decimal point are usually called decimal numbers. When the only places used are to the left of the decimal point (1 or greater), they are not typically called decimal numbers.

Zeros in Decimal Numbers

Zeros to the left of whole numbers don't have any effect on the value of the number. (033 and 33 are the same number.) Zeros between whole numbers and the decimal point, however, mark places and cannot be dropped. (330 is three hundred thirty. Without the 0, it would be thirty-three.) Zeros on the right side of the decimal point are generally significant. A 0 between the decimal point and the first digit on the right marks the tens place. (.033 is thirty-three thousandths. Without the 0, it would be .33 or thirty-three hundredths.) A 0 is often placed on the left of a decimal number, so there is less confusion about where the decimal point is to be placed, but that 0 does not affect the number (0.33 is still thirty-three hundredths.)

Zeros to the right of decimal numbers present a different case. A decimal number ending in 0 is the same value as it would be without the number. 0.330 is the same as 0.33, except that the 0 indicates a higher degree of precision. 0.33″ is the same distance as 0.330″, except the 0 at the end makes it three hundred thirty thousandths, instead of thirty-three hundredths.

In working with numbers, there are many times when a number is rounded off. For example, if you multiply 0.15×0.25, you will get 0.0375. You started with accuracy to the hundredths place, but your product is written to the ten thousandths place. If you are only working with hundredths, your product should be accurate to the hundredths place. 0.0375 rounded off to the hundredths place is 0.04. (Rounding off is covered in the next unit.) If you add a 0 to the end of your product, it signifies that the answer is accurate to the thousandths place.

Decimals in Use

Decimal fractions, or decimals, are widely used by machinists. Many of the work-pieces that machinists produce are manufactured with very tight tolerances. This requires dimensions that may be as precise as 0.0001″ (one ten thousandth of an inch) or even smaller. Using common fractions of an inch would mean fractions like 1/8,192″. Working with decimals is much simpler.

Our money system is another example of the use of the decimal system. One dime is 0.1 dollar and one penny is 0.01 dollar.

Shop Talk

Properly and quickly communicating values in the machine shop is a vital part of being a machinist. In manufacturing practice, the spoken unit is based on one thousandths of an inch, 0.001″. When communicating in the shop, all other numbers follow from this, breaking away from the usual technical terms of math.

For example, 0.346 inch is stated as "three hundred forty-six thousandths" in the shop. The same communication procedure extends to smaller numbers, where 0.0001 inch, one ten-thousandth, is stated as "one tenth" in the shop. When numbers get extremely small, the foundational unit may switch over to millionths of an inch, 0.000001 inch. This unit is common in surface roughness callouts.

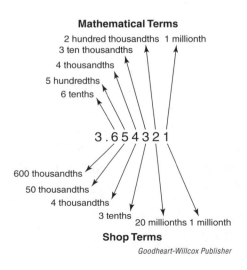

Goodheart-Willcox Publisher

Example 14-1

In math, everything is based on 1. In a machine shop, everything is based on 0.001″ (one thousandth of an inch). The following are examples of this difference.

0.1000
- In math, the technical term is "one tenth."
- In the shop, this is stated as "one hundred thousandths." (If everything is based on 0.001″, then there are 100 "thousandths" of an inch.)

0.0100
- In math, the technical term is "one hundredth."
- In the shop, this is stated as "ten thousandths." (If everything is based on 0.001″, then there are 10 "thousandths" of an inch.)

0.0010
- In math, the technical term is "one thousandth."
- In the shop, this is stated as "one thousandth" or "one thou." (Thousandths of an inch are often abbreviated as "thou" in the shop.)

0.0001
- In math, the technical term is "one ten-thousandth."
- In the shop, this is stated as "one tenth." (If everything is based on 0.001″, then there is one tenth of one thousandth of an inch.)

0.000010
- In math, the technical term is "one hundred-thousandth."
- In the shop, this is stated as "ten millionths" or "ten microinches." You may see this written as 10 μin.

0.000001
- In math, the technical term is "one millionth."
- In the shop, this is stated as "one millionth" or "one microinch." You may see this written as 1 μin.

Example 14-2

Following are some examples of how machinists communicate values in a machine shop. Remember that "thousandths" are also likely to be abbreviated as "thou" among machinists.

- **10.1234** = Ten inches, one hundred twenty-three thousandths, and four tenths (Alternatively: Ten inches, one hundred twenty-three thou, and four tenths)
- **6.8732** = Six inches, eight hundred seventy-three thousandths, and two tenths
- **0.264** = Two hundred sixty-four thousandths
- **0.0952** = Ninety-five thousandths and two tenths
- **0.04729** = Forty-seven thousandths, two tenths, and ninety millionths
- **1.632401** = One inch, six hundred thirty-two thousandths, four tenths, and one millionth

Unit 14 Review

Name _____   Date _____   Class _____

Fill in the blanks in the following review questions.

1. The word *decimal* means based on _____.

2. A _____ point is a period used to separate whole numbers from numbers less than 1.

3. Numbers less than 1 can be decimal _____, like 1/10, 1/100, and 1/1000.

4. Decimal fractions are most often referred to simply as _____.

5. Zeros to the left of whole numbers do not have any effect on the _____ of the number. Zeros on the right side of the decimal point are generally _____.

6. Decimal fractions, or decimals, are widely used by _____. This requires dimensions that may be as precise as _____.

Practice

Label the position of each digit in the following numbers according to their decimal place values. Use mathematical terms.

1. 0 . 2 5

2. 1 . 3 5

3. 0 . 4 7 8 5 6 3

4. 5 . 0 0 6 1 2 4 5

Label the position of each digit in the following numbers according to their decimal place values. Use shop terms.

5. 0 . 3 9 2 1

6. 0 . 4 7 8 5 6

7. 5.0 0 6 1 2 4

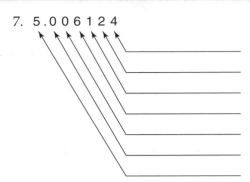

Write the decimal numbers in words. Use mathematical terms.

8. 0.12

_____

9. 0.558

_____

10. 0.43145

_____

11. 5.87

_____

12. 9.955

_____

13. 7.030

_____

14. 0.0023

_____

15. 1.240

_____

Write the decimals in words. Use shop terms.

16. 0.9

_____

17. 4.6

_____

18. 0.3

_____

19. 0.8

_____

20. 0.38

_____

Name _____   **Date** _____   **Class** _____

21. 5.09

_____

22. 0.16

_____

23. 0.63

_____

24. 4.185

_____

25. 0.461

_____

26. 0.902

_____

27. 1.284

_____

28. 3.4874

_____

29. 0.1087

_____

30. 2.4721

_____

31. 7.2069

_____

32. 0.67211

_____

33. 4.27492

_____

34. 0.38109

_____

35. 9.17238

_____

36. 0.412981

_____

37. 3.099186

_____

38. 7.714302

_____

39. 0.213507

_____

Write the following mathematical terms with numerals.

40. Six-tenths

_____

41. Seventy-seven hundredths

_____

42. Three hundred thousandths

_____

43. Fifteen and twenty-five thousandths

_____

44. Thirty and forty ten thousandths

_____

45. Twenty-five hundred thousandths

_____

46. Four and seventy-five hundredths

_____

47. Seven hundred fifty-two thousandths

_____

Write the following shop terms with numerals.

48. One hundred forty-nine thousandths

_____

49. Two thousandths

_____

50. Four hundred thousandths

_____

51. Eleven thou

_____

52. Nine tenths

_____

Name _____   **Date** _____   **Class** _____

53. Ninety-one thou

_____

54. Six thousandths and one tenth

_____

55. Three inches and eight tenths

_____

56. Eight inches and one thousandth

_____

57. Thirty-seven thousandths

_____

58. One inch, three hundred twenty-two thousandths

_____

59. Three inches, forty-three thou

_____

60. Six hundred thirty-four thousandths

_____

61. Ten inches, seventeen thousandths, and one tenth

_____

62. Nineteen thousandths and four tenths

_____

63. Ninety-eight thousandths and six tenths

_____

64. Four hundred eighty-nine thousandths and eight tenths

_____

65. Two inches, four hundred thousandths, and one tenth

_____

66. Three inches, twenty-seven thousandths, and nineteen millionths

_____

67. Five inches, one hundred thirty-six thousandths, five tenths, and two millionths

_____

68. Fourteen inches, twenty-one thousandths, three tenths, and thirty-four millionths

    _____

69. Twenty inches, nine hundred thousandths, and nine millionths

    _____

70. One tenth and twelve millionths

    _____

71. Five inches, two hundred seventeen thousandths, two tenths, and thirty-six millionths

    _____

72. Seven inches, one hundred twenty-five thousandths, and six millionths

    _____

73. One inch, eighty-one thousandths, three tenths, and forty-four millionths

    _____

74. Sixty-four microinches

    _____

75. Eight microinches

    _____

76. Sixteen microinches

    _____

77. Four microinches

    _____

UNIT 15

Converting between Common Fractions and Decimal Fractions

Objectives

Information in this unit will enable you to:

- Explain how to convert decimal fractions to common fractions.
- Describe how to convert common fractions to decimal fractions.
- Illustrate how to round decimals and explain why it might be necessary to do so.
- Explain how converting a decimal to a nearer fractional equivalent differs from converting a decimal to a common fraction.

Converting Decimal Fractions to Common Fractions

A decimal fraction can be converted to a common fraction by using the digits of the decimal fraction as the numerator and the last place value of the decimal fraction as the denominator.

Example 15-1

Convert 0.24 to a common fraction. The digits of the decimal are 24, so that becomes the numerator. The last place value of the decimal is hundredths, so the denominator is 100.

$$0.24 = \frac{24}{100}$$

Whenever a decimal is converted to a common fraction, the denominator will always have 10 as the base unit.

Example 15-2

$$0.5 = \frac{5}{10}$$

$$0.48 = \frac{48}{100}$$

$$0.937 = \frac{937}{1,000}$$

Converting Common Fractions to Decimal Fractions

To convert a common fraction to a decimal fraction, divide the numerator by the denominator.

Example 15-3

$$\frac{4}{5} = 5\overline{)4}$$

Place a decimal point to the right of the dividend and another decimal point directly above the first, where the quotient will be written. The second decimal point will help keep the decimal point in the quotient (the answer) in the proper place.

$$5\overline{)4.}\overset{\textstyle .}{}$$

Add a 0 to the right of the decimal point in the dividend and do the division.

$$5\overline{)4.0}\overset{\textstyle .8}{}$$

$$\frac{4}{5} = .8 \text{ or } 0.8$$

Continue adding 0s to the right of the decimal point of the dividend as necessary to complete the division to the degree of accuracy required.

Rounding Decimals

Some common fractions produce decimals with many more places than the degree of accuracy suggested by the common fraction. For example, 13/16 yields a decimal fraction of 0.8125. That is accurate to the ten thousandths place, while the original fraction was only accurate to sixteenths. Such results are usually rounded off to the appropriate degree of accuracy, depending on the application. Delete all of the numerals beyond the degree of accuracy needed. If the digit immediately to the right of the degree of accuracy required is 5 or more, increase the last digit that will be used by 1. If the digit immediately to the right of the required degree of accuracy is 4 or less, the last digit remains unchanged and remaining numbers are discarded.

Example 15-4

Round 0.8125 to the nearest hundredth.

> The digit in the hundredths place is 1. The next digit to the right is 2, so simply discard the 2 and the 5 (remember the rule for 4 or less). The 1 remains unchanged, and the rounded answer is 0.81.

Round 0.625 to the nearest hundredth.

> The digit in the hundredths place is 2. The next digit is 5, so increase the 2 to 3. The rounded answer is 0.63.

Some common fractions convert to unending decimals. These also should be rounded to the appropriate degree of accuracy.

Example 15-5

Convert 2/3 to a decimal fraction.

$$
\begin{array}{r}
.6666 \\
3\overline{)2.0000} \\
-18 \\
\hline
20 \\
-18 \\
\hline
20 \\
-18 \\
\hline
20 \\
-18 \\
\hline
2
\end{array}
$$

To round this number to the nearest thousandth, delete the last 6 and add 1 to the 6 in the thousandths place. The rounded answer is 0.667.

Nearer Fractional Equivalent

A chart of fractional equivalents can be used when it is necessary to convert a decimal to its **nearer fractional equivalent**, which is the common fractional equivalent that most closely matches the decimal fraction. The nearer fractional equivalent is not the same as converting the decimal to a common fraction. In converting the decimal to a common fraction, the decimal is mathematically converted to a common fraction of exactly the same value. When a machinist refers to the nearer fractional equivalent, it means the fraction that might be read from a tape measure that is closest to the decimal number.

To use the chart, find the common fractions that are "just less than" and "just greater than" the decimal fraction. Choose the line on the chart that is closest to the fraction.

Decimal Conversion Chart

Fraction		Inches	mm	Fraction		Inches	mm
	1/64	.01563	.397		33/64	.51563	13.097
1/32		.03125	.794	17/32		.53125	13.494
	3/64	.04688	1.191		35/64	.54688	13.891
1/16		.0625	1.588	9/16		.56250	14.288
	5/64	.07813	1.984		37/64	.57813	14.684
3/32		.09375	2.381	19/32		.59375	15.081
	7/64	.10938	2.778		39/64	.60938	15.478
1/8		.12500	3.175	5/8		.62500	15.875
	9/64	.14063	3.572		41/64	.64063	16.272
5/32		.15625	3.969	21/32		.65625	16.669
	11/64	.17188	4.366		43/64	.67188	17.066
3/16		.18750	4.763	11/16		.68750	17.463
	13/64	.20313	5.159		45/64	.70313	17.859
7/32		.21875	5.556	23/32		.71875	18.256
	15/64	.23438	5.953		47/64	.73438	18.653
1/4		.25000	6.350	3/4		.75000	19.050
	17/64	.26563	6.747		49/64	.76563	19.447
9/32		.28125	7.144	25/32		.78125	19.844
	19/64	.29688	7.541		51/64	.79688	20.241
5/16		.31250	7.938	13/16		.81250	20.638
	21/64	.32813	8.334		53/64	.82813	21.034
11/32		.34375	8.731	27/32		.84375	21.431
	23/64	.35938	9.128		55/64	.85938	21.828
3/8		.37500	9.525	7/8		.87500	22.225
	25/64	.39063	9.922		57/64	.89063	22.622
13/32		.40625	10.319	29/32		.90625	23.019
	27/64	.42188	10.716		59/64	.92188	23.416
7/16		.43750	11.113	15/16		.93750	23.813
	29/64	.45313	11.509		61/64	.95313	24.209
15/32		.46875	11.906	31/32		.96875	24.606
	31/64	.48438	12.303		63/64	.98438	25.003
1/2		.50000	12.700	1		1.00000	25.400

Goodheart-Willcox Publisher

Unit 15 Review

Name _____ Date _____ Class _____

Fill in the blanks in the following review questions.

1. A decimal fraction can be converted to a common fraction by using the digits of the decimal fraction as the _____ and the last place value of the decimal fraction as the _____.

2. Whenever a decimal is converted to a common fraction, the _____ will always have 10 as the base unit.

3. To convert a common fraction to a _____ fraction, divide the numerator by the denominator.

4. Some common fractions produce decimals with many more places than the _____ of _____ suggested by the common fraction.

5. Some common fractions convert to _____ decimals, which should be _____ to the appropriate degree of accuracy.

6. A chart of fractional equivalents can be used when it is necessary to convert a decimal to its _____ fractional equivalent. This is not the same as converting the decimal to a common fraction.

Practice

Convert the decimals to common fractions.

1. 0.25 _____
2. 0.3333 _____
3. 0.875 _____

4. 0.359 _____
5. 0.004 _____
6. 2.45 _____

7. 8.600 _____
8. 0.12 _____
9. 0.47 _____

10. 0.06 _____
11. 0.125 _____
12. 0.214 _____

13. 0.4165 _____
14. 0.3125 _____
15. 0.010 _____

16. 0.375 _____
17. 0.5126 _____
18. 0.075 _____

19. 1.128 _____
20. 0.4375 _____
21. 0.0625 _____

22. 0.64 _____

23. 3.1 _____

24. 1.36 _____

25. 4.184 _____

26. 6.325 _____

27. 2.55 _____

28. 0.1875 _____

29. 3.254 _____

30. 0.2824 _____

Convert the common fractions to decimals.

31. $\frac{3}{4}$ _____

32. $\frac{3}{8}$ _____

33. $\frac{5}{25}$ _____

34. $\frac{7}{16}$ _____

35. $\frac{5}{40}$ _____

36. $2\frac{1}{3}$ _____

37. $4\frac{4}{5}$ _____

38. $\frac{1}{2}$ _____

39. $\frac{1}{3}$ _____

40. $\frac{1}{5}$ _____

41. $\frac{5}{8}$ _____

42. $\frac{2}{3}$ _____

43. $\frac{7}{8}$ _____

44. $\frac{4}{10}$ _____

45. $\frac{2}{7}$ _____

46. $\frac{9}{25}$ _____

47. $\frac{9}{16}$ _____

48. $\frac{1}{32}$ _____

49. $\frac{3}{64}$ _____

50. $\frac{7}{20}$ _____

51. $\frac{3}{10}$ _____

52. $\frac{9}{40}$ _____

53. $1\frac{1}{4}$ _____

54. $3\frac{1}{2}$ _____

55. $5\frac{2}{3}$ _____

56. $2\frac{1}{8}$ _____

57. $6\frac{3}{4}$ _____

58. $7\frac{3}{5}$ _____

59. $3\frac{1}{4}$ _____

60. $8\frac{97}{100}$ _____

Name _____   **Date** _____   **Class** _____

Convert the common fractions and round the results to the indicated degree of accuracy based on mathematical terms.

61. $\frac{1}{2}$ to the nearest tenth _____

62. $\frac{7}{8}$ to the nearest hundredth _____

63. $\frac{7}{9}$ to the nearest hundredth _____

64. $\frac{5}{12}$ to the nearest thousandth _____

65. $\frac{15}{27}$ to the nearest hundredth _____

66. $\frac{5}{6}$ to the nearest tenth _____

67. $4\frac{1}{3}$ to the nearest tenth _____

68. $2\frac{3}{5}$ to the nearest hundredth _____

69. $1\frac{5}{9}$ to the nearest hundred thousandth _____

70. $6\frac{1}{14}$ to the nearest tenth _____

71. $\frac{22}{7}$ to the nearest hundredth _____

Convert the common fractions and round the results to the indicated degree of accuracy based on shop terms.

72. $\frac{1}{3}$ to the nearest thousandth _____

73. $\frac{1}{4}$ to the nearest tenth _____

74. $\frac{3}{8}$ to the nearest hundredth _____

75. $\frac{3}{16}$ to the nearest tenth _____

76. $\frac{1}{32}$ to the nearest ten thousandth _____

77. $\frac{1}{16}$ to the nearest hundredth _____

78. $2\frac{1}{4}$ to the nearest tenth _____

79. $3\frac{2}{3}$ to the nearest hundredth _____

80. $4\frac{1}{8}$ to the nearest tenth _____

81. $1\frac{3}{25}$ to the nearest tenth _____

82. $2\frac{3}{64}$ to the nearest ten thousandth _____

83. $5\frac{3}{32}$ to the nearest ten thousandth _____

84. $\frac{14}{3}$ to the nearest ten thousandth _____

Use the decimal conversion chart on page 138 to find the nearer fractional equivalent of the decimals.

85. .3152 _____   86. .62543 _____   87. .92210 _____

88. 2.23444 _____   89. 6.43688 _____   90. .8912 _____

91. 0.44 _____   92. 0.412 _____   93. 1.0149 _____

94. 5.2781 _____   95. 4.4911 _____   96. 8.89125 _____

UNIT 16

Adding Decimal Fractions

Objectives

Information in this unit will enable you to:

- Demonstrate how to add decimals.
- Explain how to check whether you have added decimals correctly.

Learning the Principles

To add decimals, write them one above the other with their decimal points lined up vertically. Place the decimal point for the sum directly below the line of decimal points in the numbers. Add the numbers the same as you would for whole numbers.

Example 16-1

$$\begin{array}{r} 4.53 \\ 7.31 \\ + 2.55 \\ \hline 14.39 \end{array}$$

If some of the numbers to be added have different decimal places, 0s can be added to the right to fill in the spaces and make the addition easier to organize.

Example 16-2

$$\begin{array}{r} 2.557 \\ 3.4\textit{00} \\ + 1.25\textit{0} \\ \hline 7.207 \end{array}$$

Adding 0s to the right of some numbers should not yield a sum with more decimal places than the greatest number of decimal places found in the numbers being added. Remember, 0s at the end of a decimal suggests an increased degree of accuracy.

Work Space/Notes

Unit 16 Review

Name _____ Date _____ Class _____

Fill in the blanks in the following review questions.

1. To add decimals, write them one above the other with their decimal points _____ up vertically.

2. Place the decimal point for the sum _____ below the line of decimal points in the numbers, and add the numbers the same as you would _____ numbers.

3. If numbers to be added have _____ decimal places, 0s can be added to the _____ to fill in the spaces and make the addition easier to organize.

4. Adding 0s should not yield a sum with more _____ _____ than the greatest number of decimal places found in the numbers being added.

Practice

Add the following numbers and check the results.

1.	0.52 + 0.25	2.	2.75 + 1.41	3.	2.04 2.45 + 6.1

1. 0.52
   + 0.25

2. 2.75
   + 1.41

3. 2.04
   2.45
   + 6.1

4. 0.33
   + 0.75

5. 0.42
   + 0.17

6. 0.84
   + 0.09

7. 1.62
   + 3.58

8. 2.19
   + 4.92

9. 0.57
   + 5.46

10. 1.0005
    + 0.9442

11. 4.0156
    + 2.0313

12. 5.1094
    + 3.672

13. 7.2
    8.2656
    + 1.625

14. 3.3438
    6.25
    + 1.625

15. 10.0938
    2.5
    + 6.9531

16. 19.05
    1.0625
    + 11.9844

17. 13.875
    17.78
    + 4.2031

18. 21.84
    9.6875
    + 5.422

19. 3.050 plus 0.03 _____

20. 65.67 plus 3.1417 _____

21. 550.25 plus 14.025 _____

22. 2.54 plus 4.06 plus 6.1 _____

23. 8.89 plus 10.16 plus 11.43 _____

24. 1.0156 plus 17.02 _____

25. 25.15 plus 23.6563 _____

26. 71.7969 plus 54.5313 _____

27. 148.813 plus 207.2362 _____

28. 314.24 plus 187.0625 _____

Add the following numbers, which are presented using shop terms, and check the results.

29. One inch, two hundred twenty-six thousandths *plus* one hundred thirty-three thousandths = _____

30. Two inches, four hundred seventy-five thousandths *plus* one inch, one hundred forty-one thousandths = _____

31. Four inches, six hundred fifty-one thou *plus* three inches, eight hundred ninety-one thou = _____

32. Seven inches, five hundred thirty-four thou *plus* one inch, nine hundred sixty-two thou = _____

33. Ten inches, three hundred forty-two thousandths, and three tenths *plus* six inches, three hundred eighty-nine thousandths, and five tenths = _____

34. Thirteen inches, four hundred twenty-eight thousandths, and six tenths *plus* one inch, five hundred seventy-three thousandths, and two tenths = _____

35. One inch, one hundred fifty-five thou, three tenths, and one millionth *plus* four inches, seven hundred sixty-five thou, six tenths, and eight millionths = _____

Name _____   **Date** _____   **Class** _____

Applications

Solve the following problems.

Day	Week 1	Week 2	Week 3	Week 4	Total
Monday	$129.50	$64.18	$131.12	$140.12	
Tuesday	$152.63	$147.29	$128.57	$137.65	
Wednesday	$143.38	$151.82	$145.91	$141.81	
Thursday	$134.13	$131.34	$150.45	$149.27	
Friday	$147.54	$139.03	$73.79	$130.76	
Total	Wk 1 Total $	Wk 2 Total $	Wk 3 Total $	Wk 4 Total $	Total $

Goodheart-Willcox Publisher

1. If you are paid the amounts shown in the table for four weeks of work, what are you paid in total for the following?

   A. Week 1: _____

   B. Week 2: _____

   C. Week 3: _____

   D. Week 4: _____

   E. Mondays: _____

   F. Tuesdays: _____

   G. Wednesdays: _____

   H. Thursdays: _____

   I. Fridays: _____

   J. Total (four weeks): _____

2. How long a piece of bar stock will be required to cut the following pieces with no allowances for kerfs: 4.375", 6.625", 4.50", and 3.667"?

   _____

3. How long must the bar stock be in question 2 if 0.09" is added to each piece for saw kerfs?

   _____

4. An aluminum flat bar is cut into five pieces. The pieces measure 1.969 inches, 1.141 inches, 2.328 inches, 4.25 inches, and 8.875 inches in length. How long was the bar? Ignore kerfs.

_____

5. If 0.125″ is added to each piece for saw kerfs, how long was the bar in the above question?

_____

6. Refer to the following drawing of a part to calculate the missing dimensions in the questions that follow.

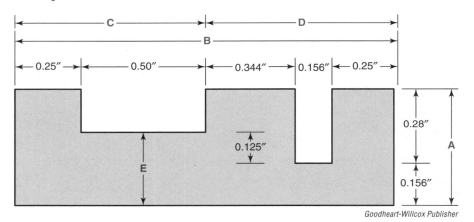

Goodheart-Willcox Publisher

Dimension A: _____

Dimension B: _____

Dimension C: _____

Dimension D: _____

Dimension E: _____

Name _____   **Date** _____   **Class** _____

7. A casting is finished by first removing 0.625″ with a hacksaw, then 0.15″ on the milling machine, and finally another 0.035″ on the grinder. How much material in total is removed in the finishing process?

_____

Refer to the axle assembly shown below for questions 8 through 10.

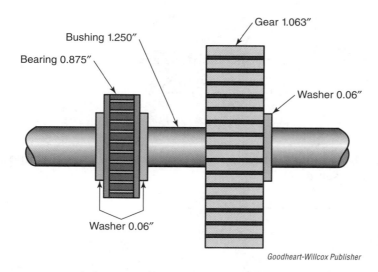

Goodheart-Willcox Publisher

8. How much axle space is taken by the bearing assembly (the bearing and the washer on each side of the bearing)?

_____

9. What is the total amount of the axle that is covered by the parts shown?

_____

10. An engineering revision changes the thickness of the washers from 0.06″ to 0.0938″. How much space is now taken up by the bearing assembly?

_____

Use this table showing the cost of the parts for the axle assembly to answer questions 11 through 13.

Part Name	Current Price	In-House Price	Cheapest Price
Axle shaft	$16.35	$15.95	
Gear	$28.65	$30.15	
Bearing	$8.42	$9.50	
Bushing	$2.60	$1.89	
Washers	$0.12 each	$0.09 each	
Total			

Goodheart-Willcox Publisher

11. What is the total cost of the assembly using current pricing?

   _____

12. If the parts were made in-house instead of purchased, what would be the total cost of the assembly?

   _____

13. If the cheapest possible price for each part were used, what would be the lowest cost of the assembly?

   _____

14. What is the overall height of the machined casting shown in the part drawing below?

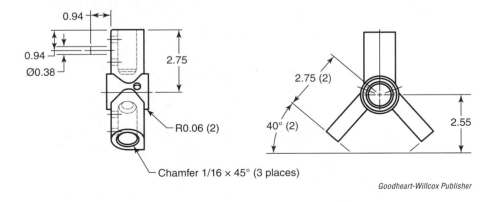

Goodheart-Willcox Publisher

   _____

Name _____   **Date** _____   **Class** _____

15. What is the outside diameter of a bushing that has an inside diameter of 1.44″ and a wall thickness of 0.295″?

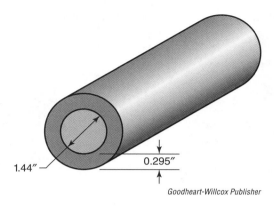

1.44″

0.295″

Goodheart-Willcox Publisher

_____

16. Gage blocks are so precisely ground they can be slid together and adhere to one another as if glued. This process is called wringing. What is the height of the following stacks given the heights of the individual gage blocks?

A. .1006″, .108″, and .09375″

_____

B. .1002″, .147″, and .45″

_____

C. .1008″, .6″, and .35″

_____

D. .1006″, .105″, .150″, and .2″

_____

E. .1001″, .121″, and .142″

_____

F. .1003″, .137″, and .55″

_____

Work Space/Notes

UNIT 17

Subtracting Decimal Fractions

Objectives

Information in this unit will enable you to:

- Demonstrate how to subtract decimals.
- Explain how to check whether you have subtracted decimals correctly.

Learning the Principles

Subtracting decimals is similar to subtracting whole numbers. Write the numbers in a column, with the decimal points lined up. Put a decimal point in the difference directly below the column of decimal points above, then subtract.

Example 17-1

$$
\begin{array}{r}
8.54 \\
- 2.53 \\
\hline
6.01
\end{array}
$$

As in addition of decimals, 0s can be added to the right, so that both numbers have the same number of decimal places.

Example 17-2

Subtract 5.5 from 13.434.

$$
\begin{array}{r}
\overset{0\ 12}{1\cancel{3}.434} \\
- \ 5.500 \\
\hline
7.934
\end{array}
$$

Work Space/Notes

Unit 17 Review

Fill in the blanks in the following review questions.

1. Subtracting decimals is similar to _____ whole numbers.

2. Write the numbers in a column with the decimal points _____ up.

3. Put a _____ point in the difference _____ below the column of decimal points above, then subtract.

4. As with the addition of decimals, 0s can be added to the _____, so that both numbers have the same number of decimal _____.

Practice

Subtract the following numbers and check your answers.

1.　0.52
　 − 0.25

2.　2.75
　 − 1.41

3.　3.625
　 − 2.44

4.　2.045
　 − 0.65

5.　11.5
　 − 8.37

6.　0.76
　 − 0.18

7.　1.02
　 − 0.15

8.　1.78
　 − 0.078

9.　3.56
　 − 1.52

10.　18.03
　 − 4.32

11.　12.125
　 − 3.1094

12.　5.2031
　 − 2.0938

13.　4.688
　 − 1.2656

14.　6.375
　 − 4.4375

15.　7.75
　 − 0.9375

16.　10.1406
　 − 5.8438

17.　9.3125
　 − 0.188

18.　22.86
　 − 15.2413

19.　50.8281
　 − 24.1306

20.　12.6164
　 − 11.1132

21.　71.354
　 − 12.051

22. 14.450 minus 9.75

_____

23. 11.5 minus 8.875

_____

24. 10.562 minus 3.7501

_____

25. 18.4303 minus 12.8612

_____

26. 0.875 minus 0.1875

_____

27. 6.912 minus 4.8294

_____

28. 101.675 minus 27.962

_____

29. 169.8908 minus 38.9752

_____

30. 327.0719 minus 109.1892

_____

31. 297.5327 minus 99.8719

_____

32. 438.219 minus 197.8294

_____

33. 92.2351 minus 19.7609

_____

Subtract the following numbers, which are given in shop terms, and check your answers.

34. Two hundred seventy thousandths from five inches, four hundred fifty thousandths

_____

35. Three inches, four hundred thousandths from twelve inches, six hundred twenty-five thousandths

_____

Name _____   **Date** _____   **Class** _____

36. Three inches, one hundred fifty thousandths from eight inches, five thousandths

   _____

37. One inch, three hundred seventy-four thou from six inches, five hundred nineteen thou

   _____

38. Four inches, six hundred eighty-nine thou, and five tenths from nine inches, one hundred thirty-four thou, and seven tenths

   _____

39. Two inches, four hundred twenty-seven thou, and two tenths from eight inches, two hundred ninety-two thou, and three tenths

   _____

40. Seven inches, eight hundred thirty-three thou, four tenths, and thirteen millionths from thirteen inches, four hundred fifty-one thou, four tenths, and sixty-four millionths

   _____

Applications

Solve the following problems.

1. What remains if 1.125″ is cut off a round bar that is 19.5″ long?

   _____

2. A one-inch-diameter steel rod is forty-eight inches long and a piece measuring twenty-two inches, eight hundred seventy-five thousandths in length (in shop terms) is cut off. How much material remains? (Ignore kerfs.)

   _____

3. If an additional piece measuring fourteen inches, two hundred fifty thou (in shop terms) is cut from the rod in question 2, how much material is left?

_____

4. A 3.75" bolt is screwed all the way through a .375" thick nut. How much of the bolt is left exposed?

_____

5. A washer is now added to the assembly of the bolt from question 4. If the thickness of the washer is two hundred eighteen thou and eight tenths (in shop terms), how much of the bolt is now exposed?

_____

6. What is your take-home pay if you are paid $750.45 and $186.14 is withheld for taxes and insurance?

_____

Use the figure to calculate the missing dimensions and answer questions 7 and 8.

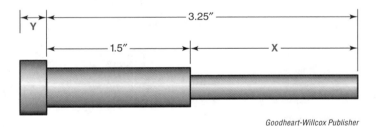

Goodheart-Willcox Publisher

7. What is dimension X in the figure?

_____

Name _____ **Date** _____ **Class** _____

8. If the overall length of the pin is five inches, four hundred twenty-one thou, and nine tenths (in shop terms), what is dimension Y?

_____

9. How thick is the bottom of the plate (dimension A) after the channel is milled in the figure below?

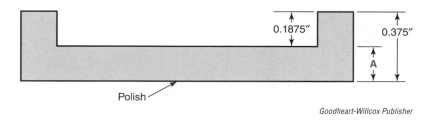

Goodheart-Willcox Publisher

_____

10. Considering question 9, if the material was 0.500″ thick before being machined to 0.375″, how much material was left as the finish allowance?

_____

11. How thick is the bottom of the plate (dimension A) in the part featured in question 9 after 0.0003″ is removed in polishing?

_____

12. What is the inside diameter of a bushing that has an outside diameter of 2.125" and a wall thickness of 0.3625"?

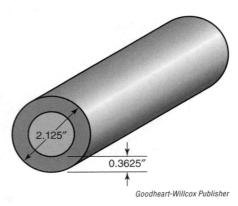

2.125"

0.3625"

Goodheart-Willcox Publisher

_____

13. If a shaft with a diameter of 1.375" is inserted into the bushing above, how much clearance will there be (difference between the diameter of the shaft and the inside diameter of the bushing)?

_____

14. If pieces 2.25" long, 2.875" long, and 6.39" long are cut from a 21"rod, how much remains of the rod?

_____

15. If 0.125" is allowed for a saw kerf on the pieces in question 14, what remains of the rod?

_____

16. A lathe removes 0.1875" of material from a diameter of 1". What size should the outside micrometer reading be after the roughing cut?

_____

Name _____   **Date** _____   **Class** _____

17. If a further forty-six thousandths and nine tenths, in shop terms, is removed from the part in question 16, what should the diameter measure after the finish cut?

_____

18. If a steel plate one inch, eight hundred twelve thou, and five tenths thick needs fifteen thou and six tenths surface ground, what is the thickness of the finished part?

_____

19. If you buy a digital caliper for $124.95 and the additional sales tax amounts to $4.99, how much change will you have left from $200?

_____

20. If you pay an additional $12.79 for express shipping of the digital caliper in question 19, how much do you have left from the $200 now?

_____

21. If you have an outside micrometer reading of (in shop terms) one inch, three hundred ninety-six thou, and the part requires a finished diameter of one inch, three hundred seventy-five thou, how much needs to be removed from the diameter to get to the required size?

_____

Use this drawing to answer problems 22 through 25.

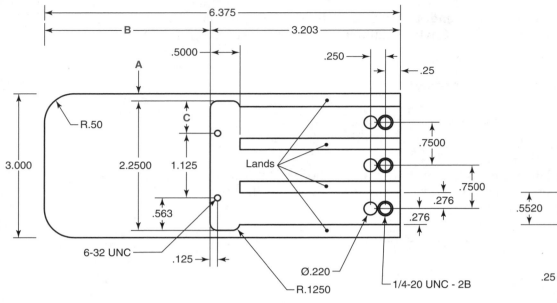
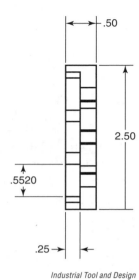

Industrial Tool and Design

22. What is the length of the two center lands?

_____

23. What is dimension A?

_____

24. What is dimension B?

_____

25. What is dimension C?

_____

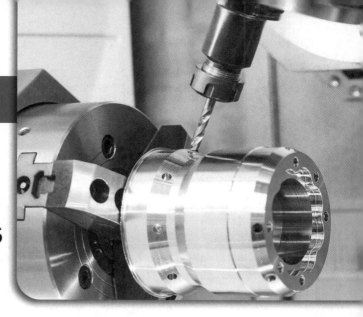

UNIT 18

Multiplying Decimal Fractions

Objectives

Information in this unit will enable you to:

- Demonstrate how to multiply decimals.
- Describe how to easily multiply decimals by 10, 100, and 1,000.

Learning the Principles

Multiplying decimal fractions is very much like multiplying whole numbers, with particular attention paid to place values and placement of the decimal point. Write the numbers above one another, putting the number with the fewest digits on the multiplier line. Start multiplication with the farthest digit to the right in the multiplier. When the digits have been multiplied, the decimal point is placed in the product. In the product line (the answer line), count the total number of digits to the right of the decimal point in the multiplier and the multiplicand and insert a decimal point at that position.

Example 18-1

$$
\begin{array}{r}
4.25 \\
\times \quad .3 \\
\hline
1.275
\end{array}
$$

There are two decimal places in the multiplicand and one in the multiplier in the example above, so in this case the decimal point in the product is placed so that there are three decimal places to the right of the decimal point.

If there is more than one digit in the multiplier, do additional rows of multiplication, the same as you would for whole numbers.

$$
\begin{array}{r}
12.6275 \\
\times \quad 0.75 \\
\hline
631375 \\
+ \ 8839250 \\
\hline
9.470625
\end{array}
$$

There are two decimal places in the multiplier and four in the multiplicand, so the decimal point in the product is placed with six places to its right. Any 0s to the far right could be dropped without changing the value of the product, but the appropriate number of 0s should be left to represent the degree of accuracy needed in the answer.

Multiplying by 10, 100, 1,000, etc.

Multiplying by 10, 100, 1,000, and so forth can be done by simply moving the decimal point to the right. To multiply by 10, move the decimal point one place; to multiply by 100, move the decimal point two places; and to multiply by 1,000, move the decimal point three places.

$$4.537 \times 10 = 45.37$$
$$4.537 \times 100 = 453.7$$
$$4.537 \times 1,000 = 4,537$$

Unit 18 Review

Name _____ Date _____ Class _____

Fill in the blanks in the following review questions.

1. Multiplying decimal fractions is very much like _____ whole numbers, with particular attention paid to place values and placement of the _____ point.

2. Write the numbers above one another, putting the number with the _____ digits on the multiplier line.

3. Start multiplication with the _____ digit to the right in the multiplier. When the digits have been multiplied, the decimal point is placed in the _____.

4. Count the total number of digits to the right of the decimal point in the _____ and multiplicand and insert a decimal point at that _____ in the product line (answer line).

5. If there is more than one _____ in the multiplier, do additional rows of multiplication, the same as you would for whole numbers.

6. To multiply by 10, move the decimal point _____ place; to multiply by 100, move the decimal point _____ places; and to multiply by 1,000, move the decimal point _____ places.

Practice

Multiply the following numbers.

1. .45
    × .5

2. .33
    × .60

3. .875
    ×   3

4. 4.375
    × .25

5. 5.6250
    ×   1.5

6. 0.51
    × 0.25

7. 0.76
    × 0.1

8. 0.84
    × 0.18

9. 0.25
    × 0.2

10. 0.85
     × 0.45

11. 1.25
     × 0.33

12. 0.125
     ×   5

13.   1.5
    $\times$ 8

14.   3.25
    $\times\,0.66$

15.   5.325
    $\times$ 7

16.   2.675
    $\times$ 0.3

17.   7.75
    $\times\,0.38$

18.   8.18
    $\times$ 2.1

19.   0.08
    $\times\,18.5$

20.   14.46
    $\times\,0.54$

21.   0.672
    $\times$ 9

22.   0.329
    $\times$ 0.5

23.   0.922
    $\times$ 10

24.   0.016
    $\times\,0.36$

25.   1.2969
    $\times$ 1000

26.   6.35
    $\times\,100$

27.   0.455
    $\times$ 4

28. $0.875 \times 0.33 =$ _____

29. $0.01 \times 3.25 =$ _____

30. $0.0003 \times 0.94 =$ _____

31. $0.875 \times 0.33 =$ _____

32. $0.6 \times 0.6 =$ _____

33. $0.04 \times 8.1 =$ _____

34. $0.0008 \times 1.25 =$ _____

35. $0.064 \times 0.6 =$ _____

Name _____   **Date** _____   **Class** _____

Multiply the following decimal numbers, which are presented in shop terms.

36. Eight hundred thou *times* one hundred thou = _____

37. Six hundred thou *times* two hundred thou = _____

38. Nine inches, four hundred thou *times* two inches, nine hundred
     thou = _____

39. One inch, six hundred twenty thou *times* three inches, seven hundred thirty
     thou = _____

40. Five inches, two hundred forty thou *times* three inches, four hundred fifty
     thou = _____

41. Three inches, eight hundred seventy-six thou *times* one inch, twenty-seven
     thou = _____

42. Four inches, ninety thou *times* three inches, two hundred twelve
     thou = _____

43. Seven inches, one hundred thirty-three thou *times* six hundred fifty-eight
     thou = _____

44. Eleven inches, nine hundred eighty-four thou, and six tenths *times* forty-six
     thou and two tenths = _____

45. Twenty inches, nine hundred thou *times* one inch, five hundred thirty-three
     thou, and one tenth = _____

46. Three inches, one hundred forty-two thou *times* one hundred seven
     thou = _____

Applications

Solve the following problems.

1. If three shims that are each .004″ thick are used on a gear shaft, what is the total thickness of the shims?

   _____

2. If four pieces of stainless steel shim stock each 0.006″ thick are required to shim a lathe tool, what is the total thickness of the shims?

   _____

3. A lathe removes .15″ in each pass. How much material is removed in four passes?

   _____

4. If a lathe tool removes 0.1875″ with each pass, how much material is removed in three passes?

   _____

5. A polishing operation covers 12.5 square inches per minute. How many square inches can be polished in 10.25 minutes?

   _____

6. How much stock is required to make nine pins that are each 6.125″ long? Allow 0.125″ saw kerf for each pin.

   _____

Name _____ **Date** _____ **Class** _____

7. If a face mill 3 inches in diameter can remove 67.5 square inches per minute, how many square inches can be removed in 3.5 minutes?

   _____

8. If there are five holes in a part and the distance between hole centers is 0.625″, what is the center-to-center distance from the first hole to the last hole?

   _____

Use this figure for problems 9 and 10.

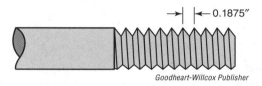

←| |←— 0.1875″

Goodheart-Willcox Publisher

9. Thread pitch is the distance from the center of one thread to the same point on the next thread. How many inches of threads would there be in 25 of the threads shown in the figure above?

   _____

10. How many inches of threads would there be in 25 of the threads shown above if the pitch is changed to .15″?

    _____

11. If there are 42 threads on a bolt with a pitch of 0.05″, how many inches of thread are on the bolt?

    _____

12. If a machining center can produce 150 parts per hour and production works a shift of 7.5 hours, how many parts can be made per shift?

_____

Use this figure for problems 13 and 14.

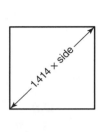
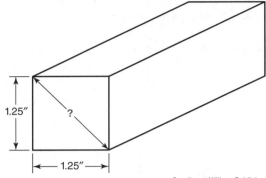

Goodheart-Willcox Publisher

13. The diagonal distance across the corners of a square is 1.414 times the length of one side. What is the distance across the corners of a square bar that has 1.25″ sides?

_____

14. What would be the distance across the corners of the square bar if 0.063″ is removed from each face of the bar?

_____

15. If you are paid $18.60 per hour and in one week you work 37 3/4 hours, what is your pay for that week? (Review converting common fractions to decimal fractions in Unit 10 if needed before solving this problem.)

_____

Name _____   **Date** _____   **Class** _____

16. If you are paid $17.76 per hour and you work 39.5 hours in one week, what is your pay for that workweek?

    _____

17. Considering question 16, if you withhold $3.56 per hour for tax and insurance, how much is withheld if you work 39.5 hours in one week?

    _____

For questions 18 through 20, calculate the number of threads inside a hole by multiplying the depth of the hole by the threads per inch (TPI).

18. If a hole is tapped 1.250" deep and has 20 threads per inch, how many threads are inside the hole?

    _____

19. If a hole is tapped 2.250" and the tap has 28 threads per inch, how many threads are inside the hole?

    _____

20. If the threaded length of a bolt is 3.5" and the TPI is 36, how many threads are on the bolt?

    _____

21. A particular bolt weighs 0.28 lb. What is the total weight of a box of 50 of these bolts if the box weighs 0.13 lb?

    _____

22. If a screw weighs 0.02 lb, what is the weight of a box of 100 of these screws?

_____

23. How much would a box of 1,000 of the screws from question 22 weigh?

_____

24. When bolts are evenly placed around the edge of a circular plate, a circle drawn through their centers is called the bolt circle. What is the length of the bolt circle in the drawing below? (The abbreviation TYP stands for "typical," meaning the dimension is typical of all the bolt locations.)

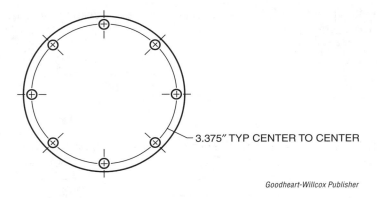

3.375" TYP CENTER TO CENTER

Goodheart-Willcox Publisher

_____

25. A shaft is turned from steel weighing .35 pounds per inch. If the shaft is 9 5/8" long, what is the weight of the steel to make it? (Review converting common fractions to decimal fractions in Unit 10 before solving this problem if needed.) Round your answer to the nearest hundredth of a pound.

_____

Dividing Decimal Fractions

Objectives

Information in this unit will enable you to:

- Identify the dividend, divisor, and quotient when dividing decimals.
- Demonstrate how to divide decimals.
- Explain what to do when a quotient is unending.
- Describe how to quickly divide decimals by 10, 100, and 1,000.

The Language of Division

The terms that were explained in Unit 5 will be used again in the explanation of how to divide decimals. The *dividend* is the number being divided. The *divisor* is the number by which the dividend is divided. The result of division is the *quotient*.

$$\begin{array}{r} 5 \leftarrow \text{Quotient} \\ \text{Divisor} \longrightarrow 6\overline{\smash{)}30} \leftarrow \text{Dividend} \end{array}$$

Learning the Principles

Dividing numbers with decimal fractions is the same as dividing whole numbers, except a decimal point must be placed in the quotient (answer). Write the dividend inside the division symbol and the divisor to the left of the symbol, and follow these steps:

1. If the divisor includes a decimal, move the decimal point to the far right of the divisor and then move the decimal point the same number of places to the right in the dividend. Place a decimal point on the quotient line directly above the decimal point in the dividend.

$$.4\overline{\smash{)}2.2.5}$$

2. Cover all but the far left digit of the dividend with your finger. Ask yourself, how many times will the divisor go into the number showing? Write that number above the digit of the dividend. If it will not fit into the number showing, leave the space blank and go to step 3.

$$4\overline{\smash{)}22.5} \qquad \text{4 won't go into 2}$$

3. Slide your finger one place to the right, so there are two digits of the dividend showing. How many times will the divisor go into that two-digit number? Write the number of times above the second digit of the dividend.

$$
\begin{array}{r}
5. \\
4\overline{)22.\underline{5}}
\end{array}
$$

4. Multiply the divisor times the number showing in the quotient (the number you found in step 2) and write it below the portion of the dividend that has been divided so far. Subtract the number you just wrote from the portion of the divided portion of the dividend. If the difference is equal to or greater than the divisor, increase the number you just wrote in the quotient and redo the multiplication and subtraction.

$$
\begin{array}{r}
5. \\
4\overline{)22.\underline{5}} \\
-\,20 \\
\hline
2
\end{array}
$$

5. Draw an arrow down from the next digit of the dividend and write that number as the next place to the right of the difference you found in step 3.

$$
\begin{array}{r}
5. \\
4\overline{)22.\underline{5}} \\
-\,20\!\downarrow \\
\hline
2\,5
\end{array}
$$

6. How many times will the divisor go into the number at the bottom (the one created through subtraction and placing the numeral from the dividend at the right)? Write the number of times as the next digit to the right in the quotient.

7. Repeat the processes of multiplying the divisor by the last numeral written in the quotient and subtracting, as in step 4. Add 0s to the right end of the dividend as necessary, so that there are enough digits to complete division to one more decimal place than the original dividend and divisor together. The dividend in this case is 2.25 and the divisor is .4, so four places are required in the quotient.

$$
\begin{array}{r}
5.6250 \\
4\overline{)22.5000} \\
-\,20\!\downarrow \\
\hline
2\,5 \\
-\,2\,4\!\downarrow \\
\hline
10 \\
-\ 8\!\downarrow \\
\hline
20 \\
-\,20\!\downarrow \\
\hline
00
\end{array}
$$

8. Round the answer to the same number of places as in the original dividend and divisor. The quotient in this case, after 0s are added, is 5.6250. When rounded to three places, the result is 5.625.

Sometimes the quotient is unending. That is the case when 1.0 is divided by 3.

$$\overset{.333}{3\overline{)1.000}}$$

When this is the case, continue the division until the quotient is one more decimal place than the desired accuracy, then round it off.

Division by 10, 100, 1,000, etc.

Dividing by 10, 100, 1,000, and so forth can be done by simply moving the decimal point to the left. To divide by 10, move the decimal point one place; to divide by 100, move the decimal point two places; and to divide by 1,000, move the decimal point three places.

Example 19-1

$$453.75 \div 10 = 45.375$$
$$453.75 \div 100 = 4.5375$$
$$453.75 \div 1,000 = .45375$$

Work Space/Notes

Unit 19 Review

Fill in the blanks in the following review questions.

1. In division, the _____ is the number being divided, the _____ is the number by which the _____ is divided, and the result is the _____ .

2. Dividing numbers with decimal fractions is the same as dividing _____ numbers, except a decimal point must be placed in the _____ (answer).

3. Write the dividend _____ the division symbol and the divisor to the _____ of the symbol.

4. Sometimes the quotient is unending. When this is the case, continue the division until the quotient is _____ more decimal place than the desired accuracy, then round it off.

5. To divide by 10, move the decimal point _____ place; to divide by 100, move the decimal point _____ places; and to divide by 1,000, move the decimal point _____ places.

Practice

Do the following division.

1. $0.55 \div 0.3 =$ _____

2. $2.75 \div 0.4 =$ _____

3. $14.25 \div 2 =$ _____

4. $9.005 \div 2.55 =$ _____

5. $5.875 \div 100 =$ _____

6. $2.5 \div 0.334 =$ _____

7. $0.375 \div 5 =$ _____

8. $1.94 \div 2 =$ _____

9. $0.83 \div 0.15 =$ _____

10. $1.86 \div 0.2 =$ _____

11. $4.89 \div 0.9 =$ _____

12. $12.32 \div 1.5 =$ _____

13. $0.76 \div 0.03 =$ _____

14. $24.8 \div 3.6 =$ _____

15. $41.9 \div 9.3 =$ _____

16. $9.2 \div 10 =$ _____

17. $342.7 \div 100 =$ _____

18. $0.052 \div 10 =$ _____

19. $8126.49 \div 1,000 =$ _____

20. 35.7 divided by 4.2 = _____

21. .875 divided by 10 = _____

22. 4 divided by 0.25 = _____

23. 78.2 divided by 0.04 = _____

24. 41.27 divided by 1.09 = _____

25. 7.8 divided by 0.2 = _____

26. 14.4 divided by 6 = _____

27. 5.84 divided by 8 = _____

28. 214.3 divided by 2.7 = _____

29. 146.718 divided by 1,000 = _____

30. 814.59 divided by 10 = _____

Name _____   **Date** _____   **Class** _____

Do the division in the following questions presented using shop terms. Round answers to the nearest tenth when necessary.

31. Seven hundred thou *divided by* ten thou = _____

32. Four hundred thou *divided by* five thou = _____

33. One inch, three hundred thou *divided by* nine hundred thou = _____

34. Three inches, seven hundred thirty thou *divided by* seven hundred twenty thou = _____

35. Eight inches, one hundred seventy thou *divided by* one inch, four hundred thou = _____

36. Five inches, nine hundred thirty-six thou *divided by* three hundred seven thou = _____

37. Two inches, eighty thou *divided by* one hundred twelve thou = _____

38. Seventeen inches, nine hundred forty-three thou *divided by* six hundred fifty-eight thou = _____

39. Thirty inches, nine hundred thou, and six tenths *divided by* three hundred twenty-two thou and two tenths = _____

40. Forty inches, one hundred seven thou *divided by* two inches, four hundred twenty-nine thou = _____

Applications

Solve the following problems.

1. A piece of bar stock 15.75" long is cut into seven equal pieces. What is the length of the pieces to the nearest thousandth of an inch?

   _____

2. A piece of bar stock 72.5" long is cut into nine equal pieces. What is the length of each piece to the nearest thousandth of an inch?

   _____

3. If you are paid $205.30 for working a 7 1/2-hour day, what is your hourly pay?

   _____

4. If you work a 40-hour workweek and are paid $1,174.45, how much are you paid per hour?

   _____

5. A 3.25-inch-long handle is to be knurled, leaving 1/10 of its length smooth at each end. What is the length of the knurling?

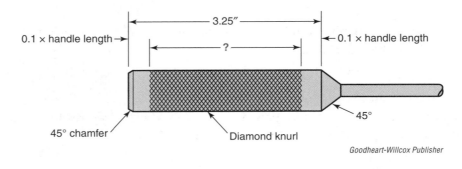

Goodheart-Willcox Publisher

   _____

Name _____ **Date** _____ **Class** _____

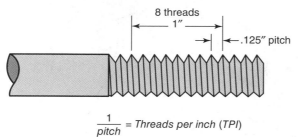

$$\frac{1}{pitch} = \textit{Threads per inch (TPI)}$$

Goodheart-Willcox Publisher

In the above example, $\frac{1}{0.125}$ = 8 TPI. Calculate the threads per inch in the following table.

	Pitch of Thread (inches)	Calculation $\left(\frac{1}{Pitch}\right)$	Threads per Inch (TPI)
6.	0.0125		_____
7.	0.025		_____
8.	0.03125		_____
9.	0.05		_____
10.	0.0625		_____
11.	0.1		_____
12.	0.2		_____
13.	0.25		_____

Goodheart-Willcox Publisher

In the metric system, to calculate the number of threads over a specified length, take the length of threads and divide by the pitch. In the following table calculate the amount of full (whole) threads.

	Pitch of Thread (mm)	Length of Threads (mm)	Calculation $\left(\frac{Length\ of\ Threads}{Pitch}\right)$	Number of Threads
14.	0.3	18		_____
15.	0.35	23.8		_____
16.	0.4	44		_____
17.	0.45	15		_____
18.	0.5	50		_____
19.	0.6	30		_____
20.	0.7	60		_____
21.	0.75	35		_____
22.	0.8	40		_____
23.	1.25	25		_____
24.	1.5	75		_____
25.	1.75	100		_____
26.	2	150		_____

Goodheart-Willcox Publisher

Refer to this drawing for problems 27 through 30.

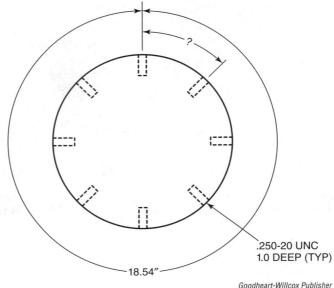

.250-20 UNC
1.0 DEEP (TYP)

18.54"

Goodheart-Willcox Publisher

27. A plate with a circumference (distance around the outside edge) of 18.54" has eight equally spaced holes drilled and tapped in its edge. What is the center-to-center distance to the nearest thousandth between the holes at the surface?

_____

28. What is the center-to-center distance to the nearest thousandth between the holes if the circumference is changed to 21.25 inches?

_____

29. To find pitch, divide 1 by the threads per inch: Pitch = $1/TPI$. What is the pitch in the callout in the drawing? (Hint: 20 = TPI)

_____

30. If the thread is changed from a coarse series to a fine series thread, and the threads per inch are now 28, what is the pitch? Round to the nearest thousandth.

_____

31. If a 6-foot length of brass rod costs $44.79, what is the cost to the nearest $0.01 of .75 feet?

_____

Name _____   **Date** _____   **Class** _____

32. If a 72" length of A2 tool steel costs $150.14, what is the cost of 1' to the nearest $0.01? (Hint: 12" = 1')

   _____

33. What is the cost of 1" of the A2 tool steel from question 32 to the nearest $0.01?

   _____

Refer to the drawing to answer questions 34 through 36.

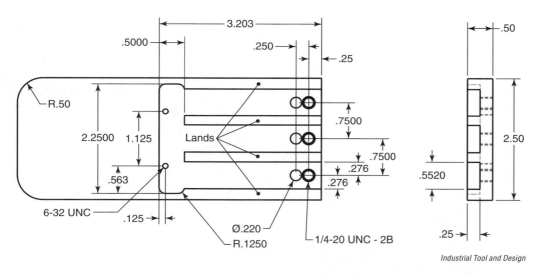

Industrial Tool and Design

34. Assuming the lands in the part drawing below are all spaced equal distances apart and all are the same thickness, what is the thickness of the lands? (Solving this problem will require multiplication, subtraction, and division.)

   _____

35. What is the pitch in the 6-32 UNC callout in the drawing? Round to the nearest tenth. (Hint: 32 = TPI)

   _____

36. If the callout is changed to a 6-40 UNF thread, what is the pitch now?

   _____

Work Space/Notes

Percent, Ratio, and Proportion

Key Terms

extremes	percent	ratio
inverse proportion	percentage	rule of three
means	proportion	

UNIT 20

Percent and Percentage

Objectives

Information in this unit will enable you to:

- Define and differentiate between the terms *percent* and *percentage*.
- Demonstrate how to change a percent to a decimal.
- Explain how to change a percent to a fraction.
- Describe the three steps involved in solving percentage problems.

Percent or Percentage

The root of the word percent is "cent," which refers to hundred, as in centipede, the hundred-legged creature. **Percent** means per hundred. The symbol for percent is %. If 4% of a batch of parts are defective, that means that 4 parts per hundred are defective.

Percentage is a variation of the word percent. Percentage is used when no specific number is used with it. One might say, "What percentage of the shipment arrived?" The answer might be, "75 percent of the shipment arrived."

Changing Percent to a Decimal

Percent may be changed to a decimal by first changing the percent to a fraction and then to a decimal.

Example 20-1

$$4\% = \frac{4}{100} = 0.04$$

Percent can be expressed as a decimal in one step by moving the decimal point two places to the left as the percent symbol is removed.

Example 20-2

$$55\% = 0.55$$

If 55% of a box of 100 bolts is removed, 0.55×100 bolts is removed. That would be 55 bolts.

Changing Percent to a Fraction

Percent is hundredths, so writing a percent as a fraction is simply a matter of removing the percent sign and writing the number as a fraction with a denominator of 100 and reducing the fraction to its lowest terms.

Example 20-3

$$40\% = \frac{40}{100}$$

$$\frac{40}{100} = \frac{4}{10}$$

$$\frac{4}{10} = \frac{2}{5}$$

$$150\% = \frac{150}{100}$$

$$\frac{150}{100} = 1\frac{50}{100}$$

$$1\frac{50}{100} = 1\frac{5}{10}$$

$$1\frac{5}{10} = 1\frac{1}{2}$$

Solving Percentage Problems

Percentage problems involve three parts: a whole, a part, and a percent. The whole is the thing that the part is a percent of. The first step in solving percentage problems is to change the percent to a decimal. Move the decimal point two places to the left and remove the percent symbol. Percentage problems often use the word "of," which is a signal to multiply.

Example 20-4

If a certain steel weighing 50 lb contains 12% chromium, what is the weight of the chromium?

Change 12% to a decimal: $12\% = .12$

Multiply .12 times the weight of the whole steel: $.12 \times 50 \text{ lb} = 6 \text{ lb}$

In this example, 50 lb is the whole and we solved for the part.

The basic formula for percentage problems is P (part) = % (percent) × W (whole). This figure is a useful aid in solving percentage problems.

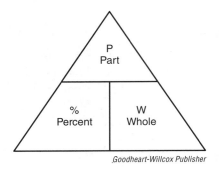

Cover the part of the triangle you are looking for and the formula to be used will be visible. To find a percent, cover the % and what is visible is P over W, so divide the part by the whole.

Example 20-5

What percentage of 60 is 15?

$$\frac{15}{60} = .25$$

Move the decimal point two places to the right and add the percent symbol: 25%

To find the part, cover the P and what is visible is % times W, so multiply the percent times the whole.

Example 20-6

What is 15% of 200?

$$15\% \times 200 = .15 \times 200 = 30$$

To find the whole, cover the W and what is visible is P over %, so divide the part by the percent.

Example 20-7

If 60% of the whole is 72, what is the whole?

$$\frac{72}{60\%} = \frac{72}{.60} = 120$$

Unit 20 Review

Name _____ **Date** _____ **Class** _____

Fill in the blanks in the following review questions.

1. The root of the word percent is _____, which refers to hundred. Percent means per _____. The symbol for percent is _____.

2. Percentage is a variation of the word percent and is used when no specific _____ is used with it.

3. Percent may be changed to a decimal by first changing it to a _____ and then to a decimal.

4. Percent can be expressed as a _____ in one step by moving the decimal point two places to the _____ as the percent symbol is removed.

5. Percent is _____, so writing a percent as a fraction is simply a matter of removing the percent sign, writing the number as a fraction with a denominator of _____, and reducing the fraction to its lowest terms.

6. Percentage problems involve three parts: a _____, a part, and a percent. The _____ is the thing the part is a percent of.

7. The basic formula for percentage problems is P (_____) = % (_____) × W (_____). The formula can be rearranged to find whichever variable is missing.

Practice

Find the whole, part, or percent to complete the table. Round your answers to the nearest tenth.

	Whole	Part	Percent
1.	56	23	
2.	140		30%
3.		50	60%
4.	14	42	
5.	110		80%
6.	200		325%
7.		62.5	120%
8.		1.01	80%
9.		48	50%
10.	356		63%
11.	206	86	
12.		100	12%
13.	1,250		35%
14.	125	350	
15.	412	18	
16.	655		95%

Goodheart-Willcox Publisher

Convert percent to fraction. Round to the lowest terms.

17. Convert 80% to a fraction.   _____

18. Convert 72% to a fraction.   _____

19. Convert 54% to a fraction.   _____

20. Convert 62.5% to a fraction.   _____

21. Convert 25% to a fraction.   _____

22. Convert 45% to a fraction.   _____

23. Convert 12% to a fraction.   _____

24. Convert $\frac{45}{100}$ to a percent.   _____

Convert fraction to percent.

25. Convert 33/100 to a percent. _____

26. Convert 7/100 to a percent. _____

27. Convert 3/50 to a percent. _____

28. Convert 4/25 to a percent. _____

29. Convert 1/20 to a percent. _____

30. Convert 1/10 to a percent. _____

Name _____ **Date** _____ **Class** _____

Applications

Solve the following problems. Round your answers to the nearest two decimal places.

1. If your wages before tax are $42,500 per year and you must pay 28% income tax, how much is your tax?

   _____

2. If you receive a 4% raise on the wages earned in problem 1, what is your new annual wage before taxes?

   _____

3. If you contribute 6% of your yearly wages in question 1 toward a 401(k) retirement plan, how much is contributed each year? (Hint: Use the wages pretax.)

   _____

4. The list price for a sine bar set is $199.95 and you receive a 15% discount. How much must you pay for the set?

   _____

5. In question 4, if a sales tax of 4% is to be collected on the purchase price, how much is the sales tax on the sine bar set?

   _____

6. If 24% of the original material is removed to produce a 6.5 lb part, what did the original blank weigh to the nearest hundredth of a pound?

   _____

You are instructed to track the percentage of rejected parts for one week. Complete the following table by computing the percentage of rejects to the nearest tenth of a percent:

	Day	Total Parts Produced	Rejected Parts	Percent Rejected
7.	Monday	52	14	_____
8.	Tuesday	61	12	_____
9.	Wednesday	55	9	_____
10.	Thursday	56	11	_____
11.	Friday	54	10	_____
12.	Total Week 1	_____	_____	

Goodheart-Willcox Publisher

After analyzing the Week 1 data, Quality Assurance discovered some measuring tools used to check the parts were not calibrated correctly, resulting in a high percentage of parts being rejected. You are instructed to track the percentage of rejected parts for a second week. Complete the following table by computing the percentage of rejects to the nearest tenth of a percent:

	Day	Total Parts Produced	Rejected Parts	Percent Rejected
13.	Monday	50	3	_____
14.	Tuesday	58	5	_____
15.	Wednesday	53	2	_____
16.	Thursday	57	4	_____
17.	Friday	59	1	_____
18.	Total Week 2	_____	_____	

Goodheart-Willcox Publisher

19. A 5.5 lb blank is turned on a lathe, removing 1.2 lb of material. What percent of the original material is removed to the nearest hundredth of a percent?

   _____

20. A 7.25-pound aluminum casting is machined on a mill, removing 0.92 pound of material. What percent of the original material is removed to the nearest hundredth of a percent?

   _____

Name _____   **Date** _____   **Class** _____

Note: The efficiency of a machine is the percentage of the input energy that is delivered as output of the machine. Use this information for questions 21 and 22.

21. A gear train is driven by 140 foot-pounds of torque. If the output is 115 foot-pounds of torque, what percentage of torque is lost in the gear train? What is the efficiency of the gear train to the nearest hundredth of a percent?

    A. _____

    B. _____

22. A small annealing oven consumes 630,000 BTUs of energy per hour and produces 580,000 BTUs per hour in the annealing chamber. What is the efficiency of the oven to the nearest tenth of a percent?

    _____

23. A belt-drive system is 85% efficient. If it is driven by a 1 1/2 horsepower electric motor, what is the output horsepower to the nearest hundredth horsepower?

    _____

24. After annual maintenance service, the belt drive from question 25 is now 88% efficient. If it is still driven by a 1 1/2 horsepower electric motor, what is the output horsepower to the nearest hundredth horsepower?

    _____

25. A conventional vertical knee mill has a power rating of 3 horsepower. After 30 minutes of continuous use, the power rating drops to 2 horsepower. What percent of horsepower is available when it drops to 2 horsepower?

    _____

26. Hydrochloric acid is sometimes used as a cleaning agent for metals. A certain job calls for 3% hydrochloric acid by weight mixed in 10 gallons of distilled water. Assuming water weighs 8.3 lb per gallon, how much acid should be added to the water?

    _____

Grade O1 oil-hardening tool steel contains four alloying elements listed in the table below. Write the number of ounces of each in a 100-pound batch of the steel. (Hint: 100 pounds is 1,600 ounces.) Round to the nearest hundredth ounce.

	Alloying Element	Weight of Element
27.	0.90% carbon	
28.	1.2% manganese	
29.	0.50% chromium	
30.	0.50% tungsten	

Goodheart-Willcox Publisher

A nickel-chromium-molybdenum steel (4340) contains three alloying elements listed in the table below. Write the number of ounces of each in a 100-pound batch. Round to the nearest hundredth ounce.

	Alloying Element	Weight of Element
31.	1.82% nickel	_____
32.	0.8% chromium	_____
33.	0.25% molybdenum	_____

Goodheart-Willcox Publisher

34. To turn a part 6.75" long, .875" is added for chucking and cutoff. How long must the blank be and what percentage to the nearest hundredth of a percent is *not* part of the finished piece?

    A. _____

    B. _____

35. Scrap steel is recycled, losing 15% due to sorting and cleaning. If 525 lb of scrap is to be recycled, how much usable steel will be produced? Round your answer to the nearest pound.

    _____

36. When aluminum is recycled, 22% is lost due to sorting, cleaning, and contamination. If 750 pounds of scrap is to be recycled, how much usable aluminum will be produced? Round your answer to the nearest pound.

    _____

UNIT 21

Ratio and Proportion

Objectives

Information in this unit will enable you to:

- Explain what a ratio is.
- Describe how gear ratios are expressed, written, and reduced.
- Explain what a proportion is and describe how proportional statements are written.

Ratio

A ratio is a comparison of two values. Both terms of a ratio must be expressed in the same units. Gears are frequently used to increase or decrease the speed of a shaft. Consider the gears in the following illustration. For the drive gear to engage all 40 teeth of the driven gear, the drive gear must make two complete revolutions. Gear ratio is expressed in terms of the number of teeth on each gear, with the driven gear given first. In this case, there are 20 teeth on the drive gear and 40 on the driven gear, so the ratio is 40 to 20.

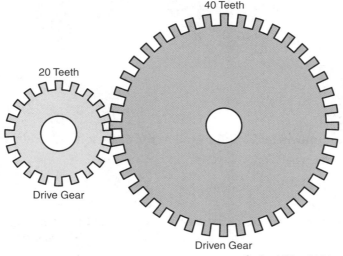

40 Teeth

20 Teeth

Drive Gear

Driven Gear

Goodheart-Willcox Publisher

Gear ratios are usually written with the first number separated from the second by a colon (:). The gear ratio in the illustration would be 40:20, read as "forty to twenty." Gear ratios are also usually reduced to show the drive gear as 1. To reduce the ratio, divide both numbers by the value of the second number. 40:20 would be reduced to 2:1. It is common for the first number in the ratio to be a fraction, usually a decimal fraction. In that case, the first number is rounded to the appropriate number of places, often hundredths.

Example 21-1

The drive gear has 22 teeth and the driven gear has 48 teeth.

The ratio is 48:22.

Divide both numbers by 22.

The ratio 2.1818:1 rounded to the nearest hundredth is 2.18:1.

In some cases, the drive gear has more teeth than the driven gear. The ratio is calculated the same way: driven gear teeth to drive gear teeth.

Example 21-2

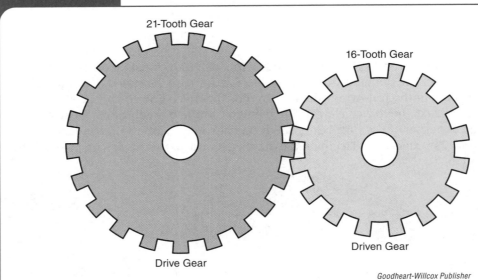

21-Tooth Gear

16-Tooth Gear

Drive Gear

Driven Gear

Goodheart-Willcox Publisher

The ratio is 16:21.

Divide both numbers by 21 and round to the nearest hundredth.

The reduced ratio is 0.76:1.

Other common examples of ratios are tax withheld to gross pay, carbon to iron in steel, and revolutions to inches of travel on a lathe. The scale on a print is also a ratio.

Proportion

A **proportion** is a statement that two ratios are equal to each other. Proportional statements are often written in the form of common fractions. The numbers or values in a proportion are called terms and are numbered left to right as follows:

First term		Second term		Third term		Fourth term
1	:	2	=	4	:	8

The first and fourth terms in a proportion are called the **extremes**. The second and third terms are called the **means**. Thus, in the proportion 1:2 = 3:6, the numbers 1 and 6 are the extremes, and the numbers 2 and 3 are the means.

The **rule of three** is a mathematical rule that allows you to solve for the fourth term of a proportion when the other three are given. A basic rule of proportions is that the product of the extremes is equal to the product of the means. In the proportion 1:2 = 3:6, the product of the extremes, $1 \times 6 = 6$, is equal to the product of the means, $2 \times 3 = 6$.

Proportion problems usually involve one unknown part and three knowns. The task is to solve for the unknown. The unknown is usually represented by a variable, which is a symbol for a value not yet known. Variables are usually represented by letters, such as x.

Given this information, we can calculate for missing information as follows:

- Unknown mean = Product of extremes divided by known mean
- Unknown extreme = Product of means divided by known extreme

There are two different rules of three, one for direct proportion and one for inverse proportion. In direct proportion, the two variables change at the same time. Stated another way, an increase in one quantity causes a corresponding increase in the other quantity, and a decrease in one quantity results in a corresponding decrease in the other quantity. Coolant used for metalworking provides an example of direct proportion: As the amount of water increases, so does the amount of concentrate used in the mixture.

Example 21-3

If the ratio of water to concentrate in a coolant is 9:1, and the amount of water used is 36 gallons, the amount of concentrate can be calculated.

$$9:1 = 36:c \quad \text{(Where } c = \text{concentrate needed)}$$

The product of the means divided by the known extreme will give us the variable c.

$$\frac{1 \times 36}{9} = 4$$

So c, the amount of concentrate, is 4 gallons.

To check our answer, be sure the product of the extremes is equal to the product of the means:

$$9 \times 4 = 1 \times 36$$
$$36 = 36$$

The answer 4 is correct. The two ratios are equal to one another.

Alternatively, we can cross multiply to figure out the amount of concentrate.

Cross
Multiplication

$$\frac{a}{b} \diagdown \frac{c}{d}$$

$$ad = bc$$

Using cross multiplication:

$$\frac{9}{1} = \frac{36}{c}$$

$$9 \times c = 1 \times 36$$

To get one c, divide both sides by 9:

$$9c = 36$$

$$\frac{9c}{9} = \frac{36}{9}$$

$$c = 4 \text{ gallons}$$

Inverse proportion is a situation in which an increase in one variable causes a decrease in the other variable, or a decrease in one variable causes an increase in the other variable. When working with gears and pulleys, transferring power from a small gear or pulley to a large gear or pulley decreases revolutions per minute, whereas transferring from a large gear or pulley to a small gear or pulley increases revolutions per minute. This will be important to remember when checking your answers.

Example 21-4

A gear train has 12 teeth on the drive gear and 30 on the driven gear. What is the speed of the driven gear if the drive gear is turning at 200 revolutions per minute (rpm)?

Step 1: Set up the proportion. The ratio of teeth on the gears is 30:12, and the ratio of rpms is x:200. Write this as a common fraction:

$$\frac{30}{12} = \frac{x}{200}$$

Step 2: Because this is an inverse proportion, the second ratio or second fraction needs to be inverted. Our calculation now looks like this:

$$30:12 = 200:x$$

Step 3: The product of the means divided by the known extreme gives us the variable x:

$$\frac{12 \times 200}{30} = x$$

$$\frac{2400}{30} = x$$

$$x = 80 \text{ rpm}$$

(Continued)

Alternatively, we can cross multiply:

$$\frac{30}{12} = \frac{200}{x}$$
$$30 \times x = 12 \times 200$$
$$30x = 2400$$

To get one x, divide both sides by 30:

$$\frac{30x}{30} = \frac{2400}{30}$$
$$x = 80$$

Transferring power from a small gear or pulley to a large gear or pulley decreases revolutions per minute, so we know our answer needs to be smaller than 200 rpm, and 80 seems reasonable.

Work Space/Notes

Unit 21 Review

Name _____ Date _____ Class _____

Fill in the blanks in the following review questions.

1. A ratio is a _____ of two values.

2. Gear ratio is expressed in terms of the number of teeth on each gear, with the _____ gear given first.

3. Gear ratios are usually written with the first number separated from the second by a _____ (:).

4. To reduce the ratio, divide both numbers by the value of the _____ number.

5. A proportion is a statement that two _____ are equal to each other.

6. Proportional statements are often written in the form of common _____.

7. Proportion problems usually involve one _____ part and three knowns. The task is to solve for the unknown.

Practice

Write the ratios for the following to the nearest hundredth.

1. A drive gear with 15 teeth and a driven gear with 45 teeth

   _____

2. A drive gear with 25 teeth and a driven gear with 60 teeth

   _____

3. A drive gear with 60 teeth and a driven gear with 25 teeth

   _____

4. A drive gear with 35 teeth and a driven gear with 15 teeth

   _____

5. A belt and pulley system with a 3" drive pulley and a 5" driven pulley

   _____

6. A threaded shaft that has nine threads per inch

   _____

7. A threaded screw that has 13 threads per inch

_____

8. A threaded bolt that has 18 threads per inch

_____

9. A metalworking coolant with 24 parts water and 1 part concentrate

_____

10. A metalworking coolant with 7.4 parts water and 1 part concentrate

_____

Solve for the unknown in the following proportions. Round your answers to the nearest hundredth.

11. $3:5 = 12:x$ _____

12. $4:8 = x:32$ _____

13. $x:21 = 25:175$ _____

14. $1.6:x = 3.5:4$ _____

15. $4.59:1 = x:1.5$ _____

16. $32:8 = 20:x$ _____

17. $60:x = 15:10$ _____

18. $x:18 = 3:9$ _____

19. $45:27 = x:3$ _____

20. $15:x = 5:12$ _____

21. $27:15 = 72:x$ _____

22. $x:9 = 27:81$ _____

Name _____ **Date** _____ **Class** _____

Applications

Solve the following problems.

Use this figure for questions 1 through 6.

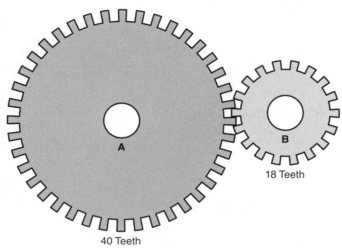

40 Teeth

18 Teeth

Goodheart-Willcox Publisher

	Gear A (Drive Gear) Teeth	Gear B (Driven Gear) Teeth	Ratio	Reduced to Lowest Terms (Nearest Hundredth)
1.	40	18	_____	_____
2.	18	40	_____	_____
3.	24	56	_____	_____
4.	57	24	_____	_____
5.	56	44	_____	_____
6.	44	57	_____	_____

Goodheart-Willcox Publisher

7. What is the gear ratio of two gears with 30 teeth on each?

_____

8. A cleaning solution is made by adding 100 ounces of caustic powder to 64 lb of water. What is the ratio of powder to water to the nearest hundredth? (1 lb = 16 oz)

_____

Use the following table for questions 9 through 20. Round to the nearest thousandth.

9,800 Grams Tool Steel	
Alloying Element	**Weight of Element**
Carbon	86 grams
Manganese	117 grams
Chromium	46 grams
Tungsten	64 grams

Goodheart-Willcox Publisher

9. What is the ratio of carbon to steel?

    _____

10. What is the ratio of manganese to steel?

    _____

11. What is the ratio of chromium to steel?

    _____

12. What is the ratio of tungsten to steel?

    _____

13. How much carbon is required to make 20,000 grams of this steel?

    _____

14. How much tungsten is required to make 20,000 grams of this steel?

    _____

15. How much manganese is required to make 20,000 grams of this steel?

    _____

Name _____ **Date** _____ **Class** _____

16. How much chromium is required to make 20,000 grams of this steel?

    _____

17. How much carbon is required to make 15,000 grams of this steel?

    _____

18. How much tungsten is required to make 15,000 grams of this steel?

    _____

19. How much manganese is required to make 15,000 grams of this steel?

    _____

20. How much chromium is required to make 15,000 grams of this steel?

    _____

Use this figure for questions 21 through 23.

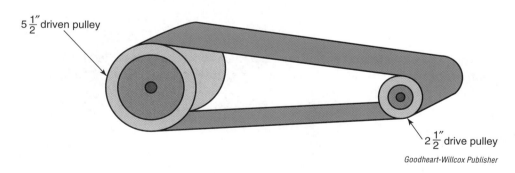

5½″ driven pulley

2½″ drive pulley

Goodheart-Willcox Publisher

21. To the nearest hundredth, what is the ratio of the drive pulley to the driven pulley in the illustration?

    _____

22. What is the speed of the driven pulley to the nearest tenth rpm if the speed of the drive pulley is 1,700 rpm?

_____

23. What size driven pulley would increase the driven pulley speed to 1,000 rpm if the speed of the drive pulley is 1,700 rpm? (Round to the nearest hundredth.)

_____

24. A gear train has 24 teeth on the drive gear and 57 teeth on the driven gear. What is the speed of the driven gear if the drive gear is turning at 630 revolutions per minute (rpm)? Round to the nearest hundredth.

_____

25. A gear train has 56 teeth on the drive gear and 44 teeth on the driven gear. What is the speed of the driven gear if the drive gear is turning at 500 revolutions per minute (rpm)? Round to the nearest hundredth.

_____

26. If the ratio of water to concentrate in a metalworking coolant is 11.5:1, and the amount of concentrate used is 1.5 gallons, how much water is needed?

_____

27. If the ratio of water to concentrate in a metalworking coolant is 15.6:1, and the amount of water used is 22 gallons, how much concentrate is needed? Round to the nearest hundredth.

_____

Linear Measure

Unit 22. Reading Rulers

Unit 23. Perimeter and Circumference

Unit 24. Tolerances

Unit 25. Vernier Measuring Instruments

Key Terms

accuracy

bilateral tolerance

circle

circumference

diameter

dimensional tolerance

graduations

linear measurement

metric system

micrometer

parallel lines

parallelogram

perimeter

pi (π)

positional tolerance

precision

radius

rectangle

square

Système International
 d'Unités (SI)

tolerance

trapezoid

US Customary System

vernier caliper

vernier scale

UNIT 22

Reading Rulers

Objectives

Information in this unit will enable you to:

- Discuss how the US Customary System works.
- Describe how the metric system works.
- Explain how to read US Customary, decimal, and metric rulers.

The US Customary System

Most of the measurements machinists encounter are **linear measurements**. That means they are measurements of distance.

The **US Customary System** of measurements is the most widely used in the United States, although use of the **metric system** is increasing. It was adapted from the English system after the Revolutionary War. The US Customary System is based on yards, feet, and inches for linear measure.

US Customary Units of Linear Measure

1 foot	12 inches
1 yard	3 feet
1 mile	5,280 feet

Goodheart-Willcox Publisher

The inch is the unit most commonly used in the machine shop. An inch can be divided into fractional parts of an inch for smaller distances. The divisions (called **graduations**) can be seen on a common ruler.

Reading a US Customary Ruler

The most common measuring instrument is the ruler, sometimes called a rule. Steel rules are used in the machine shop. Rulers can have three kinds of graduations. The graduations on a US Customary ruler divide inches into halves, quarters, eighths, sixteenths, thirty-seconds, and sometimes sixty-fourths. Each graduation level divides the next larger graduation in half. For example, 1/16 is half of 1/8, and 1/32 is half of 1/16. The whole-inch graduations are the longest on the ruler. The 1/2-inch graduations are the next longest, the 1/4-inch graduations are next, then the 1/8-inch graduations, and so forth.

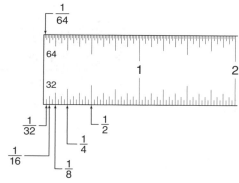

Goodheart-Willcox Publisher

When reading any measurement, with a ruler or otherwise, fractions should be reduced to their lowest terms. If you measure 12/16″, it is written or read as 3/4″.

Reading a Decimal Ruler

In the machine shop, more work is done in decimal inches than in common fractional inches. That is, the inches are divided into decimal fractions. A decimal-inch ruler normally has graduations of 1/10″ along one edge and 1/50″ along the other edge. Graduations of 1/100″ are too tightly spaced together to read accurately, so other instruments are used to measure hundredths of an inch or less.

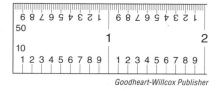

Goodheart-Willcox Publisher

The Metric System

The metric system of measurements has evolved over the last few centuries. In 1960, the metric system was officially named **Système International d'Unités** (or *SI* for short.) Today all industrialized nations except the United States use the metric system. Products that are made or serviced in the United States are often exported throughout the world, so it is important for an American machinist to be able to use the metric system.

Length measurements in the metric system are based on the meter. A meter is just over 3 inches longer than a yard.

The metric system uses powers of 10 to signify numbers that are too large or too small to be practical for use. These numbers are written with a Greek or Latin prefix to signify what power of 10 the base number is multiplied by. It is common in writing large metric numbers to leave a space after every third place, rather than a comma.

Text	Symbol	Multiplier
tera	T	1 000 000 000 000
giga	G	1 000 000 000
mega	M	1 000 000
kilo	k	1 000
hecto	h	100
deca	da	10
none	*none*	1
deci	d	0.1
centi	c	0.01
milli	m	0.001
micro	μ	0.000 001
nano	n	0.000 000 001
pico	p	0.000 000 000 001

Goodheart-Willcox Publisher

Example 22-1

3 000 meters is written as 3 kilometers; 6/1 000 meters is written as 6 millimeters.

When the number is written with symbols, the symbol for the multiplier is written as a prefix to the symbol for the unit (m for meter).

Example 22-2

3 000 meters is written as 3 km; 6/1 000 meters is written 6 mm.

Reading a Metric Ruler

Metric rulers often have one edge graduated in millimeters and the other edge graduated in half millimeters. Millimeters (mm) are the units most commonly encountered in the machine shop when using the metric system. One inch is 25.4 millimeters.

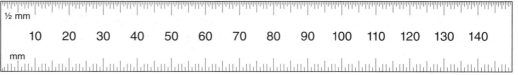

Goodheart-Willcox Publisher

Unit 22 Review

Name _____ **Date** _____ **Class** _____

Fill in the blanks in the following review questions.

1. Most of the measurements machinists encounter are _____ measurements, which means they are measurements of _____.

2. The _____ _____ _____ is the most widely used system of measurement in the United States. It is based on yards, feet, and inches for linear measure.

3. The _____ is the unit most commonly used in the machine shop.

4. Today, all industrialized nations except the United States use the _____ system.

5. Length measurements in the metric system are based on the _____.

Practice

1. Write the measurements indicated by each letter.

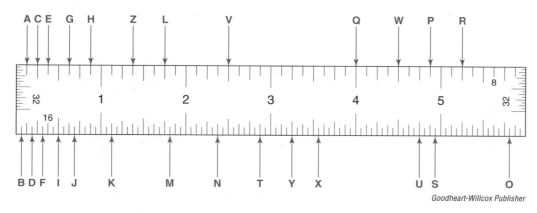

Goodheart-Willcox Publisher

A: _____   B: _____   C: _____

D: _____   E: _____   F: _____

G: _____   H: _____   I: _____

J: _____   K: _____   L: _____

M: _____   N: _____   O: _____

P: _____   Q: _____   R: _____

S: _____   T: _____   U: _____

V: _____   W: _____   X: _____

Y: _____   Z: _____

2. Write the measurements indicated by each letter.

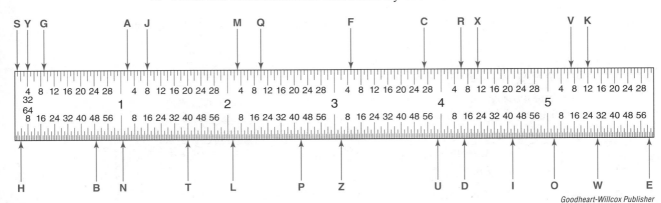

Goodheart-Willcox Publisher

A: _____   B: _____

C: _____   D: _____

E: _____   F: _____

G: _____   H: _____

I: _____   J: _____

K: _____   L: _____

M: _____   N: _____

O: _____   P: _____

Q: _____   R: _____

S: _____   T: _____

U: _____   V: _____

W: _____   X: _____

Y: _____   Z: _____

Name _____   **Date** _____   **Class** _____

3. Write both the fractional and decimal measurements indicated by each letter.

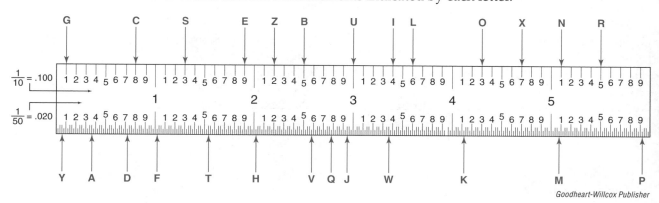

Goodheart-Willcox Publisher

A fractional: $\frac{8}{50} = \frac{4}{25}$ B fractional: _____   C fractional: _____
A decimal: 0.16   B decimal: _____   C decimal: _____

D fractional: _____   E fractional: _____   F fractional: _____
D decimal: _____   E decimal: _____   F decimal: _____

G fractional: _____   H fractional: _____   I fractional: _____
G decimal: _____   H decimal: _____   I decimal: _____

J fractional: _____   K fractional: _____   L fractional: _____
J decimal: _____   K decimal: _____   L decimal: _____

M fractional: _____   N fractional: _____   O fractional: _____
M decimal: _____   N decimal: _____   O decimal: _____

P fractional: _____   Q fractional: _____   R fractional: _____
P decimal: _____   Q decimal: _____   R decimal: _____

S fractional: _____   T fractional: _____   U fractional: _____
S decimal: _____   T decimal: _____   U decimal: _____

V fractional: _____   W fractional: _____   X fractional: _____
V decimal: _____   W decimal: _____   X decimal: _____

Y fractional: _____   Z fractional: _____
Y decimal: _____   Z decimal: _____

4. Write the measurements indicated by each letter.

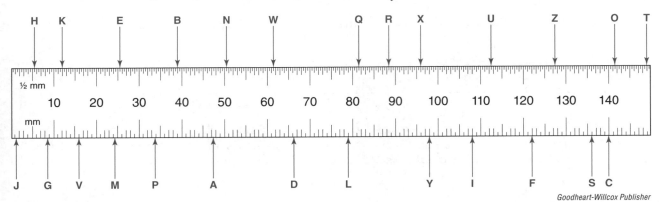

A: _____   B: _____   C: _____

D: _____   E: _____   F: _____

G: _____   H: _____   I: _____

J: _____   K: _____   L: _____

M: _____   N: _____   O: _____

P: _____   Q: _____   R: _____

S: _____   T: _____   U: _____

V: _____   W: _____   X: _____

Y: _____   Z: _____

UNIT 23

Perimeter and Circumference

Objectives

Information in this unit will enable you to:

- Explain what perimeter is and describe how to determine the perimeter of squares, rectangles, parallelograms, and trapezoids.
- Explain what circumference is and describe how to determine the circumference of a circle.
- Describe how to determine the circumference of compound shapes.

Perimeters of Four-Sided Shapes

The **perimeter** of a shape is the distance around its boundary or outside edges. A **square** is a four-sided shape in which each side is the same length and all corners are 90 degrees. Each of the sides of a square is equal to every other side, so the perimeter of a square is 4 times the length of one side. This may be stated as the formula $P = 4s$, where P is perimeter and s is the length of one side.

Example 23-1

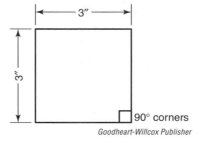

Goodheart-Willcox Publisher

What is the perimeter of the square?

$$P = 4s$$
$$P = 4 \times 3''$$
$$P = 12''$$

A **rectangle** is a four-sided shape in which opposite sides are equal and all four corners are 90 degrees. The perimeter of a rectangle is 2 times the length plus 2 times the width. This may be stated as the formula $P = 2l + 2w$, where P is the perimeter, l is the length, and w is the width.

Example 23-2

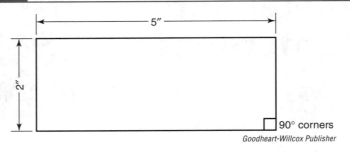

Goodheart-Willcox Publisher

What is the perimeter of the rectangle?

$$P = 2l + 2w$$
$$P = 2 \times 5'' + 2 \times 2'' = 10'' + 4''$$
$$P = 14''$$

A **parallelogram** is a four-sided shape in which opposite sides are equal and parallel, but the corners are not 90 degrees. **Parallel lines** are always the same distance apart, regardless of how long they are. The formula for the perimeter of a parallelogram is the same as the formula for the perimeter of a rectangle.

Example 23-3

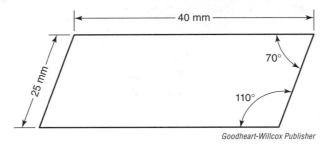

Goodheart-Willcox Publisher

What is the perimeter of the parallelogram?

$$P = 2l + 2w$$
$$P = 2 \times 40 \text{ mm} + 2 \times 25 \text{ mm} = 80 \text{ mm} + 50 \text{ mm}$$
$$P = 130 \text{ mm}$$

A **trapezoid** is a four-sided shape in which two sides are parallel, but not equal. The other two sides are not parallel. The angles between the base side of a trapezoid and the sides that adjoin the base may be equal or they may be unequal. If the base angles are equal, it is an equilateral trapezoid. To find the perimeter of a trapezoid, add the lengths of all four sides.

Example 23-4

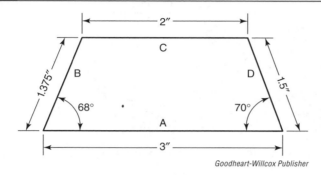

Goodheart-Willcox Publisher

What is the perimeter of the trapezoid?

$$P = A + B + C + D$$
$$P = 3'' + 1.375'' + 2'' + 1.5''$$
$$P = 7.875''$$

Circumference

Circumference is the perimeter of a **circle**. To compute circumference it is necessary to understand a few other terms relating to circles. **Diameter** is the straight-line distance from one side of a circle's circumference to the other through the center point. It is the longest chord of the circle. Diameter is often represented by the $\varnothing$ symbol in manufacturing. The **radius** of a circle is the straight-line distance from its center point to its circumference in any direction. The radius is one-half the diameter. The circumference of a circle is found by multiplying its diameter by a constant, a number that never changes. The constant used to calculate diameter and many other properties of circles is 3.14159.... It is a never-ending decimal, so it is usually rounded off to the degree of accuracy needed. This constant is so frequently used in mathematics that it has been given a name, the Greek letter π, pronounced "pie" and sometimes spelled out as **pi**. The circumference of a circle is $\pi \times$ the diameter of the circle or 2 times π times the radius. The formulas for finding the circumference of a circle are $C = \pi d$ and $C = 2\pi r$.

Example 23-5

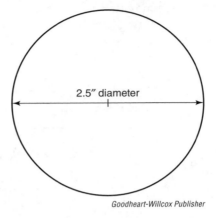

Goodheart-Willcox Publisher

What is the circumference of the circle to the nearest 0.001"?

$$C = \pi d$$
$$C = 3.1416 \times 2.5''$$
$$C = 7.854''$$

Example 23-6

Goodheart-Willcox Publisher

What is the circumference of the circle to the nearest 0.001"?

$$C = 2\pi r$$
$$C = 2 \times 3.1416 \times 1.25''$$
$$C = 7.854''$$

Compound Shapes

Not everything in the machine shop is a single shape. Often a shape is made up of a combination of the shapes described here. For example, the corners of a rectangular piece may be rounded. When the shapes are combined to make a more complex shape, the perimeter can usually be found by applying more than one of the formulas given here.

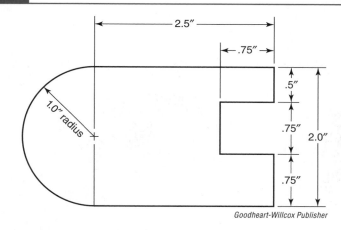

Goodheart-Willcox Publisher

This piece is made up of a 2.5″ × 2.0″ rectangle, with a 0.75″ square cut out of one end and half of a circle with a 1.0″ radius added to the other end.

To find its perimeter, find the perimeter of the rectangle and subtract the left end.

$$2 \times 2.5'' + 2 \times 2.0'' - 2.0'' = 7.0''$$

Find 2 sides of the square that was cut out of the right end.

$$2 \times 0.75'' = 1.5''$$

Add 1/2 of the circumference of the circle.

$$(1/2)C = 2\pi \times 1.0'' \div 2$$
$$= 6.283'' \div 2$$
$$= 3.14''$$

Add these dimensions.

$$7.0'' + 1.5'' + 3.14'' = 11.64''$$

Work Space/Notes

Unit 23 Review

Name _____ **Date** _____ **Class** _____

Fill in the blanks in the following review questions.

1. The _____ of a shape is the distance around its boundary or outside edges.

2. A _____ is a four-sided shape in which each side is the same length and all corners are 90 degrees. The formula for calculating the perimeter of a square is stated as $P =$ _____, where P is _____ and s is the _____ of one side.

3. A _____ is a four-sided shape in which opposite sides are equal and all four corners are 90 degrees. The formula for calculating the perimeter of a rectangle is stated as $P =$ _____ + _____, where P is the perimeter, l is the _____, and w is the _____.

4. A _____ is a four-sided shape in which opposite sides are equal and parallel but the corners are not 90 degrees. The formula for the perimeter of a parallelogram is the same as the formula for the perimeter of a _____.

5. A _____ is a four-sided shape in which two sides are parallel but not equal and the other two sides are not parallel. The angles between the base side and the sides adjoining the base may be _____ (in an equilateral trapezoid) or unequal. To find the perimeter of a trapezoid, _____ the lengths of all four sides.

6. _____ is the perimeter of a circle. The formulas for circumference of a circle are $C =$ _____ and $C =$ _____.

Practice

Find the perimeter or circumference of the following shapes to the nearest thousandth using the formulas explained in this unit.

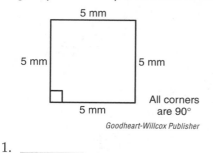

Goodheart-Willcox Publisher

1. _____

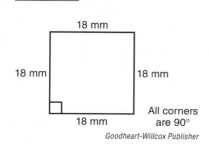

Goodheart-Willcox Publisher

3. _____

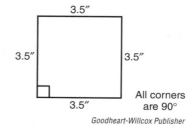

Goodheart-Willcox Publisher

2. _____

4. _____

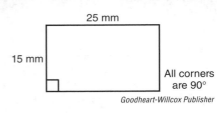

1.5″

0.75″

All corners
are 90°
Goodheart-Willcox Publisher

25 mm

15 mm

All corners
are 90°
Goodheart-Willcox Publisher

5. _____

6. _____

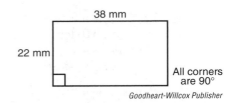

3.75″

1.25″

All corners
are 90°
Goodheart-Willcox Publisher

38 mm

22 mm

All corners
are 90°
Goodheart-Willcox Publisher

7. _____

8. _____

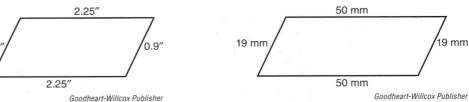

2.25″

0.9″   0.9″

2.25″
Goodheart-Willcox Publisher

50 mm

19 mm   19 mm

50 mm
Goodheart-Willcox Publisher

9. _____

10. _____

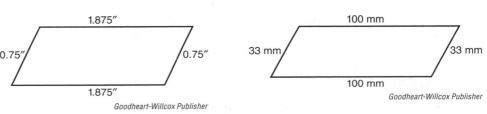

1.875″

0.75″   0.75″

1.875″
Goodheart-Willcox Publisher

100 mm

33 mm   33 mm

100 mm
Goodheart-Willcox Publisher

11. _____

12. _____

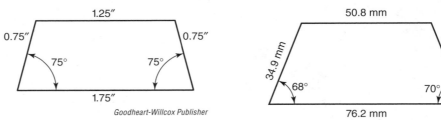

1.25″

0.75″   0.75″

75°   75°

1.75″
Goodheart-Willcox Publisher

50.8 mm

34.9 mm   38.1 mm

68°   70°

76.2 mm
Goodheart-Willcox Publisher

13. _____

14. _____

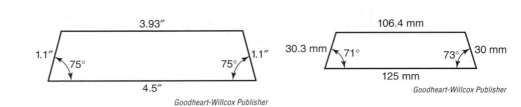

3.93″

1.1″   1.1″

75°   75°

4.5″
Goodheart-Willcox Publisher

106.4 mm

30.3 mm   71°   73°   30 mm

125 mm
Goodheart-Willcox Publisher

15. _____

16. _____

Name _____ **Date** _____ **Class** _____

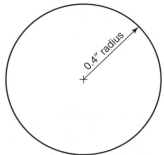

Goodheart-Willcox Publisher

Goodheart-Willcox Publisher

17. _____

18. _____

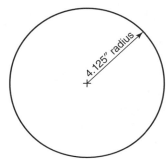

Goodheart-Willcox Publisher

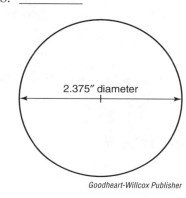

Goodheart-Willcox Publisher

19. _____

20. _____

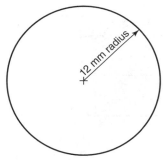

Goodheart-Willcox Publisher

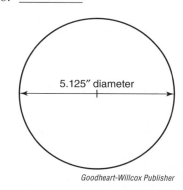

Goodheart-Willcox Publisher

21. _____

22. _____

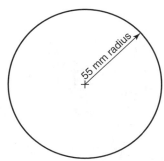

Goodheart-Willcox Publisher

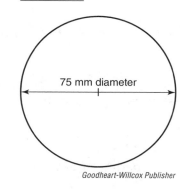

Goodheart-Willcox Publisher

23. _____

24. _____

Applications

Refer to the following part drawing for problems 1 and 2.

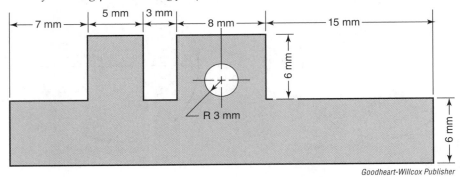

Goodheart-Willcox Publisher

1. What is the perimeter of the part? _____

2. What is the circumference of the hole/circle pictured? _____

Refer to the following illustration for problems 3 through 5.

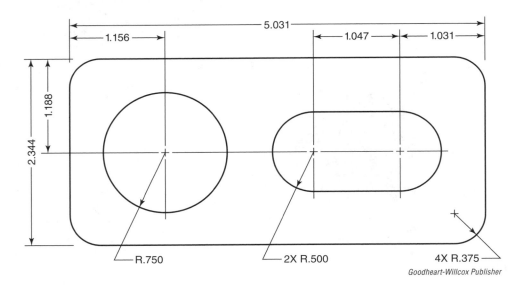

Goodheart-Willcox Publisher

3. What is the perimeter of the part to the nearest .001"? _____

4. What is the circumference of the round hole to the nearest .001"? _____

5. What is the perimeter of the slot to the nearest .001"? _____

Name _____   **Date** _____   **Class** _____

The pitch circle of a spur gear is an imaginary line drawn through the teeth halfway from the roots of the teeth to their tops. The pitch diameter is the diameter of the pitch circle. The length of the rack is the length of the pitch circle. What is the length of rack for spur gears with the pitch diameters listed in the table that follows, to the nearest .001"?

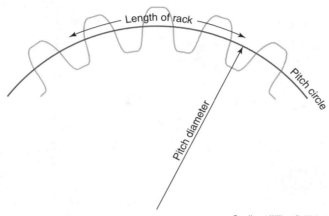

Goodheart-Willcox Publisher

	Pitch Diameter	Length of Rack
6.	3.25"	_____
7.	25.4 mm	_____
8.	2.5"	_____
9.	50 mm	_____
10.	5.125"	_____
11.	42 mm	_____

Goodheart-Willcox Publisher

Calculate the perimeters for the following objects using the dimensions given.

12. A rectangle measuring 6" by 3"   _____

13. A rectangle measuring 75 mm by 25 mm   _____

14. A square with sides of 30 mm   _____

15. A square with sides of 3.75"   _____

16. A parallelogram with sides of 25 mm and 50 mm _____

17. A parallelogram with sides of 1.5″ and 2.75″ _____

What is the perimeter of each part shown in questions 18 through 21?

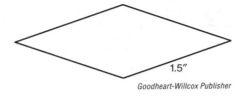

1.5″

Goodheart-Willcox Publisher

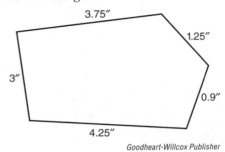

3.75″

1.25″

3″

0.9″

4.25″

Goodheart-Willcox Publisher

18. _____

19. _____

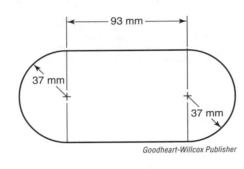

93 mm

37 mm

37 mm

Goodheart-Willcox Publisher

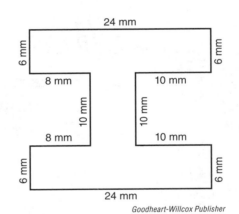

24 mm

6 mm

6 mm

8 mm

10 mm

10 mm

10 mm

8 mm

10 mm

6 mm

6 mm

24 mm

Goodheart-Willcox Publisher

20. _____

21. _____

22. What is the perimeter of the part in the drawing below to the nearest millimeter?

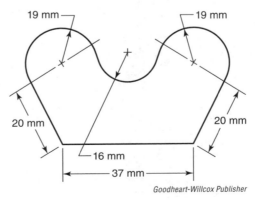

19 mm

19 mm

20 mm

20 mm

16 mm

37 mm

Goodheart-Willcox Publisher

UNIT 24

Tolerances

Objectives

Information in this unit will enable you to:

- Explain what precision and accuracy are and discuss why taking precise, accurate measurements is important.
- Describe what tolerance is and how allowable limits are determined.

Precision

Precision has to do with repeatability; if the same process is repeated with precision multiple times, the results will be the same or very close to the same.

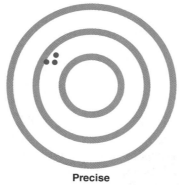

Precise

Goodheart-Willcox Publisher

The target shots shown in the figure are fairly precise, because they all hit the target in almost the same spot. A precise measurement is one that is carefully made with a reliable measuring instrument, so that if the measurement is taken several times, it will always be the same.

Accuracy

Accuracy has to do with how close a process or procedure comes to yielding the intended result. If a series of shots are made accurately, they will all hit the target near the bull's-eye.

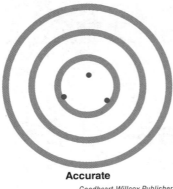

Accurate

Goodheart-Willcox Publisher

The target shots in this figure are fairly accurate, because they are all near the bull's-eye, but they are not very precise. When a dimension is calculated by performing math operations on other dimensions, the accuracy of the results may depend on how numbers are rounded off.

The target shots in the third figure are both precise and accurate, because they are closely grouped and they are close to the intended result.

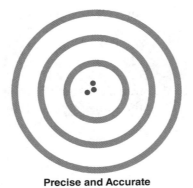

Precise and Accurate

Goodheart-Willcox Publisher

Tolerance

Tolerance is the allowable difference between design dimension and actual dimension. Machined parts cannot be perfectly precise and accurate. That is, if 10 parts are machined the same, there will be small differences in their dimensions. Those differences might be the result of human error in reading a scale, minute variations in how the machine actually performed, or tiny differences in air temperature or the composition of the metal. The job of the machinist is to keep those differences within acceptable limits. The allowable limits define the tolerances for that part. For example, a part is to be turned to a diameter of 1.450″, but it is acceptable if it is .002″ bigger or .002″ smaller. The specified dimension for that part would be 1.450″ +/− .002″. If the part is turned to 1.448″, it is within the tolerance. If it is turned to 1.447″, it is out of tolerance.

There are two types of tolerances. **Positional tolerance** has to do with the location or position of a feature, such as where a hole is drilled. **Dimensional tolerance** refers to the size or dimensions of a part or feature. The tolerance for the size of a hole is a dimensional tolerance.

Tolerances are not always the same for plus tolerance and minus tolerance. Consider a shaft that needs to slide through a hole. If the fit is too loose, the shaft will wobble. If the shaft is too big, it will not fit in the hole. The shaft might be intentionally slightly smaller than the hole but with a greater minus tolerance than plus tolerance. It might have a specified diameter of 1.425″ +.001/−.005. This is called a **bilateral tolerance**.

Another variation of the bilateral tolerance is the equal bilateral tolerance. An equal bilateral tolerance allows equal variation in both directions from the design size. An example would be 1.423″ +.003/−.003. The allowable diameter for that shaft is anything between 1.420″ and 1.426″. Expressed this way, it would be called limit dimensioning, with the lower limit (minimum) expressed first and the upper limit (maximum) second. On a print, the maximum dimension is placed above the minimum dimension: 1.426″/1.420″.

The third way you may see a tolerance expressed is as a unilateral tolerance. A unilateral tolerance is a way to express tolerance by using only minus or plus variation from a given size in one direction. Using the example of the shaft, it could be expressed as 1.426″ +0/−.006 or, alternatively, 1.420″ +.006/−0.

Work Space/Notes

Unit 24 Review

Name _____ **Date** _____ **Class** _____

Fill in the blanks in the following review questions.

1. Precision has to do with _____. If the same process is repeated with precision multiple times, the results will be the same or very close to the same.

2. A precise measurement is one that is carefully made with a _____ measuring instrument, so that if the measurement is taken several times, it will always be the same.

3. _____ has to do with how close a process or procedure comes to yielding the intended result.

4. Tolerance is the _____ difference between design dimension and actual dimension.

5. Machined parts cannot be perfectly _____ and _____. The job of the machinist is to keep those differences within acceptable limits.

6. There are two types of tolerance. _____ tolerance has to do with the location or position of a feature. _____ tolerance refers to the size or dimensions of a part or feature.

Applications

Use the following drawing to answer questions 1 through 4.

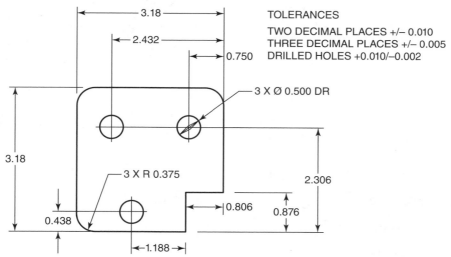

Goodheart-Willcox Publisher

1. Referring to the drawing and given the nominal size, calculate the tolerance and the minimum and maximum allowable dimensions for the lettered dimensions.

A. Nominal size: <u>3.18"</u>

    Tolerance: _____

    Minimum: _____

    Maximum: _____

B. Nominal size: <u>2.432"</u>

    Tolerance: _____

    Minimum: _____

    Maximum: _____

C. Nominal size: <u>0.750"</u>

    Tolerance: _____

    Minimum: _____

    Maximum: _____

D. Nominal size: <u>3 X 0.500" DR</u>

    Tolerance: _____

    Minimum: _____

    Maximum: _____

E. Nominal size: <u>3 X R0.375"</u>

    Tolerance: _____

    Minimum: _____

    Maximum: _____

F. Nominal size: <u>3.18"</u>

    Tolerance: _____

    Minimum: _____

    Maximum: _____

G. Nominal size: <u>0.438"</u>

    Tolerance: _____

    Minimum: _____

    Maximum: _____

H. Nominal size: <u>1.188"</u>

    Tolerance: _____

    Minimum: _____

    Maximum: _____

I. Nominal size: <u>0.806"</u>

    Tolerance: _____

    Minimum: _____

    Maximum: _____

J. Nominal size: <u>0.876"</u>

    Tolerance: _____

    Minimum: _____

    Maximum: _____

K. Nominal size: <u>2.306"</u>

    Tolerance: _____

    Minimum: _____

    Maximum: _____

Name _____   **Date** _____   **Class** _____

2. If the drilled-hole callout was changed to a reamed-hole callout with a tolerance of +.0005/−.0000, what would the new minimum and maximum allowable dimensions be?

   Reamed hole = .500″

   Tolerance: _____

   Minimum: _____

   Maximum: _____

3. If the tolerance is changed to ±.005″ for dimensions with two decimal places, what would the new minimum and maximum allowable dimensions be?

   Nominal size: _____

   Tolerance: _____

   Minimum: _____

   Maximum: _____

4. If the tolerance for the radius dimension is changed to ±.015″, what would the new minimum and maximum allowable dimensions be?

   Nominal size: _____

   Tolerance: _____

   Minimum: _____

   Maximum: _____

5. Referring to the drawing of a welding jig on page 236, write the nominal size, tolerance, and minimum and maximum allowable dimensions for the lettered dimensions.

   A. Nominal size: _____

        Tolerance: _____

        Minimum: _____

        Maximum: _____

   B. Nominal size: _____

        Tolerance: _____

        Minimum: _____

        Maximum: _____

   C. Nominal size: _____

        Tolerance: _____

        Minimum: _____

        Maximum: _____

   D. Nominal size: _____

        Tolerance: _____

        Minimum: _____

        Maximum: _____

E. Nominal size: _____

    Tolerance: _____

    Minimum: _____

    Maximum: _____

F. Nominal size: _____

    Tolerance: _____

    Minimum: _____

    Maximum: _____

G. Nominal size: _____

    Tolerance: _____

    Minimum: _____

    Maximum: _____

H. Nominal size: _____

    Tolerance: _____

    Minimum: _____

    Maximum: _____

I. Nominal size: _____

    Tolerance: _____

    Minimum: _____

    Maximum: _____

J. Nominal size: _____

    Tolerance: _____

    Minimum: _____

    Maximum: _____

K. Nominal size: _____

    Tolerance: _____

    Minimum: _____

    Maximum: _____

L. Nominal size: _____

    Tolerance: _____

    Minimum: _____

    Maximum: _____

M. Nominal size: _____

    Tolerance: _____

    Minimum: _____

    Maximum: _____

N. Nominal size: _____

    Tolerance: _____

    Minimum: _____

    Maximum: _____

6. If the tolerances are changed as follows, what are the new allowable dimensions?

$$.xx \pm .01$$

$$.xxx \pm .002$$

$$.xxxx \pm .0005$$

A. Nominal size: _____

    a. Tolerance: _____

    b. Minimum: _____

    c. Maximum: _____

B. Nominal size: _____

    a. Tolerance: _____

    b. Minimum: _____

    c. Maximum: _____

Name _____ **Date** _____ **Class** _____

C. Nominal size: _____

   a. Tolerance: _____

   b. Minimum: _____

   c. Maximum: _____

D. Nominal size: _____

   a. Tolerance: _____

   b. Minimum: _____

   c. Maximum: _____

E. Nominal size: _____

   a. Tolerance: _____

   b. Minimum: _____

   c. Maximum: _____

F. Nominal size: _____

   a. Tolerance: _____

   b. Minimum: _____

   c. Maximum: _____

G. Nominal size: _____

   a. Tolerance: _____

   b. Minimum: _____

   c. Maximum: _____

H. Nominal size: _____

   a. Tolerance: _____

   b. Minimum: _____

   c. Maximum: _____

I. Nominal size: _____

   a. Tolerance: _____

   b. Minimum: _____

   c. Maximum: _____

J. Nominal size: _____

   a. Tolerance: _____

   b. Minimum: _____

   c. Maximum: _____

K. Nominal size: _____

   a. Tolerance: _____

   b. Minimum: _____

   c. Maximum: _____

L. Nominal size: _____

   a. Tolerance: _____

   b. Minimum: _____

   c. Maximum: _____

M. Nominal size: _____

   a. Tolerance: _____

   b. Minimum: _____

   c. Maximum: _____

N. Nominal size: _____

   a. Tolerance: _____

   b. Minimum: _____

   c. Maximum: _____

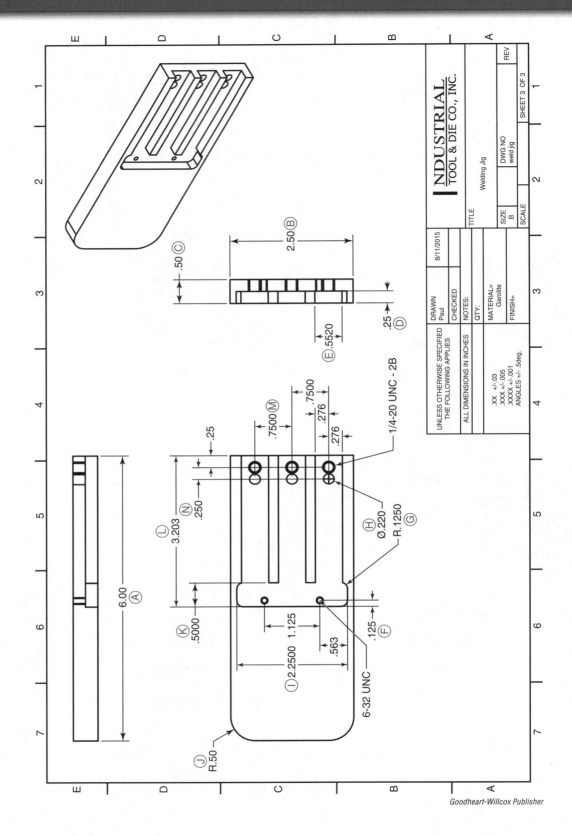

INDUSTRIAL
TOOL & DIE CO., INC.

TITLE
Welding Jig

DWG NO
weld jig

REV

SHEET 3 OF 3

SIZE
B

SCALE

DRAWN
Paul

CHECKED

NOTES:

QTY:

MATERIAL=
Garolite

FINISH=

8/11/2015

UNLESS OTHERWISE SPECIFIED
THE FOLLOWING APPLIES

ALL DIMENSIONS IN INCHES

XX +/-.03
XXX +/-.005
XXXX +/-.001
ANGLES +/- .5deg.

2.50 (B)

.50 (C)

.25 (D)

(E) .5520

.7500 (M)

.7500

.276

.276

.25

.250 (N)

3.203 (L)

Ø.220 (H)

R.1250 (G)

1/4-20 UNC - 2B

.5000 (K)

2.2500 (I)

1.125

.563

.125 (F)

6-32 UNC

6.00 (A)

R.50 (J)

UNIT 25

Vernier Measuring Instruments

Objectives

Information in this unit will enable you to:

- Explain what an inch-based micrometer is and describe how it is used.
- Describe how a vernier micrometer differs from an inch-based micrometer, and explain how to read it.
- Discuss the characteristics of a metric vernier micrometer.
- Explain what a vernier caliper is.
- Discuss how to read 25-division inch, 50-division inch, and metric vernier calipers.

Inch-Based Micrometer

The **micrometer** is the most often used precision measuring instrument in the machinist's tool chest. A micrometer caliper is often called simply a "micrometer" or a "mike." The basic parts of the micrometer are shown in the figure. The thimble and sleeve have mating threads with 40 threads per inch on the inch-based micrometer. For each full turn of the thimble, it moves 1/40 of an inch or .025" along the sleeve. The sleeve has graduation of .025", usually with every fourth graduation or .1" numbered. The thimble has 25 graduations, with each representing .001".

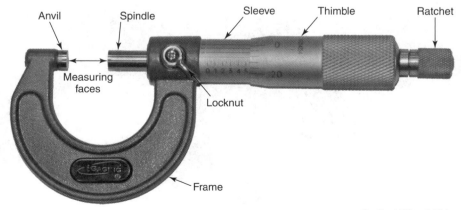

Goodheart-Willcox Publisher

As the thimble is turned to retract the spindle, graduations on the sleeve are uncovered. Until you are well practiced, it is easiest to keep track of the micrometer reading as follows. This explanation refers to the drawing of a micrometer reading below. The 2, representing two hundred thousandths or 0.200″, is the highest number showing on the sleeve, so write that down. There are two graduations showing beyond the 2, representing twenty-five thousandths or 0.025″ each, so add them together and write down 0.050″. Find the largest number on the thimble to align with the reading line or just below the reading line on the sleeve. In this case that number is 5, representing five thousandths or 0.005″, so write down 0.005″. Add the numbers written down to find the reading of the micrometer caliper.

Micrometer Caliper Reading

Sleeve

0 1 2

5 — Thimble

Reading line

0.200″
0.050″
+ 0.005″
0.255″

Vernier Micrometer

In 1631, a French mathematician named Pierre Vernier invented a method of measuring with much more precision than had been possible before. He added a second scale to the primary measuring scale, called a **vernier scale**. This second scale has slightly smaller graduations than the primary scale, so that only one graduation on the secondary scale can be aligned with any graduation on the primary scale. On a vernier micrometer, the vernier scale is on the sleeve where the graduations can be aligned with the graduations on the thimble.

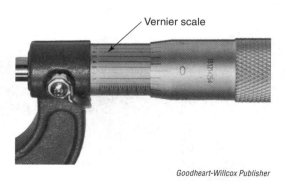

Vernier scale

The vernier scale adds one decimal place to the precision that can be obtained with a non-vernier micrometer. It is possible to measure with 0.0001" precision with an inch-based vernier micrometer. Virtually all machinist's micrometers include a vernier scale. To read a vernier micrometer, add the vernier reading to the procedure explained previously. The reading of a vernier micrometer shown in the drawing below is done as follows: The sleeve shows 0.2" plus one graduation of 0.025". The thimble graduations have passed the 0.003" mark, but have not reached 0.004", so use 0.003". Lastly, find the vernier graduation that most closely aligns with one of the thimble graduations. In this example it is the 4, so add 0.0004". The micrometer reading is 0.2284".

Customary micrometers come in a variety of sizes, all of which have measuring ranges of 1 inch. The micrometers come in sizes of 0–1 inches, 1–2 inches, 2–3 inches, etc. When taking a mike reading, it is important that the micrometer is calibrated and clean (use a soft cloth to remove any marks and debris between the anvil and spindle), and that the size of the micrometer you are using is noted. If you are using a 1–2 inch micrometer, 1 inch needs to be added to the final reading.

Inch-Based Micrometer Reading

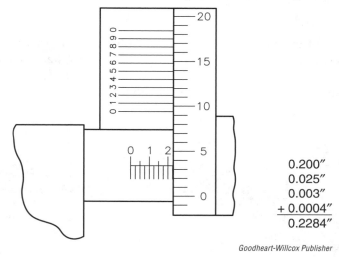

$$
\begin{array}{r}
0.200'' \\
0.025'' \\
0.003'' \\
+\ 0.0004'' \\
\hline
0.2284'' \\
\end{array}
$$

Metric Vernier Micrometer

A metric micrometer is very similar to an inch-based micrometer, except the sleeve has graduations of 0.5 mm and the thimble has graduations of 0.01 mm. The vernier scale has graduations of 0.001 mm or 0.002 mm. When the vernier scale is graduated in .001 mm increments, usually only every other graduation has a number.

Metric micrometers come in a variety of sizes, all of which have measuring ranges of 25 mm. The micrometers come in sizes of 0–25 mm, 25–50 mm, 50–75 mm, etc. It is again important that the micrometer is calibrated and clean, and that you note the size of the micrometer. If you are using a 25–50 mm micrometer, for example, 25 mm may need to be added to the final reading. Some metric micrometers start the sleeve reading at the appropriate reading, such as 25 instead of 0, for example.

Metric Micrometer Reading

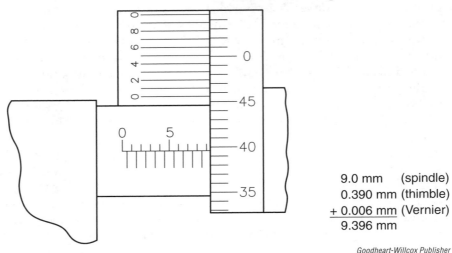

9.0 mm   (spindle)
0.390 mm (thimble)
+ 0.006 mm (Vernier)
9.396 mm

Vernier Caliper

A **vernier caliper** is another device used for accurate measurements in the machine shop. A vernier caliper, sometimes called simply a "caliper," has a fixed jaw attached to a graduated scale and a movable jaw with a vernier scale. The vernier scale of an inch caliper is graduated with either 25 or 50 graduations per inch.

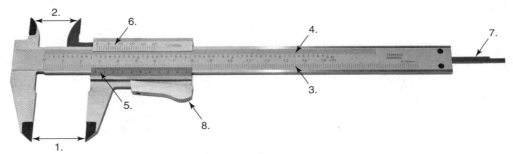

1. **Outside jaws**: used to measure outside dimensions, length, width, or thickness of an object
2. **Inside jaws**: used to measure inside diameter of an object
3. **Metric Main Scale**: graduated in mm with cm numbered
4. **Inch Main Scale**: graduated in 0.25" increments with whole inches and 0.1" numbered
5. **Metric vernier**: gives measurements to 0.1 mm
6. **Inch vernier**: 25 graduations on this model. Gives measurements to 0.01"
7. **Depth probe**: used to measure depth of holes
8. **Retainer**: locks movable parts in place to hold a reading. Many calipers have a set screw to hold a reading.

Reading a 25-Division Inch-Based Vernier Caliper

The first step in reading the caliper is to observe whether the vernier scale has 25 divisions or 50 divisions. When the vernier scale has 25 graduations, read the number of inches on the main beam to the 0 on the vernier scale and record that. Note how many tenths there are between the the last full inch and the 0 on the vernier scale. Tenths are numbered on the scale. Count the number of fortieths (0.025″) there are past the last tenth and record that. Find the graduation on the vernier scale that aligns with one of the marks on the main scale. Count the number of divisions from the 0 to that graduation. Each division on the vernier equals 0.001″. Record the vernier reading and add the readings you recorded.

25-Division Vernier Caliper Reading

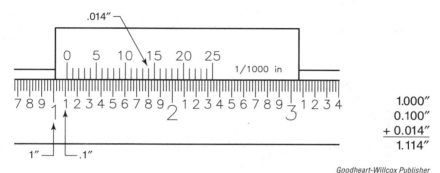

1.000″
0.100″
+ 0.014″
1.114″

Goodheart-Willcox Publisher

Reading a 50-Division Inch-Based Vernier Caliper

On a caliper with a 50-division vernier scale, each division on the main scale equals 0.05″. Every second division equals 0.10″ and is numbered. Each division on the vernier equals 0.001″.

50-Division Vernier Caliper Reading

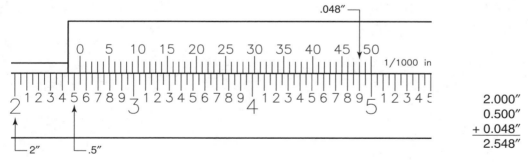

2.000″
0.500″
+ 0.048″
2.548″

Goodheart-Willcox Publisher

Reading a Metric Vernier Caliper

The divisions on the main beam of a metric vernier caliper are 1 mm. The divisions on the vernier scale are 0.02 mm each. To read the caliper, record the number of millimeters on the main scale to the left of the vernier scale 0. Record that number. Find the vernier mark that aligns with a mark on the main scale. The numbered graduations on the vernier represent tenths of a millimeter. The graduations between the numbers represent 0.02 mm each. Record the vernier reading and total the readings.

Metric Vernier Caliper Reading

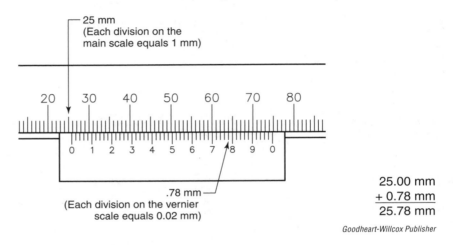

25 mm
(Each division on the main scale equals 1 mm)

.78 mm
(Each division on the vernier scale equals 0.02 mm)

25.00 mm
+ 0.78 mm
25.78 mm

Goodheart-Willcox Publisher

Unit 25 Review

Name _____ Date _____ Class _____

Fill in the blanks in the following review questions.

1. The _____ is the most often used precision measuring instrument in the machinist's tool chest.

2. The _____ and _____ have mating threads with 40 threads per inch on the inch-based micrometer. For each full turn of the _____, it moves 1/40 of an inch or 0.025" along the sleeve.

3. The _____ has graduations of 0.025", usually with every fourth graduation or 0.1" numbered. The _____ has 25 graduations, each representing 0.001".

4. A second scale to the primary measuring scale, called a _____ scale, is a method of measuring with much more precision.

5. On a vernier micrometer, the vernier scale is on the sleeve where the graduations can be aligned with the graduations on the _____. The vernier scale adds _____ decimal place to the precision that can be obtained with a non-vernier micrometer.

6. A metric micrometer is very similar to an inch-based micrometer, except the sleeve has graduations of _____ and the thimble has graduations of _____.

7. The vernier scale has graduations of _____ or _____.

8. A vernier _____ is another device used for accurate measurements in the machine shop. The vernier scale of an inch _____ is graduated with either 25 or 50 graduations per inch.

Applications

Indicate the value of each inch-based micrometer reading.

1. _____

2. _____

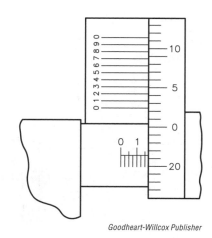

Goodheart-Willcox Publisher

Goodheart-Willcox Publisher

3. _____

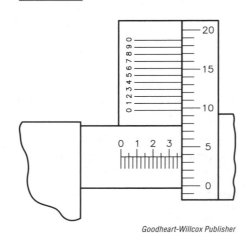

Goodheart-Willcox Publisher

4. _____

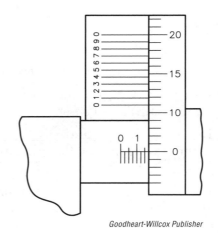

Goodheart-Willcox Publisher

5. _____

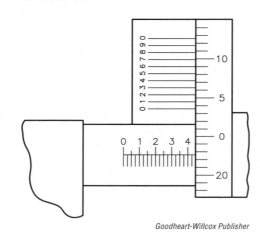

Goodheart-Willcox Publisher

6. _____

Goodheart-Willcox Publisher

Indicate the value of each metric micrometer reading.

7. _____

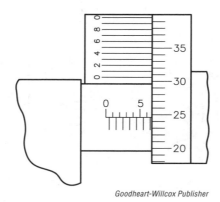

Goodheart-Willcox Publisher

8. _____

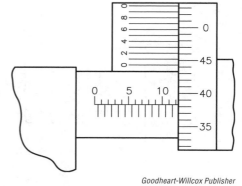

Goodheart-Willcox Publisher

Name _____   **Date** _____   **Class** _____

9. _____

Goodheart-Willcox Publisher

10. _____

Goodheart-Willcox Publisher

11. _____

Goodheart-Willcox Publisher

12. _____

Goodheart-Willcox Publisher

Fill in each micrometer measurement given the details of the inch-based micrometer reading.

	Size of Micrometer	Reading on Barrel Scale	Reading on Thimble Scale	Reading on Vernier Scale	Measurement of the Micrometer?
13.	0–1″	0.750″	0.012″	0.0002″	_____
14.	0–1″	0.925″	0.007″	0.0003″	_____
15.	1–2″	0.325″	0.015″	0.0005″	_____
16.	3–4″	0.450″	0.003″	0.0009″	_____
17.	2–3″	0.875″	0.021″	0.0006″	_____
18.	4–5″	0.125″	0.023″	0.0001″	_____
19.	1–2″	0.225″	0.001″	0.0007″	_____
20.	5–6″	0.675″	0.010″	0.0004″	_____
21.	0–1″	0.550″	0.016″	0.0008″	_____
22.	1–2″	0.025″	0.017″	0.0000″	_____

Goodheart-Willcox Publisher

Fill in each micrometer measurement given the details of the metric micrometer reading.

	Size of Micrometer	Reading on Barrel Scale	Reading on Thimble Scale	Reading on Vernier Scale	Measurement of the Micrometer?
23.	0–25 mm	4.5 mm	0.46 mm	0.001 mm	_____
24.	0–25 mm	18 mm	0.12 mm	0.004 mm	_____
25.	25–50 mm	22.5 mm	0.22 mm	0.005 mm	_____
26.	50–75 mm	3 mm	0.31 mm	0.008 mm	_____
27.	75–100 mm	12.5 mm	0.42 mm	0.002 mm	_____
28.	50–75 mm	7 mm	0.09 mm	0.001 mm	_____
29.	100–125 mm	20.5 mm	0.02 mm	0.003 mm	_____
30.	125–150 mm	1.5 mm	0.37 mm	0.009 mm	_____
31.	25–50 mm	0.5 mm	0.28 mm	0.007 mm	_____
32.	50–75 mm	16.5 mm	0.19 mm	0.006 mm	_____

Goodheart-Willcox Publisher

Indicate the value of each inch-based caliper reading.

33. _____

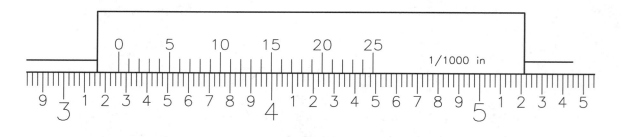

Goodheart-Willcox Publisher

34. _____

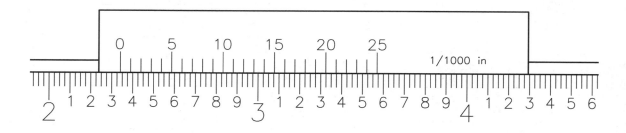

Goodheart-Willcox Publisher

Name _____   **Date** _____   **Class** _____

35. _____

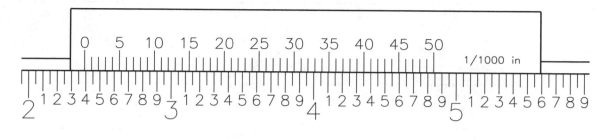

Goodheart-Willcox Publisher

36. _____

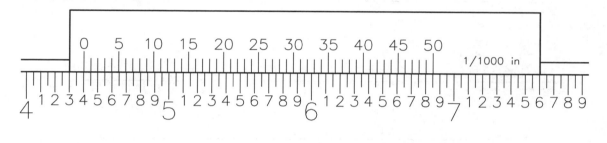

Goodheart-Willcox Publisher

37. _____

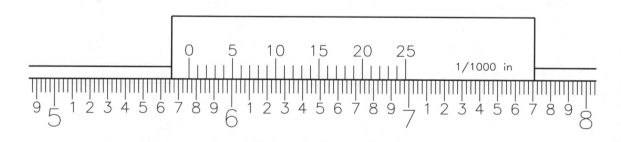

Goodheart-Willcox Publisher

Indicate the value of each metric caliper reading.

38. _____

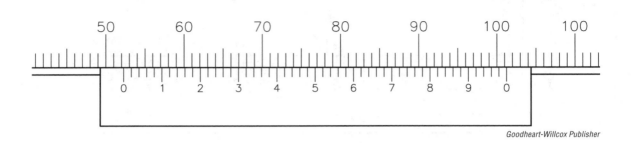

39. _____

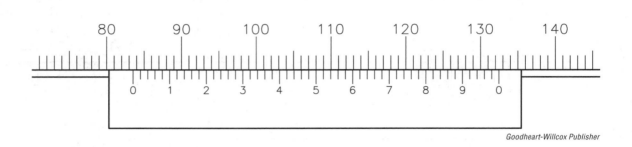

40. _____

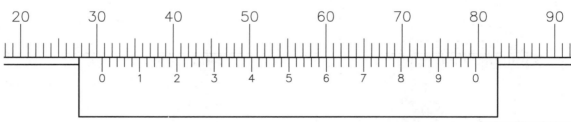

SECTION 6

Formulas

Key Terms

addendum	formula	rpm
circular pitch	ftr	speed
dedendum	major diameter	tailstock offset
diametral pitch	p	taper
equation	pitch circle	tpi
feed	pitch diameter	

UNIT 26

Equations and Formulas

Objectives

Information in this unit will enable you to:

- Differentiate between equations and formulas and explain how each is used.
- Understand how to rearrange equations.

Equations and Formulas

An **equation** is a mathematical statement that two things are equal. For example, $3 \times 2 = 6$ is an equation. Whatever is on the left side of the equal sign (=) is of the same value as whatever is on the right side.

A **formula** is a special kind of equation that includes variables. In the equation $x = y + 4$, the values of x and y can vary. If a value of 2 is assigned to y, we can solve the equation.

$$x = y + 4$$

Write the formula with 2 substituted for the y

$$x = 2 + 4$$
$$x = 6$$

Equations can be rearranged to find the value of a variable that is combined with other values. We can make any mathematical changes to the equation, as long as we do the same thing on both sides of the equal sign.

Example 26-1

Using the formula $a = b + 6$, if $a = 10$ solve for b.

Write the formula: $a = b + 6$

Subtract 6 from each side of the equation.

$$
\begin{array}{r}
a = b + 6 \\
\underline{-6 \quad\ -6} \\
a - 6 = b
\end{array}
$$

Substitute the assigned value for a.

$$10 - 6 = b$$
$$b = 4$$

Example 26-2

Using the formula $M = \frac{72}{t}$, where $M = 24$, what is t?

Write the formula: $M = \frac{72}{t}$

Substitute the known value for M.

$$24 = \frac{72}{t}$$

Multiply both sides by t.

$$24 \times t = 72$$

Divide both sides by 24.

$$t = \frac{72}{24}$$

$$t = 3$$

Check your answer by substituting the solution back into the original equation.

$$24 = \frac{72}{3}$$

Machinists frequently use formulas for calculating thread count and depth, finding the surface area of a part, computing dimensions, machining tapers, and determining the proper feeds and speeds for various machining operations.

Work Space/Notes

Unit 26 Review

Fill in the blanks in the following review questions.

1. An equation is a _____ statement that two things are equal. Whatever is on the _____ side of the equal sign (=) is of the same value as whatever is on the _____ side.

2. A formula is a special kind of _____ that includes variables.

3. Equations can be rearranged to find the value of a _____ that is combined with other values.

4. We can make any _____ changes to the equation, as long as we do the same thing on both sides of the equal sign.

Practice

Rearrange the equations so that x stands alone on one side of the equal sign.

1. $x - 4 = 17$

   _____

2. $35 + y = x - 5$

   _____

3. $\frac{x}{3} = 14$

   _____

4. $22 = 3x + 1$

   _____

5. $x + a = 31$

   _____

6. $2x - a = y$

   _____

7. $\frac{1}{3}x = a + 3$

   _____

8. $\frac{xa}{4} = y$

   _____

9. $yx + 5 = 4a$

   _____

10. $8 = x - 14 - y$

    _____

Solve the following to the nearest two decimal places.

11. What is a if $a = .667h$ and $h = 5$? _____

12. What is x if $12x - a = 108$ and $a = 12$? _____

13. What is x if $\frac{x}{4} + c = 6$ and $c = 4$? _____

14. What is a if $16 = 6a - b$ and $b = 8$? _____

15. What is y if $y - \frac{25}{a} = 4$ and $a = 5$? _____

16. What is t if $84 + 8b = 4t$ and $b = 4$? _____

17. Find b when $a = 4.5\,b$ and $a = 1.25$. _____

18. Find c when $c = \pi d$ and $d = 2.375$. _____

19. Find x when $x = \frac{t}{3}$ and $t = 25$. _____

20. Find r when $\frac{1}{2y} = r + 3$ and $y = 7.55$. _____

21. Given the formula $a = \frac{1}{2}bh$, what is a when $b = 2.25$ and $h = 3.667$? _____

22. What is the perimeter of a rectangle with a length of 2.375" and a width of 1.25"? Use the formula $P = 2L + 2W$, where P is perimeter, L is length, and W is width.

_____

UNIT 27

Thread Formulas

Objectives

Information in this unit will enable you to:

- Discuss the use of standard threads in the Unified Thread Standard (UTS).
- Describe how to find the pitch of a thread when threads per inch (tpi) is known.
- Explain how to determine thread height.
- Understand how to calculate the major diameter of a gage number thread.
- Explain how to find a tap drill of correct size.
- Use formulas to check the pitch diameter of a thread.

Thread Formulas

Standard threads in the United States and Canada are specified according to the UTS (Unified Thread Standard). In this system, there are three series of threads: UNC (Unified National Coarse), UNF (Unified National Fine), and UNEF (Unified National Extra Fine). Coarse threads have fewer threads per inch or **tpi**.

The Unified Thread Standard also specifies the **major diameter** (largest diameter) of the thread in fractions of an inch. That is the dimension from the peak of a thread to the peak of a thread on the opposite side of the bolt or shaft. The coarseness of the threads is specified as the number of threads per inch.

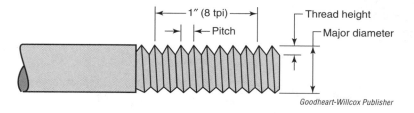

Goodheart-Willcox Publisher

The specification includes an indication of the thread series (UNC, UNF, or UNEF), LH if it is a left-hand thread, and possibly a tolerance specification. The parts of a UTS thread specification are shown in the following figure.

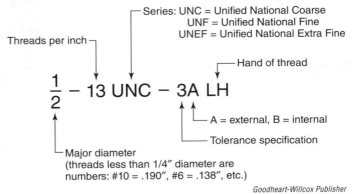

Metric threads are specified by their own rules. A letter *M* indicates that it is a metric thread, the diameter is specified in millimeters, and the coarseness is specified by the pitch in millimeters. The following figure shows the parts of a metric thread specification.

A couple of formulas are especially useful to the machinist when working with threads. To find the pitch of a thread, or **p**, when the tpi is known, use the formula:

$$p = \frac{1}{\text{tpi}}$$

where *p* is the pitch and *tpi* is the threads per inch.

Example 27-1

Using the formula above, what is the pitch of the thread shown in the following figure?

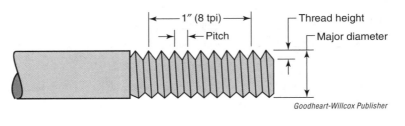

$$p = 1/8'' \text{ or } .125''$$

The major diameter of a gage number thread (threads less than 1/4″ diameter) can be calculated as follows:

$$\text{Major } \varnothing = \text{gage \#} \times 0.013 + 0.06$$

Example 27-2

What is the major diameter of a thread #8 – 32, using the above formula?

$$\text{Major } \varnothing = \text{gage \#} \times 0.013 + 0.06$$

Substitute 8 for the gage #.

$$\text{Major } \varnothing = 8 \times 0.013 + 0.06$$

Major $\varnothing$ of the #8 thread = 0.164″.

The ratio of thread height to thread pitch is the same for all UTC and ISO (metric) threads. The thread height is .866 times the pitch. The formula for thread height is $H = .866 \times p$, where H is thread height and p is pitch.

Example 27-3

What is the height of the thread on a 5/16 – 18 UNC bolt?

Convert 18 tpi to pitch with the formula:

$$p = \frac{1}{\text{tpi}}$$
$$p = .056″$$

Use the formula for thread height.

$$H = .866 \times p$$

Substitute .056″ for p.

$$H = .866 \times .056$$
$$H = .048″$$

Another common situation in the machine shop is having to find a tap drill of the correct size before tapping a hole. Most shops will have a tap drill chart for this purpose. However, if one is not readily available, the size can be calculated. To determine tap drill sizes for the Unified Thread Standard or metric system (or any thread with a thread angle of 60°), use the following formula:

$$\text{Major diameter of thread} - \text{Thread pitch} = \text{Tap drill size}$$

Note: This will yield an approximate 75% thread engagement.

Example 27-4

What is the tap drill size needed for an M10 × 1.5 thread using the above formula?

Tap drill size = Major diameter of the thread − Thread pitch

The major diameter is 10 and the pitch is 1.5.

Tap drill size = 10 − 1.5

Tap drill size needed = 8.5 mm.

Formulas for Checking Pitch Diameter

The pitch diameter of a thread can be measured accurately using a micrometer and three precision wires of equal diameter. Two of the wires are placed in contact with the thread on one side, and the third wire is placed opposite, as illustrated in the drawing below.

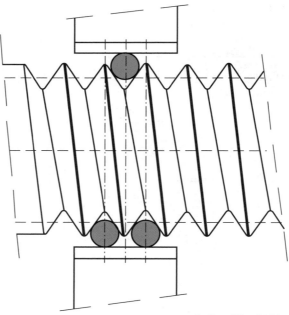

Goodheart-Willcox Publisher

The formula to calculate the micrometer reading for Unified Thread Standard (inches) is:

$$\text{Micrometer reading} = \text{Pitch diameter} - \frac{0.86603}{\text{tpi}} + 3w$$

The formula to calculate the micrometer reading for metric threads (mm) is:

$$\text{Micrometer reading} = \text{Pitch diameter} - 0.86603 \times \text{pitch} + 3w$$

Where w = wire size used for measurement.

Pitch diameter limits (max and min) need to be looked up in a reference book or chart for these calculations. (Your instructor can help with this.)

Note: The formulas do not account for lead angle error upon measurement. This is usually acceptable unless an extremely accurate measurement is required.

To calculate the correct wire size (w) to use for taking your measurement, use the formulas below:

$$\text{The smallest wire size that can be used for a given thread} = \frac{0.56}{\text{tpi}}$$

$$\text{The largest wire size that can be used for a given thread} = \frac{0.9}{\text{tpi}}$$

Example 27-5

Calculate the micrometer reading for a 7/16 – 20 UNF 2A thread. The pitch diameter limits are 0.3995″ and 0.4037″.

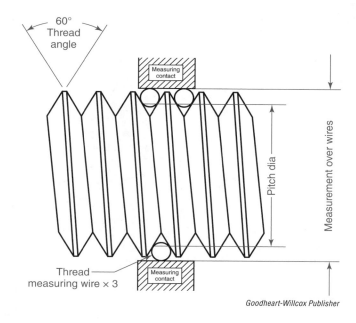

Goodheart-Willcox Publisher

First calculate the range of wire sizes that can be used to take a measurement.

$$\text{Minimum wire size} = \frac{0.56}{\text{tpi}} \qquad\qquad \text{Maximum wire size} = \frac{0.9}{\text{tpi}}$$

$$\text{Minimum wire size} = \frac{0.56}{20} \qquad\qquad \text{Maximum wire size} = \frac{0.9}{20}$$

$$\text{Minimum wire size} = 0.028 \qquad\qquad \text{Maximum wire size} = 0.045$$

A standard set of pitch diameter measuring wires would have wires of 0.029″, 0.032″, 0.040″, and 0.045″ that could be used for measuring this thread. For this example, 0.029″ wires will be used. Remember that all three wires must be of equal size.

Now calculate the micrometer reading using the 0.029″ wires.

$$\text{Minimum micrometer reading} = \text{Minimum pitch diameter} - \frac{0.86603}{\text{tpi}} + 3w$$

$$\text{Maximum micrometer reading} = \text{Maximum pitch diameter} - \frac{0.86603}{\text{tpi}} + 3w$$

(Continued)

For the 7/16 – 20 UNF Class 2A thread, the minimum pitch diameter is 0.3995″ and the maximum pitch diameter is 0.4037″. The tpi = 20, and w = 0.029″.

$$\text{Minimum micrometer reading} = 0.3995 - \frac{0.86603}{20} + 3 \times 0.029$$

$$\text{Minimum micrometer reading} = 0.4432''$$

$$\text{Maximum micrometer reading} = 0.4037 - \frac{0.86603}{20} + 3 \times 0.029$$

$$\text{Maximum micrometer reading} = 0.4474''$$

Unit 27 Review

Name _____ **Date** _____ **Class** _____

Fill in the blanks in the following review questions.

1. Standard threads in the United States and Canada are specified according to the UTS (_____ _____ _____).

2. In this system, there are _____ series of threads.

3. Coarse threads have _____ threads per inch (tpi).

4. The Unified Thread Standard also specifies the _____ _____ (largest diameter) of the thread in fractions of an inch.

5. Metric threads are specified by their own rules. A letter _____ indicates that it is a metric thread.

6. The _____ is specified in millimeters.

7. The coarseness is specified by the _____.

Applications

Label the parts of the bolt specifications.

$$\overset{1.}{\frac{7}{16}} - \overset{2.}{14} \; \overset{3.}{\text{UNC}} - \overset{4.}{2\text{A}} \times \overset{5.}{2\frac{1}{2}}$$

Goodheart-Willcox Publisher

1. _____

2. _____

3. _____

4. _____

5. _____

$$\overset{6.\;7.}{\text{M8}} \times \overset{8.}{1.0} \times \overset{9.}{20}$$

Goodheart-Willcox Publisher

6. _____

7. _____

8. _____

9. _____

Compute the answers to the following questions.

10. To the nearest .001", what is the thread height of a UTC thread with pitch of .5555"?

_____

11. Compute the height of 7/8 – 14 UNF threads to three decimal places.

_____

12. Compute the height of M12 × 1.75 threads to two decimal places.

_____

Calculate the major diameter of the gage number threads listed below. The major diameter can be calculated using the following formula:

$$\text{Major } \varnothing = \text{Gage \#} \times 0.013 + 0.06$$

	Size of Thread	Major Diameter (inches)
13.	#0	_____
14.	#1	_____
15.	#2	_____
16.	#3	_____
17.	#4	_____
18.	#5	_____
19.	#6	_____
20.	#8	_____
21.	#10	_____
22.	#12	_____

Goodheart-Willcox Publisher

Name _____ **Date** _____ **Class** _____

In the table below, given the tpi, calculate the pitch of the threads, minimum wire size, and maximum wire size. Use the following formulas and round to the nearest tenth (4 decimal places) where necessary.

$$\text{Pitch } (p) = \frac{1}{\text{tpi}}$$

$$\text{Minimum wire size} = \frac{0.56}{\text{tpi}} \qquad \text{Maximum wire size} = \frac{0.9}{\text{tpi}}$$

	Threads Per Inch (TPI)	Pitch	Minimum Wire Size	Maximum Wire Size
23.	4			
24.	4 ½			
25.	6			
26.	7			
27.	8			
28.	10			
29.	11			
30.	12			
31.	13			
32.	14			
33.	16			
34.	18			
35.	20			
36.	22			
37.	24			
38.	26			
39.	27			
40.	28			
41.	30			
42.	32			
43.	36			
44.	40			
45.	44			
46.	48			
47.	50			
48.	56			
49.	64			
50.	72			
51.	80			

Goodheart-Willcox Publisher

Calculate the required tap drill size for the following metric threads with an approximate 75% full thread.

Tap drill size = Major diameter of thread − Thread pitch

	Major Diameter (mm)	Pitch (mm)	Tap Drill Size Required
52.	M1.6	0.35	_____
53.	M2	0.4	_____
54.	M2.5	0.45	_____
55.	M3	0.5	_____
56.	M3.5	0.6	_____
57.	M4	0.7	_____
58.	M5	0.8	_____
59.	M6	1.0	_____
60.	M8	1.25	_____
61.	M8	1.0	_____
62.	M10	1.5	_____
63.	M10	1.25	_____
64.	M12	1.75	_____
65.	M12	1.25	_____
66.	M14	2	_____
67.	M14	1.5	_____
68.	M16	2	_____
69.	M16	1.5	_____
70.	M18	2.5	_____
71.	M18	1.5	_____
72.	M20	2.5	_____
73.	M20	1.5	_____
74.	M22	2.5	_____
75.	M22	1.5	_____
76.	M24	3.0	_____
77.	M24	2.0	_____
78.	M27	3.0	_____
79.	M27	2.0	_____
80.	M30	3.5	_____
81.	M30	2.0	_____
82.	M33	3.5	_____
83.	M33	2.0	_____
84.	M36	4.0	_____
85.	M36	3.0	_____

Goodheart-Willcox Publisher

Name _____ **Date** _____ **Class** _____

Use the formulas that follow to answer questions 86 through 88.

Minimum micrometer reading = Minimum pitch diameter $- \dfrac{0.86603}{\text{tpi}} + 3w$

Maximum micrometer reading = Maximum pitch diameter $- \dfrac{0.86603}{\text{tpi}} + 3w$

86. Calculate the minimum and maximum micrometer reading for a 3/8 – 24 UNF 2A thread. The pitch diameter limits are 0.3430″ and 0.3468″, and the wire size to use is 0.024″.

    _____

87. Calculate the minimum and maximum micrometer reading for a 1/2 – 13 UNC 2A thread. The pitch diameter limits are 0.4435″ and 0.4485″, and the wire size to use is 0.045″.

    _____

88. Calculate the minimum and maximum micrometer reading for a 3/4 – 16 UNF 2A thread. The pitch diameter limits are 0.7029″ and 0.7079″, and the wire size to use is 0.040″.

    _____

Work Space/Notes

UNIT 28

Taper Formulas

Objectives

Information in this unit will enable you to:

- Identify the letter symbols used with tapers and explain the meaning of each.
- Describe how to calculate an unknown dimension of a taper when other dimensions are known.
- Explain how to calculate the amount of tailstock offset on a taper.

Tapers

Machinists are often called upon to turn **tapers** on the lathe. The following table shows common symbols for various aspects of tapers. Note that in this context, tpi stands for taper per inch rather than threads per inch.

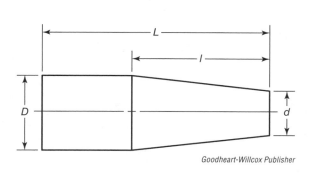

Goodheart-Willcox Publisher

Letter Symbol	Meaning
tpi	Taper per inch
tpf	Taper per foot
D	Diameter of large end
d	Diameter of small end
l	Length of taper
L	Total length of piece

Goodheart-Willcox Publisher

The formulas to calculate an unknown dimension of a taper when other dimensions are known are given in the following table. The formulas to find D and d involve parentheses inside parentheses. When this is encountered, begin with the math operation that is enclosed in the inner parentheses first, then perform the operation enclosed in the outer parentheses, and finally perform any operation(s) not enclosed in parentheses.

Unknown	Known	Formula
tpi	tpf	$tpi = tpf/12$
tpf	tpi	$tpf = tpi \times 12$
tpf	D, d, l	$tpf = \dfrac{(D - d) \times 12}{l}$
D	d, l, tpf	$D = d + \left(l\left(\dfrac{tpf}{12} \right) \right)$
d	D, l, tpf	$d = D - \left(l\left(\dfrac{tpf}{12} \right) \right)$
l	D, d, tpf	$l = 12\left(\dfrac{D - d}{tpf} \right)$

Goodheart-Willcox Publisher

Example 28-1

Compute D in the figure below.

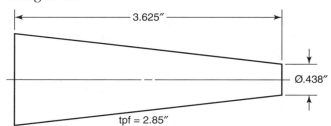

Goodheart-Willcox Publisher

Use the following formula.

$$D = d + \left(l\left(\frac{tpf}{12} \right) \right)$$

Substitute the known values in the formula.

$$D = .438 + \left(3.625\left(\frac{2.85}{12} \right) \right)$$

Reduce the fraction in the inner parentheses to a decimal.

$$D = .438 + (3.625 \times .2375)$$

Do the multiplication indicated in the inner parentheses.

$$D = .438 + .8609$$

Do the addition and round the answer to the degree of accuracy desired, which is three decimal places in this case, the maximum degree of accuracy of the other dimensions.

$$D = 1.299$$

Tailstock Offset for Turning a Taper

External tapers may be turned between centers by offsetting the tailstock as shown in the figure.

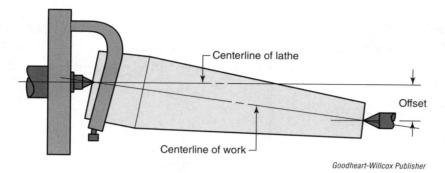

Centerline of lathe

Offset

Centerline of work

Goodheart-Willcox Publisher

The machinist must calculate the amount of **tailstock offset**. When the taper per inch (tpi) and the total length of the workpiece (*L*) are known, the following formula can be used to compute the amount of offset:

$$\text{Offset} = \frac{L \times \text{tpi}}{2}$$

Example 28-2

What is the offset for a workpiece 10″ long with a .25″ taper per inch?

Use the following formula.

$$\text{Offset} = \frac{L \times \text{tpi}}{2}$$

Then, substitute given dimensions for tpi and *L*.

$$\text{Offset} = \frac{10 \times .25}{2}$$

Do the math operations to reduce the answer to a decimal.

$$\text{Offset} = 1.25″$$

When the dimensions of the tapered section are known but the taper per inch is not known, use the following formula.

$$\text{Offset} = \frac{L \times (D - d)}{2 \times l}$$

Example 28-3

What is the offset for a workpiece that is 8″ long with a large diameter of 1.25″ tapering 4″ to .75″?

Use the following formula.

$$\text{Offset} = \frac{L \times (D - d)}{2 \times l}$$

Substitute the known dimensions in the formula.

$$\text{Offset} = \frac{8 \times (1.25 - .75)}{2 \times 4}$$

Perform the math operation inside the parentheses.

$$\text{Offset} = \frac{8 \times .5}{8}$$

Reduce the answer to a decimal.

$$\text{Offset} = .50''$$

Unit 28 Review

Name _____ **Date** _____ **Class** _____

Fill in the blanks in the following review questions.

1. Machinists are often called upon to turn _____ on the lathe.

2. The meaning of the abbreviation tpf is _____ _____ _____.

3. When working with parentheses inside parentheses, begin with the math operation that is enclosed in the _____ parentheses first.

4. External tapers may be turned between centers by offsetting the _____. The machinist must calculate the amount of _____ _____.

Applications

Compute the missing information in the table below. Round to the nearest thousandth.

	l	Taper per foot (tpf)	Taper per inch (tpi)	*D*	*d*
1.	5.0″	_____	0.10″	0.875″	_____
2.	_____	0.375″	_____	4.25″	3.75″
3.	3.75″	0.5″	_____	0.625″	_____
4.	_____	_____	0.06″	1.5″	1.0″
5.	_____	0.25″	_____	0.375″	0.25″
6.	2.687″	1.741″	_____	1.75″	_____
7.	4.25″	_____	_____	2.5″	1.125″
8.	_____	_____	0.052″	1.5″	0.875″
9.	3.5″	_____	_____	1.25″	0.5″
10.	6.0″	0.75″	_____	3.5″	_____
11.	5.75″	_____	0.042″	_____	0.9″
12.	4.0″	0.625″	_____	_____	0.75″
13.	6.25″	_____	_____	3.0″	1.75″
14.	4.75″	0.6″	_____	_____	0.675″
15.	8.0″	_____	0.05″	_____	0.725″

Goodheart-Willcox Publisher

Compute the tailstock offset for the indicated tapers to the nearest thousandth of an inch or hundredth of a millimeter.

16. $D = 1.125''$, $d = .8125''$, $l = 3''$, $L = 5.5''$ _____

17. $L = 250$ mm, $l = 145$ mm, $D = 30$ mm, $d = 15$ mm _____

18. $tpi = .05''$, $L = 7.25''$ _____

19. $tpf = .5''$, $L = 9''$ _____

20. What is the diameter of the large end of the taper on the end mill shown in the figure?

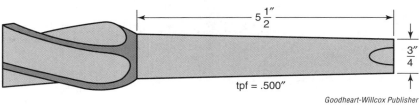

Goodheart-Willcox Publisher

_____

Name _____   **Date** _____   **Class** _____

Compute the missing information for each of the images below. Round to the nearest thousandth.

21.

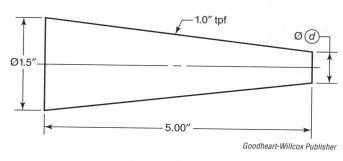

Goodheart-Willcox Publisher

_____

22.

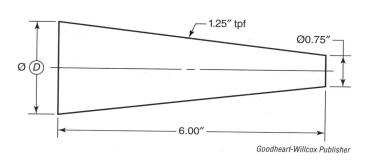

Goodheart-Willcox Publisher

_____

23.

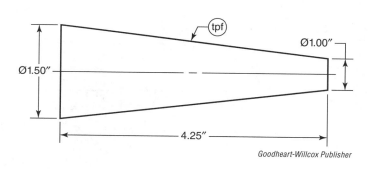

Goodheart-Willcox Publisher

_____

24.

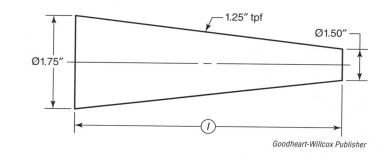

Goodheart-Willcox Publisher

_____

25.

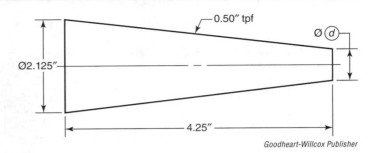

0.50″ tpf

Ø ⓓ

Ø2.125″

4.25″

Goodheart-Willcox Publisher

26.

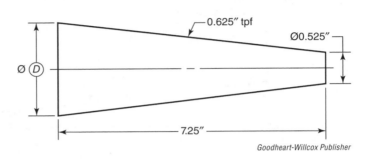

0.625″ tpf

Ø0.525″

Ø ⓓ

7.25″

Goodheart-Willcox Publisher

27.

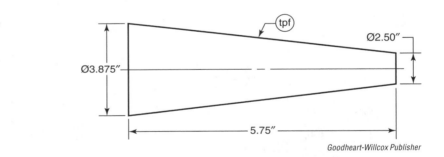

ⓣⓟⓕ

Ø2.50″

Ø3.875″

5.75″

Goodheart-Willcox Publisher

28.

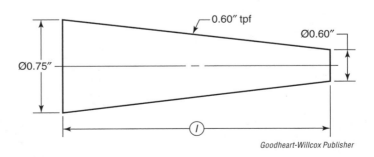

0.60″ tpf

Ø0.60″

Ø0.75″

ⓛ

Goodheart-Willcox Publisher

UNIT 29

Speeds and Feeds

Objectives

Information in this unit will enable you to:

- Explain the difference between speeds and feeds.
- Describe how to calculate cutting speed.
- Discuss how to calculate the feed rate on a milling machine.

Speeds and feeds, also called feeds and speeds, are actually two different things. They are often grouped together, because combined they play a major role in the quality of finish a machining operation produces, the time required to do that operation, and its cost. **Speed** in this context refers to the speed difference between the cutting tool and the surface of the workpiece. *Cutting speed* may be the speed at which the workpiece revolves against a stationary tool or the speed at which a cutting tool revolves against a stationary workpiece. **Feed** refers to the speed with which the cutter is advanced along the workpiece.

Calculating Cutting Speed

Cutting speed is expressed in feet per minute (fpm), or meters per minute (m/min). Cutting speed varies considerably, depending on the material being machined, the rigidity of the machine, and the geometry of the cutting tool. The ideal speed is the same for a revolving workpiece as it is for a revolving cutter.

There are several sources of recommended cutting speeds. Tables of speeds may be found on the Internet and in reference books for machinists. Manufacturers and suppliers of materials publish speeds for each of the materials they supply. Cutting speed tables usually suggest a range of speeds and the machinist uses the speed within that range that produces the best results, trying the middle of the range first. The table shown here suggests speeds for several common types of materials.

Material	High-Speed Steel Cutter		Carbide Cutter	
	fpm	m/min	fpm	m/min
SAE 1020 – low-carbon steel	80–120	25–40	300–400	90–120
SAE 1050 – high-carbon steel	60–100	18–30	180–220	55–65
Stainless steel	100–120	30–40	240–300	75–90
Aluminum	400–700	120–220	1,000–1,400	300–450
Brass	200–600	60–180	1,000–2,000	300–600

Goodheart-Willcox Publisher

Speed calculations are generally done for revolving workpieces (e.g., lathes) or revolving tools (e.g., milling machine cutters), so to set the machine up requires determining the **rpm** (revolutions per minute) of the revolving piece. The formula for calculating rpm is:

$$rpm = \frac{fpm \times 12}{\pi D}$$

In this formula, D is the diameter of the revolving piece in inches.

Note: Feet per minute (fpm) is often shown as surface feet per minute (sfm) on cutting speed charts.

Example 29-1

Calculate the mid-range rpm to turn a 2" diameter piece of stainless steel on the lathe with a carbide cutter. The average or mid-range fpm can be found by adding the slowest and fastest speeds, then dividing by 2.

$$\frac{(240+300)}{2} = \text{the average fpm of 270}$$

First, use the following formula.

$$rpm = \frac{fpm \times 12}{\pi D}$$

Substitute known information in the formula.

$$rpm = \frac{270 \times 12}{\pi \times 2}$$

$$rpm = \frac{3,240}{6.28}$$

Round rpm to nearest whole number.

$$rpm = 516$$

(Somewhere between 500 and 520 rpm would be a good place to start.)

A simplified formula for calculating rpm can also be used. If we take $\frac{12}{\pi}$ and calculate it, we get 3.81972. Rounding the answer to make it easier to remember, we can use 3.82 in the formula.

$$rpm = \frac{fpm \times 3.82}{D}$$

Example 29-2

Calculate the mid-range rpm to turn a 2″ diameter piece of stainless steel on the lathe with a carbide cutter. The average or mid-range fpm can be found by adding the slowest and fastest speeds, then dividing by 2.

$$\frac{(240 + 300)}{2} = \text{the average fpm of } 270$$

Use the simplified formula.

$$\text{rpm} = \frac{\text{fpm} \times 3.82}{D}$$

Substitute known information in the formula.

$$\text{rpm} = \frac{270 \times 3.82}{2}$$

$$\text{rpm} = \frac{1031.4}{2}$$

Round rpm to nearest whole number.

$$\text{rpm} = 516$$

The formula to calculate rpm in the metric system is:

$$\text{rpm} = \frac{\text{m/min} \times 1000}{\pi D}$$

Example 29-3

Calculate the mid-range rpm to turn a 50 mm diameter piece of stainless steel on the lathe with a carbide cutter. The average or mid-range fpm can be found by adding the slowest and fastest speeds, then dividing by 2.

$$\frac{(75 + 90)}{2} = \text{the average m/min of } 82.5$$

Use the formula.

$$\text{rpm} = \frac{\text{m/min} \times 1000}{\pi D}$$

Substitute known information in the formula.

$$\text{rpm} = \frac{82.5 \times 1000}{\pi \times 50}$$

$$\text{rpm} = \frac{82,500}{157.08}$$

Round rpm to nearest whole number.

$$\text{rpm} = 525$$

Calculating Feed Rate

The rate at which a cutting tool can be advanced into the workpiece (the speed at which the workpiece moves on a milling machine or the speed at which the tool holder moves on a lathe) depends on the material being machined, the number of cutting surfaces (teeth/flutes) on the cutter, the speed of the cutting edge(s), the coolant used, and the finish desired. Calculating the feed rate on a lathe is fairly simple, because there is only one cutting tool. The feed rate for a milling machine is more complex because different types of mills can accept different rates of material removal and the number of teeth doing the cutting can vary greatly. Tables are available that suggest a range of feed rates per revolution of the tool per tooth for various materials. The feed rate per tooth per revolution of the tool is expressed as inches per tooth per revolution using the symbol ftr. The table below shows suggested ftr ranges for common types of materials. As with speed calculations, the middle of the ftr range is a good starting point.

Type of Cutter	Free Cutting Steel		Alloy Steel		Brass		Aluminum	
	Inches per tooth	mm per tooth	Inches per tooth	mm per tooth	Inches per tooth	mm per tooth	Inches per tooth	mm per tooth
End mill	.005–010	.13–.25	.003–.007	.08–.18	.007–.015	.18–.38	.009–.022	.22–.55
Face mill	.008–.020	.20–.50	.005–.012	.13–.30	.012–.030	.30–.75	.016–.040	.40–1.02
Shell end mill	.007–.015	.18–.38	.004–.009	.10–.22	.010–.022	.25–.55	.012–.030	.30–.75
Slab mill	.004–.008	.10–.20	.001–.004	.03–.10	.006–.012	.15–.30	.008–.017	.20–.43
Side cutter	.005–.011	.13–.28	.003–.007	.08–.18	.008–.016	.20–.40	.010–.020	.25–.50
Saw	.003–.005	.08–.13	.001–.003	.03–.08	.004–.007	.10–.18	.006–.010	.15–.25

Goodheart-Willcox Publisher

To find the suggested rate at which the workpiece should travel in inches per minute, use the formula:

$$F = ftr \times T \times rpm$$

In this formula, F = feed in inches per minute, ftr = feed per tooth per revolution (from the table), T = number of teeth or flutes, and rpm = revolutions per minute. Where a range is given, the average can be found by adding the lowest number and the highest number together, then dividing that sum by 2.

Example 29-4

Calculate the feed for a 3″ shell end mill with 10 flutes (teeth) milling aluminum.

Calculate the speed (rpm) using the formula for speed.

$$\text{rpm} = \frac{\text{fpm} \times 12}{\pi D}$$

$$\text{rpm} = \frac{550 \times 12}{\pi \times 3}$$

$$\text{rpm} = \frac{6{,}600}{9.4}$$

$$\text{rpm} = 702 \text{ (speed can be rounded to nearest whole number)}$$

Now use the formula for feed:

$$F = \text{ftr} \times T \times \text{rpm}$$

$$F = .021 \times 10 \times 702$$

$$F = 147.42 \text{ inches per minute}$$

Calculating Spindle Speed

Other common cutting tools a machinist will use in the shop are drills, reamers, and taps. Hole-making is a class of machining operations that are specifically used to create a hole, or holes, into a workpiece. Most parts include a hole of some sort, so being able to calculate the correct speed and feed for these tools is essential.

Drills, reamers, and taps all enter the workpiece axially. A drill bit creates either a blind hole or a through hole with a diameter equal to that of the tool. A reamer enlarges an existing hole to the diameter of the tool. Reaming removes a minimal amount of material and is performed after drilling to obtain a more accurate diameter and a higher quality internal finish. A tap cuts internal threads into an existing hole. The existing hole is typically drilled by the required tap drill size that will accommodate the desired tap.

Calculating the spindle speed or the *rpm* is done in the same way as other cutting tools.

$$\text{rpm} = \frac{\text{fpm} \times 12}{\pi D} \quad \text{or} \quad \text{rpm} = \frac{\text{fpm} \times 3.82}{D}$$

Charts similar to the following providing speed and feed rates are common in machine shops.

Feeds and Speeds for HSS Drills, Reamers, and Taps

Material	Drills		Reamers		Taps - FPM by Threads per inch			
	FPM	Feed	FPM	Feed	3–7 1/2	8–15	16–24	25–up
Aluminum	200–250	M	150–160	M	50	100	150	200
Aluminum bronze	60	M	40–45	M	12	25	45	60
Brass	200–250	H	150–160	H	50	100	150	200
Bronze, common	200–250	H	150–160	H	40	80	100	150
Bronze, phosphor, 1/2 hard	175–180	M	130–140	M	25	40	50	80
Bronze, phosphor, soft	200–250	H	150–160	H	40	80	100	150
Cast iron, soft	140–150	H	100–110	H	30	60	90	140
Cast iron, medium soft	80–110	M	50–65	M	25	40	50	80
Cast iron, hard	45–50	L	67–75	L	10	20	30	40
Cast iron, chilled*	15	L	8–10	L	5	5	10	10
Cast steel	40–50*	L	70–75	L	20	30	40	50
Copper	70	L	45–55	L	40	80	100	150
Drop forgings (steel)	60	M	40–45	M	12	25	45	60
Duralumin	200	M	150–160	M	50	100	150	200
Everdur	60	L	40–45	L	20	30	40	50
Machinery steel	110	H	67–75	H	35	50	60	85
Magnet steel, soft	35–40	M	20–25	M	20	40	50	75
Magnet steel, hard*	15	L	10	L	5	10	15	25
Manganese steel, 7%–13%	15	L	10	L	15	20	25	30
Manganese copper, 30% Mn.*	15	L	10–12	L	…	…	…	…
Malleable iron	85–90	H	…	H	20	30	40	50
Mild steel, .20–.30 C	110–120	H	75–85	H	40	55	70	90
Molybdenum steel	55	M	35–45	M	20	30	35	45
Monel metal	50	M	35–38	M	8	10	15	20
Nickel, pure*	75	L	40	L	25	40	50	80
Nickel steel, 3 1/2%	60	L	40–45	L	8	10	15	20
Screw stock, C.R.	110	H	75	H	20	30	40	50
Spring steel	20	L	12–15	L	10	10	15	15
Stainless steel	50	M	30	M	8	10	15	20
Stainless steel, C.R.*	20	L	15	L	8	10	15	20
Steel, .40–.50 C	80	M	8–10	M	20	30	40	50
Tool, SAE, and forging steel	75	H	35–40	H	25	35	45	55
Tool, SAE, and forging steel	50	M	12	M	15	15	25	25
Tool, SAE, and forging steel*	15	L	10	L	8	10	15	20
Zinc alloy	200–250	M	150–175	M	50	100	150	200

*Use specially constructed heavy-duty drills.
Note: Carbon steel tools should be run at speeds 40% to 50% of those recommended for high-speed steel.
Spiral point taps may be run at speeds 15% to 20% faster than regular taps.

Goodheart-Willcox Publisher

Example 29-5

Calculate the mid-range rpm for a Ø0.25″ drill bit cutting aluminum.

From the cutting speed chart, aluminum has a range of 200–250 fpm.

$$\frac{(200 + 250)}{2} = \text{the average fpm of 225}$$

Use the formula.

$$\text{rpm} = \frac{\text{fpm} \times 12}{\pi D}$$

Substitute known information in the formula.

$$\text{rpm} = \frac{225 \times 12}{\pi \times 0.25}$$

$$\text{rpm} = \frac{2700}{0.785}$$

$$\text{rpm} = 3{,}439 \text{ (rounded to nearest whole number)}$$

Example 29-6

Calculate the rpm for a Ø0.5″ reamer cutting stainless steel.

From the cutting speed chart, stainless steel has an fpm of 30.

Use the formula.

$$\text{rpm} = \frac{\text{fpm} \times 12}{\pi D}$$

Substitute known information in the formula.

$$\text{rpm} = \frac{30 \times 12}{\pi \times 0.5}$$

$$\text{rpm} = \frac{360}{1.57}$$

$$\text{rpm} = 229 \text{ (rounded to nearest whole number)}$$

Example 29-7

Calculate the rpm for a 5/16 – 18 UNC tap cutting malleable iron.

From the cutting speed chart, malleable iron has an fpm of 40.

Use the formula.

$$rpm = \frac{fpm \times 12}{\pi D}$$

Substitute known information in the formula.

$$rpm = \frac{40 \times 12}{\pi \times 0.3125}$$

$$rpm = \frac{480}{0.982}$$

rpm = 489 (rounded to nearest whole number)

The feed for drills, reamers, and taps varies depending on the material being cut, the size of the cutting tool being used, and other machining conditions. The following table contains guidelines that can be utilized when drilling different materials.

Drill Feed per Revolution (IPR) by Diameter of Drill					
Reference Symbol	Under ⅛	⅛ to ¼	¼ to ½	½ to 1	Over 1″
L - Low	.001	.002	.003	.005	.006
M - Medium	.0015	.003	.006	.010	.012
H - High	.0025	.005	.010	.020	.025

Note: All diameters of reamer use a feed equal to two or three times that recommended for drills.

Goodheart-Willcox Publisher

To determine the feed rate for a drill, find the reference symbol in the feed column from the cutting speed chart. The symbol will be an L, M, or H, where L represents a low feed, M a medium feed, and H a high feed.

After determining the symbol, the size of the tool needs to be considered. Using the rows representing L, M, and H, and the columns representing the size of the tool, a suggested IPR (inches per revolution) can be found.

Feed on the lathe is typically measured in IPR, so the value can be used as is. However, on a mill, feed rate is typically IPM (inches per minute). To calculate IPM, use the following:

$$Feed\ (IPM) = IPR \times rpm$$

Example 29-8

Calculate the feed for a #3 drill (Ø0.213) drilling copper on a mill.

First calculate the rpm. From the cutting speed chart, copper has an fpm of 70. Use the formula.

$$\text{rpm} = \frac{\text{fpm} \times 12}{\pi D}$$

Substitute known information in the formula.

$$\text{rpm} = \frac{70 \times 12}{\pi \times 0.213}$$

$$\text{rpm} = \frac{840}{0.669}$$

rpm = 1,256 (rounded to nearest whole number)

Now use the formula for feed. The IPR from the table is 0.002″.

$$\text{Feed (IPM)} = \text{IPR} \times \text{rpm}$$

Substitute known information in the formula.

$$\text{Feed (IPM)} = 0.002 \times 1256$$

$$\text{Feed (IPM)} = 2.512$$

The feed rate for a reamer is typically two to three times faster than that for a drill. The application, rigidity of the setup, type of material being reamed, amount of stock to be removed, and the finish required will all determine which feed gives better results.

Example 29-9

Calculate the feed for a Ø0.25″ reamer cutting manganese steel 7%–13% on a mill.

First calculate the rpm. From the cutting speed chart, manganese steel 7%–13% has an fpm of 10. Use the formula.

$$\text{rpm} = \frac{\text{fpm} \times 12}{\pi D}$$

Substitute known information in the formula.

$$\text{rpm} = \frac{10 \times 12}{\pi \times 0.25}$$

$$\text{rpm} = \frac{120}{0.785}$$

rpm = 153 (rounded to nearest whole number)

(Continued)

Now use the formula for feed for a drill. The IPR from the table is 0.002″.

$$\text{Feed (IPM)} = \text{IPR} \times \text{rpm}$$

Substitute known information in the formula.

$$\text{Feed (IPM)} = 0.002 \times 153$$
$$\text{Feed (IPM)} = 0.306$$

Now double the answer.

$$\text{Feed (IPM)} = 0.612$$

Now triple the answer.

$$\text{Feed (IPM)} = 0.918$$

The reamer should have a feed rate of between 0.612 and 0.918 IPM.

The feed rate for a tap can be calculated via a simple formula.

$$\text{F (IPM)} = \frac{\text{rpm}}{\text{tpi}}$$

In this formula, F = feed in inches per minute, rpm = revolutions per minute, and tpi = threads per inch.

Example 29-10

Calculate the feed for a Ø0.375 – 16 UNC tap, tapping spring steel on a mill.
First calculate the rpm. From the cutting speed chart, spring steel has an fpm of 15. Use the formula.

$$\text{rpm} = \frac{\text{fpm} \times 12}{\pi D}$$

Substitute known information in the formula.

$$\text{rpm} = \frac{15 \times 12}{\pi \times 0.375}$$
$$\text{rpm} = \frac{180}{1.178}$$

rpm = 153 (rounded to nearest whole number)

Now use the formula for feed, where tpi is 16.

$$\text{F (IPM)} = \frac{\text{rpm}}{\text{tpi}}$$

Substitute known information in the formula.

$$\text{F (IPM)} = \frac{153}{16}$$

Feed (IPM) = 9.5625 (rounded to nearest ten-thousandth of an inch)

Unit 29 Review

Name _____ **Date** _____ **Class** _____

Fill in the blanks in the following review questions.

1. In the context of _____ and _____, speed refers to the speed difference between the cutting tool and the surface of the workpiece.

2. _____ _____ may be the speed at which the workpiece revolves against a stationary tool or the speed at which a cutting tool revolves against a stationary workpiece.

3. _____ refers to the speed with which the cutter is advanced along the workpiece.

4. _____ _____ is expressed in feet per minute (fpm) or meters per minute (m/min).

5. Cutting speed varies considerably, depending on the _____ being machined, the rigidity of the machine, and the geometry of the cutting tool.

6. Cutting speed tables usually suggest a _____ of speeds, and the machinist uses the speed within that range that produces the best results, trying the middle of the range first.

7. The feed rate per tooth per revolution of the tool is expressed as inches per tooth per revolution using the symbol _____.

8. As with speed calculations, the _____ of the ftr range is a good starting point.

Applications

Compute mid-range speeds for the following. Use the table in this unit and round your answer to the nearest 10 rpm.

1. Turning a 5" diameter piece of high carbon steel on a lathe with a carbide cutting tool.

   _____

2. Turning a 1.5" diameter piece of aluminum on a lathe with an HSS cutting tool.

   _____

3. Milling low-carbon steel with a 6″ diameter HSS side cutter.

    _____

4. Milling stainless steel with a 2 1/2″ diameter carbide slab cutter.

    _____

5. Drilling low-carbon steel with a 3/4″ HSS twist drill.

    _____

Compute mid-range speeds for the following. Use the table in this unit and round your answer to the nearest whole number.

6. Milling low-carbon steel with a 12 mm diameter HSS end mill.

    _____

7. Turning a 40 mm diameter piece of high-carbon steel on a lathe with a carbide cutting tool.

    _____

8. Milling aluminum with a 100 mm diameter carbide insert face mill.

    _____

Name _____ **Date** _____ **Class** _____

9. Turning a 75 mm diameter piece of brass on a lathe with an HSS cutting tool.

    _____

10. Milling stainless steel with a 25 mm carbide end mill.

    _____

*Compute the mid-range feed rates for the following milling machine operations. Use the ftr
table in this unit and the stated speeds. Round your answer to the nearest 1 inch per minute.*

11. Alloy steel; 5″ diameter HSS face mill with 7 teeth; speed = 150 sfm.

    _____

12. Brass; 6″ diameter HSS shell end mill with 10 teeth; speed = 1,500 sfm.

    _____

13. Free cutting steel; 3″ diameter HSS slitting saw with 72 teeth; speed = 250 sfm.

    _____

14. Aluminum; 3/8″ diameter HSS end mill with 2 flutes; speed = 600 sfm.

    _____

15. Brass; 2″ diameter carbide end mill with 3 flutes; speed = 1,500 sfm.

    _____

16. Aluminum; 1″ diameter carbide end mill with 3 flutes; speed = 1,000 sfm.

    _____

17. Aluminum; 3″ diameter face mill carbide inserts with 5 teeth; speed = 1,200 sfm.

    _____

18. Alloy steel; 0.625″ diameter carbide end mill with 4 flutes; speed = 270 sfm.

    _____

19. Free cutting steel; 2.5″ diameter HSS slitting saw with 30 teeth; speed = 100 sfm.

    _____

20. To the nearest tenth of a minute, how long will it take to machine 720 square inches of cast iron with a 2 1/2″ diameter × 3″ wide slab mill cutter with 6 teeth? The cutting speed is 90 sfm. The feed rate per tooth is .005 in/min.

    _____

Name _____ **Date** _____ **Class** _____

In the table below, calculate the rpm and feed rate for each tool. Calculate the rpm to the nearest whole number using the average fpm where necessary. Calculate the feed to the nearest ten-thousandth of an inch where necessary.

	Material to Be Machined	Cutting Tool	Tool Ø in Inches	Decimal Equivalent	RPM	Feed
21.	Copper	Drill	3/16	_____	_____	_____
22.	Brass	Reamer	1/8	_____	_____	_____
23.	Aluminum	Tap	7/16 – 20 UNF	_____	_____	_____
24.	Mild steel, 0.2–0.3 C	Drill	#7	_____	_____	_____
25.	Stainless steel	Reamer	21/32	_____	_____	_____
26.	Cast steel	Tap	7/8 – 9 UNC	_____	_____	_____
27.	Zinc alloy	Drill	"Z"	_____	_____	_____
28.	Molybdenum steel	Reamer	19/64	_____	_____	_____
29.	Bronze, common	Tap	1/4 – 28 UNF	_____	_____	_____

Goodheart-Willcox Publisher

Work Space/Notes

UNIT 30

Spur Gear Calculations

Objectives

Information in this unit will enable you to:

- Describe the physical characteristics of a spur gear.
- Define and knowledgeably discuss the terminology associated with spur gears.
- Explain how to calculate all of the information necessary to cut a spur gear.

Spur Gears Defined

There are many types of gears. The simplest type is the spur gear, which has teeth running straight across the perimeter of the gear. The size, pitch, and shape of the teeth are critical because most gears must mesh with another gear, usually of a different diameter.

Spur Gear Nomenclature

Most of the terms associated with gear specifications have corresponding symbols that are used in gear formulas. The **pitch circle** is an imaginary circle that passes through the approximate mid-height of the gear teeth. When two gears are meshed, their pitch circles just touch each other. The **pitch diameter** is the diameter of the pitch circle.

The thickness of a tooth is measured at the pitch circle. **Diametral pitch** is the number of teeth per inch on the pitch circle. The **circular pitch** is the distance measured on the pitch circle between the same points on adjacent teeth.

The **addendum** is the distance a tooth extends above the pitch circle and the **dedendum** is the distance a tooth extends below the pitch circle. Teeth of a mating gear do not usually extend all the way to the bottom of the dedendum, so there is a small clearance between the tops of the teeth and the bottom of the tooth cutout. An imaginary circle at the bottom of the depth to which a mating tooth reaches is the working depth circle. The working depth is slightly less than the whole depth.

Spur gear parts, their symbols, and their definitions are shown in the following drawing and table.

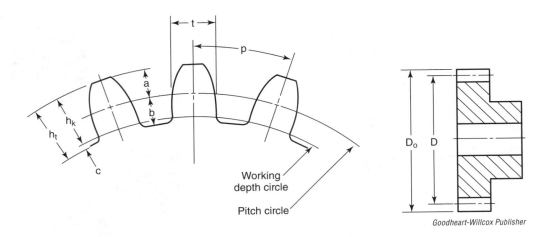

Goodheart-Willcox Publisher

Spur Gear Parts and Formulas

Gear Part	Symbol	Definition
Number of teeth on the gear	N	The total number of teeth formed along the pitch circle.
Pitch circle		An imaginary circle located approximately half the distance from the roots and tops of gear teeth. It is tangent to the pitch circle of the mating gear.
Pitch diameter	D	The diameter of the pitch circle.
Diametral pitch	P	The number of teeth per inch of pitch diameter.
Circular pitch	p	The distance measured on the pitch circle between similar points on adjacent teeth.
Tooth thickness	t	Thickness of the tooth at the pitch circle. The dimension used in measuring tooth thickness with a vernier gear tooth caliper.
Addendum	a	The distance the tooth extends above the pitch circle.
Dedendum	b	The distance the tooth extends below the pitch circle.
Working depth	h_k	The sum of the addendums of the two mating gears.
Whole depth	h_t	Total depth of a tooth space, equal to the addendum (a) plus dedendum (b), or to the depth to which each tooth is cut.
Clearance	c	The difference between the working depth and the whole depth of a gear tooth. The amount by which the dedendum on a given gear exceeds the addendum of the mating gear.
Center distance	C	Distance between the centers of two mating gears.
Outside diameter	D_o	Diameter or size of gear blank.
Pressure angle	θ	The angle of pressure between contacting teeth of mating gears. It represents the angle at which the forces from the teeth on one gear are transmitted to the mating teeth of another gear. Pressure angles of 14 1/2°, 20°, and 25° are standard. However, the 20° is replacing the older 14 1/2°.

Goodheart-Willcox Publisher

Spur Gear Formulas

It is possible to calculate all of the information necessary to cut a spur gear by using the formulas in the following chart.

Formulas for Spur Gear Calculations

To Find	Rule	Formula
Diametral pitch	Divide π by the circular pitch.	$P = \pi / p$
Circular pitch	Divide π by the diametral pitch.	$p = \pi / P$
Pitch diameter	Divide the number of teeth by the diametral pitch.	$D = N / P$
Outside diameter	Add 2 to the number of teeth and divide the sum by the diametral pitch.	$D_o = \dfrac{N + 2}{P}$
Number of teeth	Multiply the pitch diameter by the diametral pitch.	$N = D \times P$
Tooth thickness	Divide 1.5708 by the diametral pitch.	$t = 1.5708 / P$
Addendum	Divide 1.0 by the diametral pitch.	$a = 1.0 / P$
Minimum dedendum	Divide 1.250 by the diametral pitch.	$b = 1.250 / P$
Working depth	Divide 2 by the diametral pitch.	$h_k = 2 / P$
Minimum whole depth	Divide 2.250 by the diametral pitch.	$h_t = 2.250 / P$
Minimum clearance	Divide 0.250 by the diametral pitch.	$c = 0.250 / P$
Center distance	Add the number of teeth in both gears and divide the sum by two times the diametral pitch.	$C = \dfrac{N_1 + N_2}{2P}$
Length of rack	Multiply the number of teeth in the rack by the circular pitch.	$L = N \times p$

Note: Use 3.1416 for pi when performing calculations.

Goodheart-Willcox Publisher

Example 30-1

Calculate the outside diameter (D_o), whole depth (h_t), addendum (a), and tooth thickness (t) for a spur gear with 60 teeth and a diametral pitch of 12.

Outside diameter

$$D_o = \frac{N + 2}{P}$$

$$D_o = \frac{60 + 2}{12} = \frac{62}{12}$$

$$D_o = 5.167''$$

Whole depth

$$h_t = \frac{2.157}{P}$$

$$h_t = \frac{2.157}{12}$$

$$h_t = .180''$$

(Continued)

Addendum

$$a = \frac{1.0}{P}$$

$$a = \frac{1.0}{12}$$

$$a = .083''$$

Tooth thickness

$$t = \frac{1.5708}{P}$$

$$t = \frac{1.5708}{12}$$

$$t = .131''$$

Unit 30 Review

Name _____ Date _____ Class _____

Fill in the blanks in the following review questions.

1. There are many types of gears. The simplest type is the _____ gear, which has teeth running straight across the perimeter of the gear.

2. The size, pitch, and shape of the _____ are critical because most gears must mesh with another gear, usually of a different diameter.

3. The _____ circle is an imaginary circle that passes through the approximate mid-height of the gear teeth.

4. When two gears are meshed, their _____ circles just touch each other.

5. The _____ _____ is the diameter of the pitch circle.

6. The _____ of a tooth is measured at the pitch circle.

7. The _____ pitch is the number of teeth per inch on the pitch circle.

8. The _____ pitch is the distance measured on the pitch circle between the same points on adjacent teeth.

9. The _____ is the distance a tooth extends above the pitch circle.

10. The _____ is the distance a tooth extends below the pitch circle.

11. The _____ _____ circle is an imaginary circle at the bottom of the depth to which a mating tooth reaches and is slightly less than the whole depth.

Applications

Complete this table for a gear with 44 teeth and an outside diameter of 6 1/2". Round dimensions to the nearest thousandth.

	Feature	Dimension or Value
1.	Diametral pitch	_____
2.	Pitch diameter	_____
3.	Addendum	_____
4.	Dedendum	_____
5.	Whole depth	_____
6.	Clearance	_____
7.	Tooth thickness	_____
8.	Circular pitch	_____
9.	Working depth	_____

Goodheart-Willcox Publisher

Complete this table for a gear with 60 teeth and a diametral pitch of 8. Round dimensions to the nearest thousandth.

	Feature	Dimension or Value
10.	Outside diameter	_____
11.	Pitch diameter	_____
12.	Addendum	_____
13.	Dedendum	_____
14.	Whole depth	_____
15.	Working depth	_____
16.	Tooth thickness	_____
17.	Circular pitch	_____
18.	Clearance	_____

Goodheart-Willcox Publisher

Complete this table for a gear with 40 teeth and a diametral pitch of 10. Round dimensions to the nearest thousandth.

	Feature	Dimension or Value
19.	Outside diameter	_____
20.	Pitch diameter	_____
21.	Addendum	_____
22.	Dedendum	_____
23.	Whole depth	_____
24.	Working depth	_____
25.	Tooth thickness	_____
26.	Circular pitch	_____
27.	Clearance	_____

Goodheart-Willcox Publisher

Powers and Roots

Unit 31. Powers

Unit 32. Roots of Numbers

Key Terms

cube of a number

exponent

index

power of a number

radical symbol

radicand

square of a number

vinculum

UNIT 31

Powers

Objectives

Information in this unit will enable you to:

- Explain what powers and exponents are in reference to multiplication, and how they are used.
- Discuss what squares and cubes are in reference to multiplication, and discuss their practical functions.
- Describe how to compute the power of a number on a standard calculator and on a scientific calculator.
- Explain how to write and use fractions with exponents in the denominator, numerator, and entire fraction.

Powers

A **power of a number** is the product of multiplying that number by itself. If 5 is multiplied by 5, the product is 25. Twenty-five is 5 to the second power. If 5 is multiplied by itself three times, the result is 125. One hundred twenty-five is the third power of 5. Rather than write out 5×5 or $5 \times 5 \times 5$, powers of numbers are usually written with an **exponent**. The exponent is written as a small numeral near the upper-right corner of the base number. Four to the second power (4×4) would be written 4^2. Two to the fifth power ($2 \times 2 \times 2 \times 2 \times 2$) would be written 2^5. This is called raising the number to the fifth power. The exponent is the power to which the number is raised.

Squares and Cubes

A square can be divided by equally spaced vertical and horizontal grid lines. Consider a square with 6″ sides. Both sides are the same size, so if the grid lines are spaced 1″ apart, there are six rows and six columns each 1″ wide. The original square is divided into six parts each way, with 36 smaller squares covering its entire area. The area of a square can be computed by multiplying the length of a side by itself ($6″ \times 6″$ in this case). Because of this mathematical principle, a number multiplied by itself (the number to the second power) is called the **square of a number**. 6^2 is called 6 squared, so the area of the original square is 36 square inches.

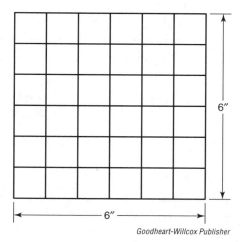

If a third dimension (depth) is added to our original square and that dimension is the same as the dimensions of our original square (6″ × 6″), a cube is formed. A cube is a solid object with all six surfaces being equal squares. If our cube is divided into six equal parts in all dimensions, it is made up of 216 small cubes measuring 1″ in all three dimensions. The volume of the cube can be found by multiplying the height times the width times the depth (6″ × 6″ × 6″ in this case). Because of this mathematical principle, a number to the third power is called the **cube of a number** or that number cubed.

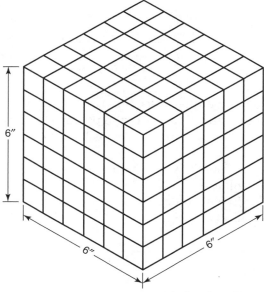

In linear measure (measuring distances) it is common to use the symbols ′ for feet and ″ for inches. However, those symbols should not be used in area or volume measure. Exponents can be used to write area or volume measure, but not with the symbols for inches or feet. To indicate an area of 28 square inches, it would be proper to write 28 sq in or 28 in^2. To indicate a volume of 28 cubic inches, it would be proper to write 28 cu in or 28 in^3.

Example 31-1

Find the volume of a cube with sides measuring 2.4 mm to the nearest .01 mm.

$$2.4^3 = 2.4 \times 2.4 \times 2.4 = 13.824$$

Round to nearest .01 mm = 13.82 mm^3.

Computing Powers with a Calculator

Calculators vary in complexity and in operating systems. The simplest calculators are handy to add, subtract, multiply, and divide, and have a $\sqrt{}$ key to extract square roots (discussed in Unit 32). To compute the power of a number on the most basic calculators, it is necessary to multiply the number once for each power.

Example 31-2

Raise 3 to the third power.

Press **3**, press **X**, press **3**, press **X**, press **3**, press **ENTER =**

The display shows:

27

Each manufacturer and model of scientific calculator uses slightly different key strokes to perform an operation like raising to powers. Consult the instructions for the calculator you are using. The sequences of keystrokes given here are for one popular scientific calculator.

Scientific calculators usually have a shift key to allow each key to perform two functions. Scientific calculators have an x^y key or a $\wedge$ key. The $\wedge$ or x^y key allows the user to enter the power to which the number is to be raised.

Example 31-3

Raise 3 to the fifth power.

Press **3**, press **∧**, press **5**, press **ENTER =**

The display shows:

243

Fractions with Exponents

Fractions may have exponents associated with either the numerator, the denominator, or the entire fraction. When the exponent applies to only the numerator or the denominator, the exponent is written at the top right of that part of the fraction.

Example 31-4

Numerator squared:

$$\frac{3^2}{4} = \frac{3 \times 3}{4} = \frac{9}{4}$$

Denominator cubed:

$$\frac{4}{5^3} = \frac{4}{5 \times 5 \times 5} = \frac{4}{125}$$

When the exponent applies to both the numerator and the denominator, the fraction is enclosed in parentheses and the exponent is written outside the parentheses. When parentheses are used in any mathematical expression, the operations inside the parentheses are performed first, then whatever is indicated outside the parentheses. If the fraction inside the parentheses can be reduced, that is done before raising to the indicated power.

Example 31-5

Entire fraction cubed:

$$\left(\frac{3}{6}\right)^3 = \left(\frac{1}{2}\right)^3 = \frac{1 \times 1 \times 1}{2 \times 2 \times 2} = \frac{1}{8}$$

Example 31-6

Fractions added, then squared:

$$\left(\frac{1}{2} + \frac{2}{3}\right)^2 = \left(\frac{3}{6} + \frac{4}{6}\right)^2 = \left(\frac{7}{6}\right)^2 = \frac{7 \times 7}{6 \times 6} = \frac{49}{36}$$

Work Space/Notes

Unit 31 Review

Name _____ **Date** _____ **Class** _____

Fill in the blanks in the following review questions.

1. The _____ of a number is the product of multiplying that number by itself.

2. Powers of numbers are usually written with an _____, a small numeral near the upper-right corner of the base number.

3. The exponent is the _____ to which the number is raised.

4. A number multiplied by itself is called the _____ of a number.

5. A number to the third power is called the _____ of a number or that number _____.

6. _____ can be used to write area or _____ measure, but not with the symbols for inches or feet.

7. When the exponent applies to only the numerator or denominator, the exponent is written at the top right of that _____ of the fraction.

8. When the exponent applies to both the numerator and denominator, the fraction is enclosed in parentheses and the exponent is written _____ the parentheses.

9. When parentheses are used in any mathematical expression, the operations inside the parentheses are performed _____.

10. If the fraction inside the parentheses can be reduced, that is done _____ raising to the indicated power.

Practice

Raise the following to the indicated power.

1. 5^2 _____

2. 4^3 _____

3. 6^5 _____

4. 1^3 _____

5. $\dfrac{2^2}{5}$ _____

6. $\dfrac{9}{3^3}$ _____

7. $\dfrac{7}{16^2}$ _____

8. $\left(\dfrac{3}{4}\right)^2$ _____

9. $\left(\dfrac{2}{7}\right)^3$ _____

10. $\left(\dfrac{3}{8}+\dfrac{1}{3}\right)^2$ _____

11. 8^7 _____

12. 7^6 _____

13. $\dfrac{1}{3^2}$ _____

14. $\dfrac{3^3}{4}$ _____

15. $\dfrac{1^2}{8}$ _____

16. $\left(\dfrac{2}{3}\right)^3$ _____

17. $\left(\dfrac{1}{5}\right)^2$ _____

18. $\left(\dfrac{4}{5}-\dfrac{1}{2}\right)^3$ _____

Write each problem using exponents, then solve each problem.

Example: $7 \times 7 \times 7$ is written as 7^3 and $= 343$

19. $6 \times 6 \times 6 \times 6 \times 6 \times 6 \times 6 \times 6$

_____

20. 10×10

_____

21. $3 \times 3 \times 3 \times 3 \times 3 \times 3 \times 3 \times 3 \times 3$

_____

22. $14 \times 14 \times 14$

_____

23. $5 \times 5 \times 5 \times 5 \times 5$

_____

24. $100 \times 100 \times 100$

_____

25. 4×4

_____

26. $2 \times 2 \times 2 \times 2 \times 2 \times 2 \times 2 \times 2$

_____

27. $12 \times 12 \times 12$

_____

28. 9×9

_____

29. $8 \times 8 \times 8 \times 8$

_____

30. $1 \times 1 \times 1 \times 1 \times 1$

_____

31. 15×15

_____

32. $76 \times 76 \times 76$

_____

33. $\frac{2}{3} \times \frac{2}{3}$

_____

34. $\frac{3}{5} \times \frac{3}{5} \times \frac{3}{5}$

_____

35. $\frac{3}{4} \times \frac{3}{4} \times \frac{3}{4} \times \frac{3}{4}$

_____

36. $\frac{1}{2} \times \frac{1}{2} \times \frac{1}{2}$

_____

37. $\frac{1}{5} \times \frac{1}{5} \times \frac{1}{5} \times \frac{1}{5} \times \frac{1}{5}$

_____

38. $\frac{6}{7} \times \frac{6}{7}$

_____

Name _____   **Date** _____   **Class** _____

Applications

Solve the following problems using exponents.

1. What is the area of the square shown in the figure below to the nearest .1 mm?

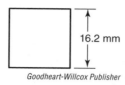

16.2 mm

Goodheart-Willcox Publisher

_____

Determine the area of each square in the table below. The length of each side is given. Round answers to the nearest thousandth when necessary.

	Side of Square	Area
2.	$\frac{1}{2}$″	_____
3.	2 inches	_____
4.	$\frac{5}{8}$″	_____
5.	25 mm	_____
6.	6.5 inches	_____
7.	$2\frac{1}{3}$″	_____
8.	41 mm	_____
9.	3.125 inches	_____
10.	12.7 mm	_____
11.	$\frac{2}{3}$″	_____

Goodheart-Willcox Publisher

12. What is the area of the part shown in the figure below to the nearest .01 in²? (Crosshatching highlights three squares.)

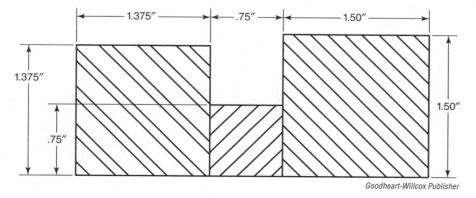

Goodheart-Willcox Publisher

_____

13. What is the volume of the cube shown in the figure below to the nearest .01 in³?

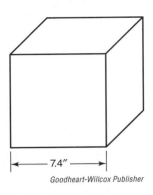

|← 7.4″ →|

Goodheart-Willcox Publisher

_____

Determine the volume of each cube in the table below. The length of each side is given. Round answers to the nearest thousandth when necessary.

	Side of Square	Volume
14.	$\frac{2''}{3}$	
15.	35 mm	
16.	$\frac{5''}{9}$	
17.	5 inches	
18.	7 mm	
19.	$5\frac{3''}{4}$	
20.	4.5 inches	
21.	24.5 mm	
22.	1.6 inches	
23.	$\frac{4''}{5}$	

Goodheart-Willcox Publisher

UNIT **32**

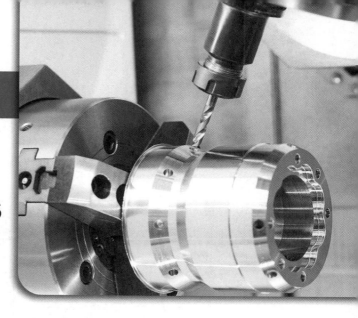

Roots of Numbers

Objectives

Information in this unit will enable you to:

- Explain what a root is and how it is written.
- Describe the difference between square roots and cube roots.
- Discuss the ways in which roots might be incorporated into fractions, and how to reduce a fraction containing a root.
- Explain how to extract a root using a calculator.

Roots

The root of a number is the opposite of the power of a number. The root is the number which, when multiplied by itself a specified number of times, yields the given number as a product. For example, the fourth root of 16 is 2 because $2 \times 2 \times 2 \times 2 = 16$. This would be written as $\sqrt[4]{16}$. The $\sqrt{}$ is called a **radical symbol**. The horizontal line is called the **vinculum**. The number written smaller and to the left (4 in this example) is the **index**. The index is written inside the check-mark part of the radical symbol. The number of which the root is to be calculated, under the vinculum (16 in this example), is called the **radicand**.

$$\overset{\text{Index} \longrightarrow}{} \sqrt[4]{16} = 2 \longleftarrow \text{Root}$$

Index → 4
Vinculum
Radical symbol
Radicand

Goodheart-Willcox Publisher

For a square root, the index 2 is not included.

Squares and Cubes

Square roots and cube roots are the roots most often encountered by machinists. The terms come from the same origins as the square and cube powers explained in Unit 31. A square that is made up of 16 smaller squares has the edges of 4 of the smaller squares along each of its sides. The square root of 16 is 4.

1	2	3	4
5	6	7	8
9	10	11	12
13	14	15	16

Goodheart-Willcox Publisher

In the same way, a cube that is made up of 64 smaller cubes has 4 of the smaller cube edges along each side of the larger cube. The cube root of 64 is 4.

Fractions Involving Roots

The rules that apply to fractions involving powers also apply to fractions involving roots. If only the numerator of the fraction is under the radical symbol, the numerator is the root of that radicand. If only the denominator is under the radical symbol, the denominator is the root of that radicand.

Example 32-1

Square root of numerator:

$$\frac{\sqrt{25}}{10} = \frac{5}{10} = \frac{1}{2}$$

Square root of denominator:

$$\frac{3}{\sqrt{16}} = \frac{3}{4}$$

If the entire fraction is enclosed by the radical symbol, both the numerator and the denominator are the roots of the indicated radicands. There are two methods that can be used to reduce the fraction. You should use the one that is simplest for the problem at hand.

Method 1: Extract both roots, then divide.

$$\sqrt{\frac{16}{4}} = \frac{4}{2} = 2$$

Method 2: Divide, then extract the root.

$$\sqrt{\frac{16}{4}} = \sqrt{4} = 2$$

Extracting Roots with a Calculator

Even the most basic calculators have a $\sqrt{\ }$ key to extract square roots. To find a square root using the $\sqrt{\ }$ key, enter the radicand (the number for which you want the square root) and then tap the $\sqrt{\ }$ key. Alternatively, on some calculators press the $\sqrt{\ }$ key, and then enter the radicand. Such basic calculators cannot usually be used to find roots with an index of 3 or more. Scientific calculators often require shifting to the 2nd function of the key pad to use the $\sqrt{\ }$ function.

Each manufacturer and model of scientific calculator uses slightly different keystrokes to perform an operation like finding roots. Consult the instructions for the calculator you are using. The sequences of keystrokes given here are for one type of scientific calculator.

Example 32-2

Find the square root of 54 to the nearest hundredth.

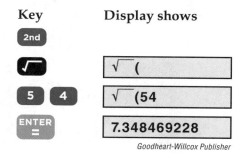

Goodheart-Willcox Publisher

Note: Some calculators do not show the parenthesis mark in the display.
Round to nearest .01 = 7.35.

Finding roots other than the square root requires the use of the $\sqrt[x]{}$ key.

Example 32-3

Find the fifth root of 122 to the nearest hundredth.

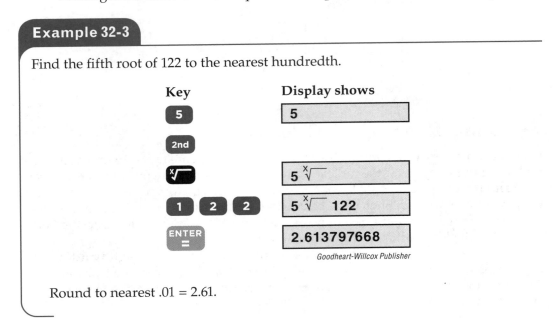

Key	Display shows
5	5
2nd	
$\sqrt[x]{}$	5 $\sqrt[x]{}$
1 2 2	5 $\sqrt[x]{}$ 122
ENTER =	2.613797668

Goodheart-Willcox Publisher

Round to nearest .01 = 2.61.

Unit 32 Review

Fill in the blanks in the following review questions.

1. The root of a number is the opposite of the _____ of a number.

2. The _____ is the number which, when multiplied by itself a specified number of times, yields the given number as a product.

3. The $\sqrt{}$ is called a _____ symbol.

4. For a square root, the index _____ is not included.

5. _____ roots and _____ roots are the roots most often encountered by machinists.

6. The _____ that apply to fractions involving powers also apply to fractions involving roots.

7. If only the _____ of the fraction is under the radical symbol, the numerator is the root of the radicand. If only the _____ is under the radical symbol, the denominator is the root of that radicand.

8. If the _____ fraction is enclosed by the radical symbol, both the numerator and the denominator are the roots of the indicated radicands.

9. There are two methods that can be used to reduce the fraction. You should use the one that is simplest for the problem at hand. Method 1 is to _____ both roots, then divide. Method 2 is to _____, then extract the root.

Practice

Find the indicated root of the following to the nearest thousandth.

1. $\sqrt{36}$ _____   2. $\sqrt{64}$ _____   3. $\sqrt[4]{81}$ _____

4. $\sqrt[3]{27}$ _____   5. $\sqrt{60}$ _____   6. $\sqrt{34}$ _____

7. $\sqrt{49}$ _____   8. $\sqrt{25}$ _____   9. $\sqrt{64}$ _____

10. $\sqrt{52}$ _____   11. $\sqrt[4]{256}$ _____   12. $\sqrt[5]{218}$ _____

13. $\sqrt{93}$ _____

14. $\sqrt[7]{321}$ _____

15. $\sqrt[9]{512}$ _____

16. $\sqrt[6]{242}$ _____

17. $\sqrt{121}$ _____

18. $\sqrt[3]{343}$ _____

19. $\sqrt[3]{100}$ _____

20. $\sqrt[5]{220}$ _____

21. $\dfrac{\sqrt{36}}{9}$ _____

22. $\dfrac{36}{\sqrt{9}}$ _____

23. $\sqrt{\dfrac{36}{9}}$ _____

24. $\dfrac{\sqrt{25}}{\sqrt[3]{64}}$ _____

25. $\sqrt[3]{330}$ _____

26. $\sqrt{10,000}$ _____

27. $\sqrt{1}$ _____

28. $\sqrt[4]{1,296}$ _____

29. $\sqrt[3]{729}$ _____

30. $\dfrac{25}{\sqrt{4}}$ _____

31. $\dfrac{\sqrt{49}}{8}$ _____

32. $\sqrt{\dfrac{64}{16}}$ _____

33. $\dfrac{\sqrt{9}}{4}$ _____

34. $\dfrac{7}{\sqrt{2}}$ _____

35. $\sqrt{\dfrac{81}{25}}$ _____

36. $\dfrac{\sqrt[3]{100}}{\sqrt{9}}$ _____

Name _____   **Date** _____   **Class** _____

Applications

Solve the following problems.

1. The square cover plate shown in the figure below has a total area of 215.5 in². What is dimension x to the nearest .01"?

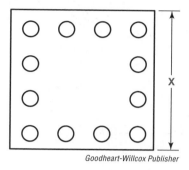

Goodheart-Willcox Publisher

_____

Determine the length of one side of each square in the table below. The area of each square is given. Round answers to the nearest thousandth when necessary.

	Side of Square	Area
2.	_____	100 in²
3.	_____	5,625 mm²
4.	_____	645 mm²
5.	_____	50 in²
6.	_____	42.25 in²
7.	_____	225 mm²
8.	_____	10.5625 in²
9.	_____	161.29 mm²
10.	_____	5.0625 in²
11.	_____	72.25 mm²

Goodheart-Willcox Publisher

12. The stud block shown in the figure below is a perfect cube and has a volume of 1.953 in³. What is the length of one side to the nearest .001″?

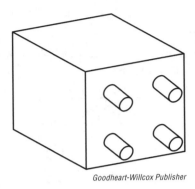

Goodheart-Willcox Publisher

_____

Determine the length of one side of each cube in the table below. The volume of each cube is given. Round answers to the nearest thousandth when necessary.

	Side of Square	Volume
13.		34.328 in³
14.		187.5 mm³
15.		11.391 in³
16.		52.734 in³
17.		5,832 mm³
18.		107.172 in³
19.		134.611 in³
20.		256.048 mm³
21.		3.375 in³
22.		125 mm³

Goodheart-Willcox Publisher

Angles and Trigonometry

Key Terms

acute angle
acute triangle
adjacent side
base of a triangle
complementary angles
equilateral triangle
height of a triangle
hypotenuse
isosceles triangle
law of cosines
law of cosines (angle version)

law of sines
minuend
minute (angular measure)
obtuse angle
obtuse triangle
opposite side
Pythagorean theorem
rays
reflex angle
right angle

right triangle
scalene triangle
second (angular measure)
soh cah toa
straight angle
subtrahend
supplementary angles
tolerance
trigonometry
vertex

UNIT 33

Units of Angular Measure

Objectives

Information in this unit will enable you to:

- Explain what rays and vertices are.
- Differentiate between degrees, minutes, and seconds, and explain how they are used when measuring angles.
- Identify and discuss different types of angles.

Rays and Vertices

An angle is the amount of opening between two lines that intersect (meet). In mathematics, the lines forming the angle are called **rays**. The point at which two lines meet is their **vertex**. The four corners of a square are the vertices of the sides. In discussing angles, the point at which the two lines forming the angle meet is the vertex of the angle. The symbol for angle is ∠.

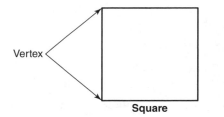

Square

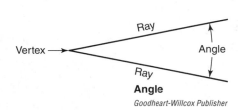

Angle

Goodheart-Willcox Publisher

Degrees, Minutes, and Seconds

The angle formed by two rays is commonly measured in degrees, **minutes**, and **seconds**. The lengths of the rays can be changed without affecting the angle they form.

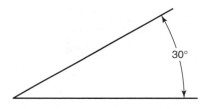

30°

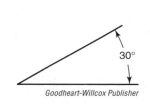

30°

Goodheart-Willcox Publisher

316

A degree is 1/360th of a circle. The symbol for indicating degrees is °. An angle equal to 1/4 of a circle may be written as 90°. For more precise measurement, an angle can be divided into 60 minutes. Each minute is 1/60th of a degree, which is 1/360th of a complete circle, so a minute is 1/21,600th of a circle. The symbol for minutes is ′. An angle of 1/2 degree greater than 90° would be written as 90° 30′. For even greater precision, a minute can be divided into 60 seconds. There are 1,296,000 seconds in a complete circle. The symbol for seconds is ″. An angle of 1/3 minute more than 90° 30′ would be written as 90° 30′ 20″ because 20 seconds is 1/3 of a 60-second minute.

Kinds of Angles

There are several kinds of angles. The most common angles are shown in the following illustration. An **acute angle** is an angle that is less than 90°. A **right angle** is 90° and is indicated by a square at the vertex. An **obtuse angle** is an angle that is more than 90° but less than 180°. A **straight angle** is 180°. A straight angle is also a straight line. A **reflex angle** is an angle that is more than 180° but less than 360°.

Angle Drawing	Angle Name	Description
	Acute angle	Less than 90°
	Right angle	90°, indicated by a square at the vertex
	Obtuse angle	More than 90° but less than 180°
	Straight angle	180°
	Reflex angle	More than 180° but less than 360°
	Complementary angles	Two angles whose measure equals 90°
	Supplementary angles	Two angles whose measure equals 180°

Goodheart-Willcox Publisher

Notice that **complementary angles** and **supplementary angles** are pairs of angles. If one line passes through another, the angles formed on opposite sides of the vertex are supplementary angles, as shown below. Supplementary angles add up to form a straight line (180°), so any angles that are based on a straight line intersecting another straight line are supplementary.

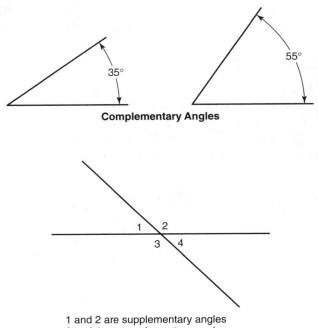

Complementary Angles

1 and 2 are supplementary angles
2 and 4 are supplementary angles
3 and 4 are supplementary angles
1 and 3 are supplementary angles

Goodheart-Willcox Publisher

When two lines cross (intersect), two pairs of opposite angles are created. The opposite angles can also be called vertical angles. Vertical angles are congruent. Vertical in this case means they share the same vertex, not up-down, which is the usual meaning. Congruent means identical in form, or simply equal.

In the above figure:

∠1 and ∠4 are opposite angles and therefore equal.

∠2 and ∠3 are opposite angles and therefore equal.

Unit 33 Review

Name _____ **Date** _____ **Class** _____

Fill in the blanks in the following review questions.

1. An angle is the amount of opening between two lines that _____ (meet). The point at which two lines meet is their _____.

2. The angle formed by two rays is commonly measured in _____, _____, and _____.

3. The symbol for indicating degrees is _____, the symbol for minutes is _____, and the symbol for seconds is _____.

4. An acute angle is an angle less than _____°, a _____ angle is 90° and is indicated by a square at the vertex, and an obtuse angle is more than _____° but less than _____°.

5. A _____ angle is 180°. A reflex angle is more than _____° but less than _____°.

6. Complementary angles are two angles whose measure equals _____°.

7. Supplementary angles add up to form a _____ angle (180°), so any angles based on two intersecting straight lines are supplementary.

Practice

Label the following according to the kind of angle.

1. _____ 32°

2. _____ 280°

3. _____ 121°

4. _____ ?

5. _____ 70° 110° ?

6. _____ ? 70°

7. _____ ?

8. _____ 60°

9. _____ 315°

10. _____ 118°

Write the following angles in words.

11. 34° 30′ 45″

_____

12. 95° 59′ 35″

_____

13. 118° 13′ 51″

_____

14. 12° 27′ 09″

_____

15. 61° 43′ 22″

_____

16. 136° 56′ 19″

_____

17. 110° 02′ 58″

_____

18. 173° 14′ 36″

_____

19. 233° 31′ 45″

_____

Write the following angles using the proper symbols for degrees, minutes, and seconds.

20. Fifty-two degrees, twenty minutes, five seconds

_____

21. Two hundred twelve degrees, fifty-eight minutes, and thirty-four seconds

_____

22. Twelve degrees, five minutes, and twenty-nine seconds

_____

23. Sixty-one degrees, thirty-four minutes, and forty-two seconds

_____

24. One hundred degrees, fifteen minutes, and seven seconds

_____

25. One hundred twenty-three degrees, forty-three minutes, and twelve seconds

_____

26. Two hundred thirty-nine degrees, fifty-one minutes, and nineteen seconds

_____

27. Three hundred seven degrees, one minute, and eight seconds

_____

Name _____   **Date** _____   **Class** _____

28. Label the following according to the kind of angle represented.

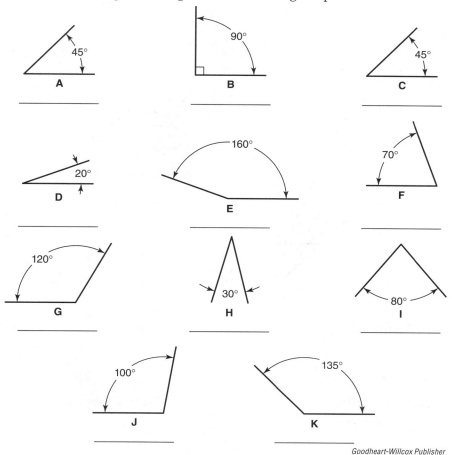

Goodheart-Willcox Publisher

Use illustrated angles A through K above to answer questions 29 through 33.

29. Which angle(s) is/are a right angle? Choose one correct answer.

   A. J

   B. B

   C. H and F

   D. None are correct.

In questions 30 through 33, which angles are complementary? Circle the correct answer.

30. A. A and C

   B. C and K

   C. I and J

   D. None are correct.

31. A. G and D

   B. A and K

   C. F and H

   D. None are correct.

32. A. E and G

   B. D and F

   C. D and I

   D. None are correct.

33. A. F and G

   B. B and J

   C. E and H

   D. None are correct.

Which angles in the following figure are supplementary? Circle one correct answer for each question, 34 through 38.

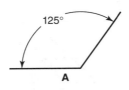

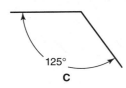

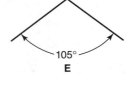

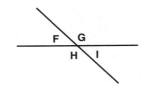

34. A. A and B
    B. B and C
    C. E and B
    D. None are correct.

35. A. C and D
    B. F and I
    C. C and A
    D. None are correct.

36. A. F and G
    B. G and I
    C. E and B
    D. All are correct.

37. A. A and D
    B. B and C
    C. G and H
    D. None are correct.

38. A. J and K
    B. A and B
    C. I and K
    D. None are correct.

Which angles are equal?

39. A. A and C
    B. A and B
    C. I and K
    D. None are correct.

40. A. J and K
    B. A and B
    C. F and I
    D. None are correct.

41. A. F and H
    B. G and H
    C. J and K
    D. None are correct.

Name _____   **Date** _____   **Class** _____

Use the following figure to answer questions 42 through 47.

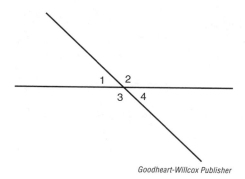

Goodheart-Willcox Publisher

42. If ∠1 and ∠2 are supplementary and ∠1 = 39°, what does ∠2 measure?

_____

43. If ∠2 and ∠4 are also supplementary, what does ∠4 measure?

_____

44. If ∠4 and ∠3 are also supplementary, what does ∠3 measure?

_____

When two lines intersect, the opposite angles are equal. If ∠4 = 42° 30′, calculate the following:

45. ∠1 = _____

46. ∠2 = _____

47. ∠3 = _____

Use the following figure to answer questions 48 and 49.

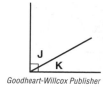

Goodheart-Willcox Publisher

48. If ∠J measures 47° 25′, what is the measurement of ∠K?

_____

49. If ∠K measures 41° 45′ 37″, what is the measurement of ∠J?

_____

50. In the figure below, what is the measurement of ∠X?

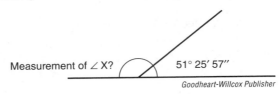

Measurement of ∠X?   51° 25′ 57″

Goodheart-Willcox Publisher

_____

51. In the figure below, what is the measurement of ∠X?

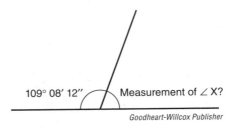

109° 08′ 12″   Measurement of ∠X?

Goodheart-Willcox Publisher

_____

Applications

Answer the following questions.

1. In the illustration of a dovetail slide, are angles A and B complementary, supplementary, or reflex?

Dovetail Slide

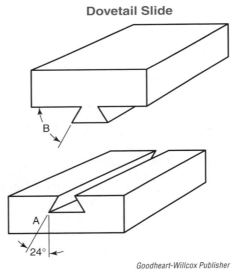

B

A

24°

Goodheart-Willcox Publisher

_____

2. What is the sum of angles A and B on the dovetail slide?

_____

3. What does angle B measure?

_____

UNIT 34

Basic Math Operations with Angles

Objectives

Information in this unit will enable you to:

- Describe how to add, subtract, multiply, and divide angles expressed in whole degrees.
- Discuss how to add, subtract, multiply, and divide angles expressed in degrees, minutes, and seconds.
- Describe how to convert degrees, minutes, and seconds to decimal angles.
- Explain what a tolerance is and discuss why it is common for drawings and specifications for machined parts to include tolerances.

Operations with Whole Degrees

Angles can be added, subtracted, multiplied, and divided with basic math operations. If the angles are expressed only as whole degrees, the operations can be performed on them just as on any whole numbers.

Example 34-1

Addition $37° + 104° = 141°$

Subtraction $58° - 15° = 43°$

Multiplication $45° \times 3 = 135°$

Division $60° \div 2 = 30°$

Adding Degrees, Minutes, and Seconds

If the angles are expressed with minutes and/or seconds, write each element in a separate column. To add angles with degrees, minutes, and seconds, add each column separately. If the seconds column total is 60 or more, add 1 minute for every 60 seconds to the minute column and subtract those seconds from the seconds column. If the minutes column total is 60 or more, add 1 to the degree column for every 60 minutes and subtract those minutes from the minutes column.

Example 34-2

Add 34° 15′ 42″ + 14° 55′ 22″.

$$
\begin{array}{rrr}
34° & 15′ & 42″ \\
+\ 14° & +\ 55′ & +\ 22″ \\
\hline
48° & 70′ & 64″
\end{array}
$$

Subtract 60″ from the seconds column and add it to the minutes column in the form of 1′.

$$
\begin{array}{rrr}
34° & 15′ & 42″ \\
+\ 14° & +\ 55′ & +\ 22″ \\
\hline
48° & 70′ & 64″ \\
 & +\ 1′ & -\ 60″ \\
\hline
48° & 71′ & 4″
\end{array}
$$

Subtract 60′ from the minutes column and add it to the degrees column in the form of 1°.

$$
\begin{array}{rrr}
34° & 15′ & 42″ \\
+\ 14° & +\ 55′ & +\ 22″ \\
\hline
48° & 70′ & 64″ \\
 & +\ 1′ & -\ 60″ \\
\hline
48° & 71′ & 4″ \\
+\ 1° & -\ 60′ & \\
\hline
49° & 11′ & 4″
\end{array}
$$

Subtracting Degrees, Minutes, and Seconds

Subtraction of angles expressed with minutes and/or seconds is very similar to addition of those angles. Write the problem in columns of degrees, minutes, and seconds. Perform the subtraction in each column separately. Start with the column representing the smallest units (the farthest to the right) and work toward the left. If the **subtrahend** (number being subtracted) is greater than the **minuend** (number being subtracted from) in any column, borrow from the column to its left to increase the minuend. Reduce the result as explained above for addition.

Example 34-3

Subtract 18° 28′ 40″ from 64° 15′ 55″.

$$
\begin{array}{rrr}
64° & 15′ & 55″ \\
-\ 18° & -\ 28′ & -\ 40″
\end{array}
$$

For the minutes, 28′ is greater than 15′, so borrow 1° from the 64° in the degree column and add it to the 15′ in the form of 60′.

$$
\begin{array}{rrr}
64° - 1° = 63° & 15′ + 60′ = 75′ & 55″ \\
-\ 18° & -\ 28′ & -\ 40″ \\
\hline
45° & 47′ & 15″
\end{array}
$$

Multiplying Degrees, Minutes, and Seconds

Multiplying angles expressed in degrees, minutes, and seconds is a matter of simple multiplication. Multiply each term (column) and then reduce the products as for addition or subtraction.

Example 34-4

Multiply 25° 25′ 12″ by 3.

$$
\begin{array}{ccc}
25° & 25′ & 12″ \\
\times\ 3 & \times\ 3 & \times\ 3 \\
\hline
75° & 75′ & 36″ \\
+\ 1° & -\ 60′ & \\
\hline
76° & 15′ & 36″ \\
\end{array}
$$

Dividing Degrees, Minutes, and Seconds

Dividing angles expressed in degrees, minutes, and seconds is done by dividing each term separately. Unlike subtraction, division should be done starting with the column representing the largest units first (degrees in the example) and moving to the right. Any remainder from the division of degrees is converted to minutes and added to the minutes column. Remainders from division of minutes is converted to seconds and added to the seconds column. Any remainder from the division of seconds is used as the numerator of a fraction with the divisor (the number by which the angle is being divided) as the denominator. If that fraction is less than 1/2, round the minutes down by dropping the remainder. If the fraction is 1/2 or greater, round up the number of seconds to the next whole number.

Example 34-5

Divide 40° 20′ 35″ by 3.

$$
\begin{array}{ccc}
13° & 20′ & 35″ \\
3\overline{)40°} & +\ 60′ & +\ 120″ \\
-\ 39 & \overline{80′} & \overline{155″} \\
1r & & \\
\end{array}
$$

$$
\begin{array}{cc}
26′ & 51\ 2/3″,\ \text{round to } 52″ \\
3\overline{)80′} & 3\overline{)155″} \\
-\ 78 & -\ 153 \\
2r & 2r \\
\end{array}
$$

$$
\begin{array}{ccc}
13° & 26′ & 52″ \\
\end{array}
$$

Converting Degrees, Minutes, and Seconds to Decimal Angles

Machinists sometimes work with decimal angles. Decimal degrees are primarily used in the metric system. CNC machines capable of four-axis and five-axis machining also measure rotation of the axis in decimal degrees. To convert an angle of degrees, minutes, and seconds to decimal form, start with the seconds. There are 60 seconds in a minute, so divide the number of seconds by 60 and add the result to the minutes. Likewise, there are 60 minutes in a degree, so divide the new number of minutes by 60 and add that to the degrees.

Example 34-6

Convert 94° 30′ 30″ to decimal form.

Divide 30″ by 60.

$$\frac{30''}{60} = .5'$$

Add .5′ to the minutes.

$$94°\ 30.5'$$

Divide 30.5′ by 60.

$$\frac{30.5'}{60} = .5083°$$

Add .5083° to the degrees.

$$94°\ 30'\ 30'' = 94.5083'$$

Converting Decimal Angles to Degrees, Minutes, and Seconds

To convert a decimal angle to degrees and minutes, start with the minutes. Multiply the decimal portion of the figure by 60. The whole number becomes the minutes. Take the remaining decimal and multiply it by 60. The resulting number becomes the seconds. Seconds can remain as a decimal if needed or rounded to the nearest whole second depending on the required accuracy.

Example 34-7

Convert 33.6875° to degrees, minutes, and seconds.

Multiply decimal by 60.

$$0.6875° \times 60 = 41.25'$$

The number 41 is the whole number and gives 41′.

The remaining decimal is 0.25.

Multiply remaining decimal by 60.

$$0.25 \times 60 = 15''$$

Combining the numbers, we get: 33.6875° = 33° 41′ 15″.

Tolerances

Measurements can never be absolutely accurate. They are only as accurate as the instrument being used to measure them. For this reason, it is common for drawings and specifications for machined parts to include **tolerances**. A tolerance describes the limits of difference between the stated dimension and the finished piece. A part might be drawn as 22° with a tolerance of ±30'. (The ± symbol is read as "plus or minus.") A dimension of 22° ±30' would mean that anything up to 30' greater or less than 22° is acceptable, or from 21° 30' to 22° 30'.

Work Space/Notes

Unit 34 Review

Name _____ **Date** _____ **Class** _____

Fill in the blanks in the following review questions.

1. Angles can be added, subtracted, multiplied, and divided with _____ math operations. If the angles are expressed only as whole degrees, the operations are performed just as on any _____ numbers.

2. If the angles are expressed with minutes and/or seconds, write each element in a separate _____. Then add or subtract each _____ separately.

3. For subtraction, start with the column representing the _____ units (the farthest to the right) and work toward the left.

4. If the _____ (number being subtracted) is greater than the _____ (number being subtracted from) in any column, borrow from the column to its left to increase the minuend.

5. Multiplying angles expressed in degrees, minutes, and seconds is a matter of _____ multiplication. Multiply each term (column) and then _____ the products as for addition or subtraction.

6. Dividing angles expressed in degrees, minutes, and seconds is done by dividing each term _____. Unlike subtraction, division should start with the column representing the _____ units first (degrees) and moving to the right.

7. To convert an angle of degrees, minutes, and seconds to _____ form, start with the seconds.

8. Measurements can never be absolutely _____.

Practice

Add the angles. Show your work.

1. 10° 22′ 17″ + 18° 9′ 30″

   _____

2. 47° 28′ 19″ + 52° 31′ 15″

   _____

3. 112° 45′ 52″ + 14° 12′ 5″

   _____

4. 66° 51′ 30″ + 15° 25′ 20″

   _____

5. 210° 45′ 7″ + 2° 31′ 55″

   _____

6. 5° 30′ 55″ + 1° 44′ 41″

   _____

7. 15° 27′ 0″ + 5° 40′ 3″

_____

8. 314° 12′ 15″ + 13° 50′ 28″

_____

9. 265° 43′ 38″ + 52° 31′ 43″

_____

10. 165° 17′ 51″ + 133° 54′ 22″

_____

11. 31° 32′ 43″ + 18° 5′ 49″ + 47° 27′ 12″

_____

12. 102° 52′ 34″ + 59° 4′ 17″ + 156° 41′ 49″

_____

Subtract the angles. Show your work.

13. 42° 40′ 25″ − 11° 21′ 11″

_____

14. 510° 59′ 45″ − 420° 42′ 42″

_____

15. 114° 28′ 47″ − 99° 21′ 45″

_____

16. 15° 12′ 32″ − 8° 35′ 10″

_____

17. 110° 23′ 14″ − 5° 54′ 25″

_____

18. 14° 47′ 55″ − 10° 55′ 58″

_____

19. 347° 12′ 20″ − 250° 48′ 36″

_____

20. 34° 10′ 35″ − 11° 0′ 21″

_____

21. 176° 23′ 31″ − 82° 52′ 33″

_____

22. 209° 1′ 54″ − 186° 24′ 38″

_____

23. 21° 2′ 4″ − 19° 54′ 17″

_____

24. 117° 13′ 28″ − 12° 5′ 45″ − 32° 44′ 9″

_____

Name _____   **Date** _____   **Class** _____

Multiply the angles. Show your work.

25. $3 \times 20° \ 15' \ 18''$

_____

27. $6 \times 15°$

_____

29. $11 \times 15° \ 8' \ 12''$

_____

31. $5 \times 50° \ 16' \ 24''$

_____

33. $10 \times 17° \ 8' \ 19''$

_____

35. $3 \times 56° \ 11' \ 27''$

_____

26. $4 \times 105° \ 14' \ 7''$

_____

28. $2 \times 57° \ 45' \ 51''$

_____

30. $3 \times 28° \ 20' \ 20''$

_____

32. $5 \times 12° \ 30'$

_____

34. $2 \times 102° \ 21' \ 54''$

_____

36. $12 \times 10° \ 30' \ 15''$

_____

Divide the angles. Show your work.

37. $16° \div 3$

_____

39. $48° \ 30' \ 39'' \div 4$

_____

41. $212° \ 58' \ 16'' \div 7$

_____

38. $55° \ 25' \ 35'' \div 5$

_____

40. $124° \ 47' \ 52'' \div 3$

_____

42. $5° \ 12' \ 10'' \div 6$

_____

43. 54° 13′ 2″ ÷ 4

_____

44. 10° 30′ 16″ ÷ 2

_____

45. 119° 30′ 48″ ÷ 3

_____

46. 206° 44′ 36″ ÷ 4

_____

47. 358° 25′ 40″ ÷ 10

_____

48. 45° 45′ 45″ ÷ 9

_____

Calculate the maximum and minimum angles and write the results in the table below.

	Angle	Tolerance	Maximum Angle	Minimum Angle
49.	45°	±2°		
50.	30°	±5°		
51.	60°	+1°⁄−0°		
52.	15° 25′	+30′⁄−0′		
53.	34° 42′ 35″	±25′		
54.	25° 30′	±30′		
55.	35° 15′ 0″	±30″		
56.	110°	±20″		
57.	90°	+0°⁄−30′		
58.	125°	±20′		
59.	75° 40′	±30″		
60.	100°	+1°⁄−2°		
61.	115° 30′ 30″	±15″		
62.	24° 20′	±1′ 30″		
63.	15° 45′	±45′		
64.	37° 20′ 15″	±15′		
65.	30° 30′	+0°⁄−40″		

Goodheart-Willcox Publisher

Name _____ **Date** _____ **Class** _____

Convert the degrees, minutes, and seconds to decimal degrees.

66. 30° 26'

_____

67. 96° 44'

_____

68. 142° 07'

_____

69. 102° 1' 39"

_____

70. 215° 16' 12"

_____

71. 37° 39' 36"

_____

72. 89° 11' 52"

_____

73. 9° 30' 45"

_____

74. 271° 17' 42"

_____

75. 338° 57' 36"

_____

Convert the decimal degrees to degrees, minutes, and seconds.

76. 129.5°

_____

77. 58.78°

_____

78. 68.31°

_____

79. 27.965°

_____

80. 210.625°

_____

81. 334.775°

_____

82. 153.752°

_____

83. 93.325°

_____

84. 27.928°

_____

85. 9.296°

_____

Applications

When bolts are evenly placed around the edge of a circular plate, a circle drawn through their centers is called the bolt circle. To calculate the angle between bolts when equally spaced (∠X in the figure below) the formula is:

$$\angle X = \frac{360°}{\# \ of \ holes/bolts}$$

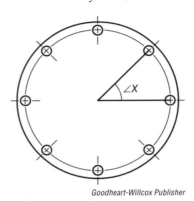

Goodheart-Willcox Publisher

Calculate the angle between bolts in the following problems. Round to the nearest second when necessary.

	# of Holes/Bolts in Bolt Circle	∠ between Holes/Bolts
1.	3	
2.	4	
3.	5	
4.	6	
5.	7	
6.	8	
7.	9	
8.	10	

Goodheart-Willcox Publisher

Name _____ **Date** _____ **Class** _____

9. What is angle X in the part drawing below?

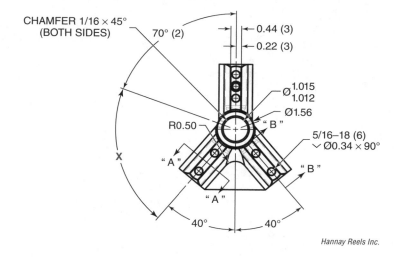

Hannay Reels Inc.

_____

10. A 22° 15′ part will be joined with a 10° 55′ part as shown in the following illustration. What will be the total angle of the assembly?

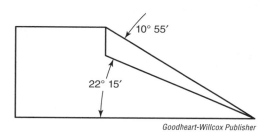

Goodheart-Willcox Publisher

_____

11. Continuing from question 10, if the 22° 15′ angle has a tolerance of ±1° and the 10° 55′ angle has a tolerance of ±45′, answer the following:

What is the total tolerance of the two dimensions combined?

_____

What is the maximum angle allowed if the tolerances are added?

_____

What is the minimum angle allowed if the tolerances are subtracted?

_____

Work Space/Notes

UNIT **35**

Triangles

Objectives

Information in this unit will enable you to:

- Explain what a triangle is.
- Identify and differentiate between different types of triangles.
- Describe how to determine the area of a triangle.

Kinds of Triangles

A triangle is a shape made up of three straight sides joined at three corners or vertices. There are several kinds of triangles, and each might be encountered in the machine shop. One way of naming triangles is based on the number of sides that are equal in length. The kinds of triangles based on the number of equal sides are shown in the following figure.

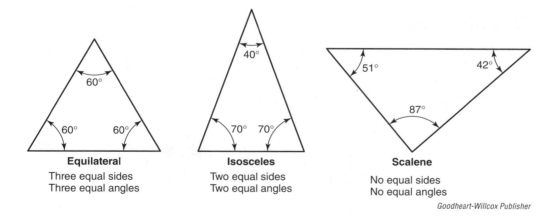

Equilateral
Three equal sides
Three equal angles

Isosceles
Two equal sides
Two equal angles

Scalene
No equal sides
No equal angles

Goodheart-Willcox Publisher

An **equilateral triangle** has three equal sides. Because the sides are all equal in length, the angles must also be equal. Each angle will always equal 60°. Notice that the sum of the three angles in an equilateral triangle is 180°. That is true for all types of triangles. The angles may be different, but they always add up to 180°.

An **isosceles triangle** has two sides that are the same length. These equal sides are called the legs. The side that is not the same (unequal) is called the base. An isosceles triangle also has two angles that are the same, called the base angles. The third angle that is different is called the vertex angle. The three angles still add up to 180°.

In a **scalene triangle**, no two sides or angles are the same. The angles do, however, still add up to 180°.

A triangle may also be named according to the largest angle it includes, as shown in the following figure.

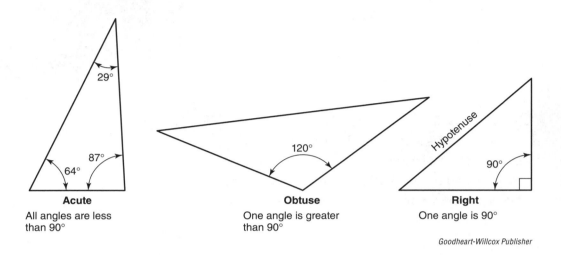

In an **acute triangle**, all of the angles are less than 90°. An equilateral triangle is an acute triangle and it is possible that an isosceles triangle is an acute triangle.

An **obtuse triangle** is one in which one angle is greater than 90°.

A **right triangle** has one 90° angle. The side opposite the 90° angle is called the **hypotenuse**.

Area of Triangles

Two new terms are necessary to find the areas of triangles: **base** and **height** (sometimes called altitude). The base can be any side of the triangle. The height is the distance from the base to the uppermost part of the triangle measured along a line perpendicular to the base, as shown in the figure.

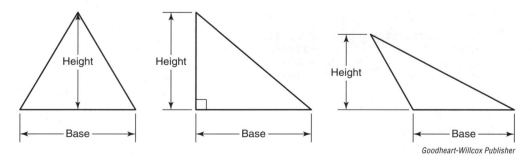

The area of a triangle is $1/2 \times$ base $\times$ height, regardless of the kind of triangle. If the triangle has a base of 36 mm and a height of 48 mm, the area would be:

$$\frac{1}{2} \times 36 \text{ mm} \times 48 \text{ mm} = 864 \text{ mm}^2$$

Unit 35 Review

Name _____ **Date** _____ **Class** _____

Fill in the blanks in the following review questions.

1. A _____ is a shape made up of three straight sides joined at three corners or vertices.

2. One way of _____ triangles is based on the number of sides equal in length.

3. A(n) _____ triangle has three equal sides. Because the sides are all equal in length, the angles must also be _____.

4. The sum of the three angles in any triangle is _____.

5. A(n) _____ triangle has two sides that are of equal length. It also has _____ angles that are the same.

6. In a(n) _____ triangle, no two sides or angles are the same.

7. In a(n) _____ triangle, all angles are less than 90°.

8. A(n) _____ triangle is one in which one angle is greater than 90°.

9. A(n) _____ triangle has one 90° angle. The side opposite the 90° angle is called the _____.

10. When calculating the area of a triangle, the _____ can be any side of the triangle, and the _____ is the distance from the base to the uppermost part of the triangle measured along a line perpendicular to the base.

Practice

Identify the kind of triangle as equilateral, isosceles, or scalene.

1.

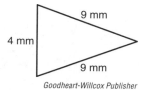

9 mm
4 mm
9 mm

Goodheart-Willcox Publisher

_____

2.

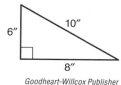

10″
6″
8″

Goodheart-Willcox Publisher

_____

3.

5 mm   5 mm
5 mm

Goodheart-Willcox Publisher

_____

4.

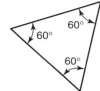

60°
60°
60°

Goodheart-Willcox Publisher

_____

5.

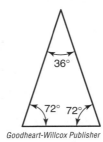

Goodheart-Willcox Publisher

_____

6.

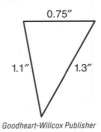

Goodheart-Willcox Publisher

_____

7.

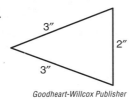

Goodheart-Willcox Publisher

_____

8.

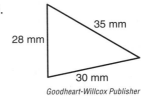

Goodheart-Willcox Publisher

_____

9.

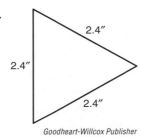

Goodheart-Willcox Publisher

_____

10.

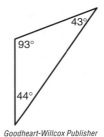

Goodheart-Willcox Publisher

_____

Identify the kind of triangle as acute, right, or obtuse.

11.

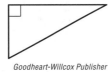

Goodheart-Willcox Publisher

_____

12.

Goodheart-Willcox Publisher

_____

13.

Goodheart-Willcox Publisher

_____

14.

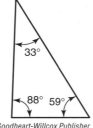

Goodheart-Willcox Publisher

_____

Name _____ **Date** _____ **Class** _____

15.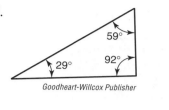

16.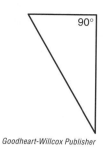

_____

_____

17.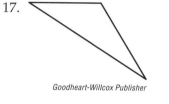

18.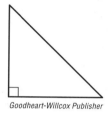

_____

_____

19.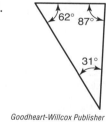

20.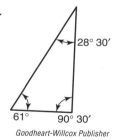

_____

_____

Calculate the areas of the triangles. Round your answers to the nearest tenth. Show your work.

21.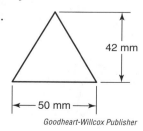

22.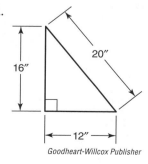

_____

_____

23.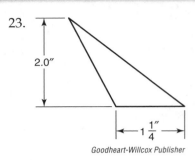

2.0″

1 1/4″

Goodheart-Willcox Publisher

_____

24.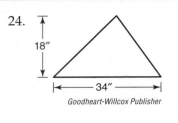

18″

34″

Goodheart-Willcox Publisher

_____

25.

28 mm

22 mm

Goodheart-Willcox Publisher

_____

26.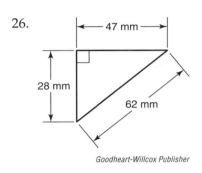

47 mm

28 mm

62 mm

Goodheart-Willcox Publisher

_____

27.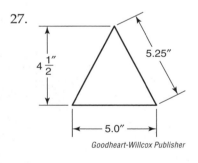

4 1/2″

5.25″

5.0″

Goodheart-Willcox Publisher

_____

28.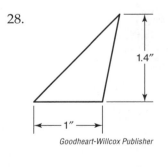

1.4″

1″

Goodheart-Willcox Publisher

_____

29.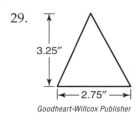

3.25″

2.75″

Goodheart-Willcox Publisher

_____

30.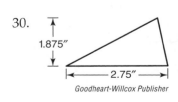

1.875″

2.75″

Goodheart-Willcox Publisher

_____

Name _____   **Date** _____   **Class** _____

31.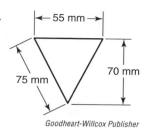

32.

_____

_____

Calculate the missing angle/angles and identify the kind of triangle as acute, right, or obtuse.

33.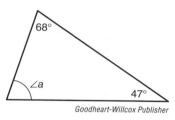

_____

_____

34.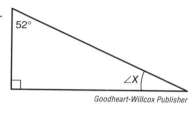

_____

_____

35.

_____

_____

36.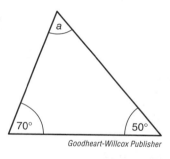

_____

_____

37.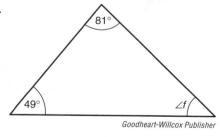

_____

_____

38.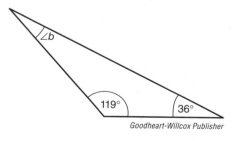

_____

_____

39.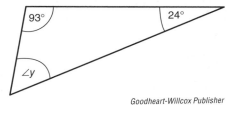

Goodheart-Willcox Publisher

_____

_____

40.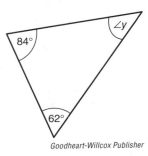

Goodheart-Willcox Publisher

_____

_____

41.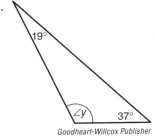

Goodheart-Willcox Publisher

_____

_____

42.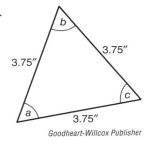

Goodheart-Willcox Publisher

_____

_____

43.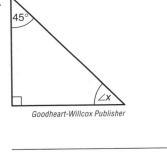

Goodheart-Willcox Publisher

_____

_____

UNIT 36

Right Triangle Trigonometry

Objectives

Information in this unit will enable you to:

- Discuss the Pythagorean theorem, including what it is used for and how it is used.
- Explain what trigonometry is and identify the three sides of a triangle.
- Identify and describe the three principal functions of trigonometry, and discuss the meaning of soh cah toa.

Pythagorean Theorem

The lengths of the sides of a right triangle are described by **Pythagorean theorem**, a rule discovered centuries ago by the Greek mathematician Pythagoras. Pythagorean theorem states that the sum of the squares of the two shortest sides (the ones next to the 90° angle) is equal to the square of the hypotenuse. Pythagorean theorem holds true only for triangles with a 90° angle.

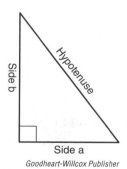

Side b · Hypotenuse · Side a

Goodheart-Willcox Publisher

This relationship is often stated as $a^2 + b^2 = c^2$, where a is the shortest side, b is the next shortest side, and c is the hypotenuse. $a^2 + b^2 = c^2$ is an equation. Whatever is on the left side of the equals sign is equal to whatever is on the right side. If the same operation is performed on both sides of an equation, that statement is still true.

Square roots and other roots are discussed in Unit 32. For this unit, it should be sufficient to know that all handheld calculators have a √ key. To find the square root of a number, enter the number and press the √ key. (Alternatively, on some calculators press the √ key, then enter the number, then press =.)

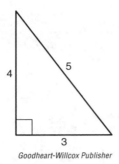

Square the lengths of the short sides.

$$3 \times 3 = 9$$
$$4 \times 4 = 16$$

Add the two squared sides.

$$9 + 16 = 25$$

Find the square root.

$$\sqrt{25} = 5 \text{ (the length of the hypotenuse)}$$

Example 36-1

By subtracting b^2 from both sides of this equation, we can leave a^2 alone on the left side.

$$a^2 = c^2 - b^2$$

In the previous figure, b represents 4 and c represents 5. Substitute those numbers to solve for a.

$$a^2 = 5^2 - 4^2$$
$$a^2 = 25 - 16$$
$$a^2 = 9$$

Find the square root of both sides.

$$a = 3$$

The three forms of the Pythagorean theorem are:

$$c^2 = a^2 + b^2$$
$$a^2 = c^2 - b^2$$
$$b^2 = c^2 - a^2$$

To isolate the variables c, a, and b in the above equations, take the square root of both sides. The formulas can now provide a direct answer when using a scientific calculator.

$$c = \sqrt{(a^2 + b^2)}$$
$$a = \sqrt{(c^2 - b^2)}$$
$$b = \sqrt{(c^2 - a^2)}$$

The unknown side of any right triangle can be found with one of these equations as long as two sides are known.

Angles and Sides

Trigonometry is the study of the ratios of sides and angles in triangles. Using three basic trigonometry functions, it is possible to calculate the length of an unknown side or the angle of an unknown corner of a triangle. Trigonometry can be used to solve problems with all types of triangles, but right triangles are the most common in the machine shop, so that is what is covered in this unit.

It is necessary to give names to the sides of a triangle, so the trigonometric formulas make sense. The side opposite the right angle in a right triangle is always the hypotenuse. One of the sides touching a given angle other than the right angle is called the **adjacent side**. The side that is across from the given angle and not touching it is called the **opposite side**.

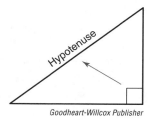

Goodheart-Willcox Publisher

The side opposite the right angle is the hypotenuse.

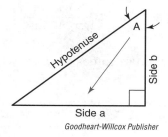

Goodheart-Willcox Publisher

Side b is adjacent to angle A.
Side a is opposite angle A.

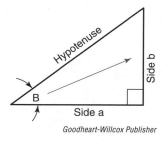

Goodheart-Willcox Publisher

Side a is adjacent to angle B.
Side b is opposite angle B.

Soh Cah Toa

There are three principal functions in trigonometry. They are sine, cosine, and tangent, which are abbreviated as sin, cos, and tan, respectively. For any angle in a triangle, these functions specify the ratios of certain sides.

Memorize the phrase "**soh cah toa.**" *Soh* represents the sine function, and the letters stand for **s**ine of an angle equals **o**pposite over **h**ypotenuse. *Cah* represents the cosine function, with the letters representing **c**osine of an angle equals **a**djacent over **h**ypotenuse. *Toa* represents the tangent function, with the letters representing **t**angent of an angle equals **o**pposite over **a**djacent.

$$\sin(\theta) = \frac{\text{opposite}}{\text{hypotenuse}} \qquad \cos(\theta) = \frac{\text{adjacent}}{\text{hypotenuse}} \qquad \tan(\theta) = \frac{\text{opposite}}{\text{adjacent}}$$

The Greek letter θ (theta) is used in mathematics as a variable to represent the unknown measure of an angle.

Each of the trigonometric functions has an inverse. The inverse is used to obtain the unknown measure of an angle. Since the sine of an angle is equal to the ratio of *opposite/hypotenuse*, the sine inverse (sometimes referred to as *arcsine*, abbreviated as *arcsin*) of the same ratio will give the measure of the angle. The inverse sine function is represented using $\sin^{-1}$. (This does not mean sine raised to the negative-one power). The inverse of cosine is $\cos^{-1}$, and the inverse of tangent is $\tan^{-1}$, alternatively expressed as *arccosine* (*arccos*) and *arctangent* (*arctan*). Pressing the shift key followed by the sin, cos, or tan key on a scientific calculator will display the inverse function.

To find the measure of an unknown angle, we can rearrange the formulas as follows:

$$\theta = \sin^{-1}\left(\frac{\text{opposite}}{\text{hypotenuse}}\right) \qquad \theta = \cos^{-1}\left(\frac{\text{adjacent}}{\text{hypotenuse}}\right) \qquad \theta = \tan^{-1}\left(\frac{\text{opposite}}{\text{adjacent}}\right)$$

Knowing which formula to use comes down to what information the problem presents. See the table below.

Information Known	Formula to Use to Determine Angle
Opposite and hypotenuse	$\theta = \sin^{-1}\left(\dfrac{\text{opposite}}{\text{hypotenuse}}\right)$
Adjacent and hypotenuse	$\theta = \cos^{-1}\left(\dfrac{\text{adjacent}}{\text{hypotenuse}}\right)$
Opposite and adjacent	$\theta = \tan^{-1}\left(\dfrac{\text{opposite}}{\text{adjacent}}\right)$

Consider the right triangle in this figure.

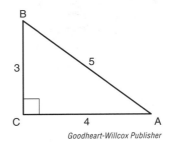

The sine of angle A equals 3 (the length of the opposite side) over 5 (the length of the hypotenuse) or 3/5. This can be stated as sin ∠A = 3/5. There are tables on the Internet and in reference books that show angles and their corresponding trigonometric functions, but that can also be done with a scientific calculator. The following are the key strokes for one type of scientific calculator to find the angle for which 3/5 is the sine: Make sure the calculator is set to degree, not radian or gradient.

The display shows:

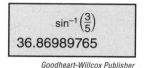

The calculator will likely output the answer in decimal degrees. Decimal degrees are the preferred unit of measurement in the metric system. Most scientific calculators have a button to convert between decimal degrees and degrees, minutes, and seconds. The key on the calculator looks like this:

Pressing the DMS (degrees, minutes, seconds) key will alternate between decimal degrees and degrees, minutes, and seconds. Using the key in this case gives an answer of ∠A = 36.870° rounded to the nearest thousandth, or 36° 52′ 12″ rounded to the nearest second.

Knowing that one angle is 90°, by using these functions and having three pieces of information about a right triangle, the length of any side can be found and either of its remaining angles can be found.

Rearranging the formulas, we can solve for the hypotenuse, opposite, and adjacent sides.

Using the sine equation:

$$\sin(\theta) = \frac{\text{opposite}}{\text{hypotenuse}} \qquad \text{opposite} = \sin(\theta) \times \text{hypotenuse} \qquad \text{hypotenuse} = \frac{\text{opposite}}{\sin(\theta)}$$

Using the cosine equation:

$$\cos(\theta) = \frac{\text{adjacent}}{\text{hypotenuse}} \qquad \text{adjacent} = \cos(\theta) \times \text{hypotenuse} \qquad \text{hypotenuse} = \frac{\text{adjacent}}{\cos(\theta)}$$

Using the tangent equation:

$$\tan(\theta) = \frac{\text{opposite}}{\text{adjacent}} \qquad \text{opposite} = \tan(\theta) \times \text{adjacent} \qquad \text{adjacent} = \frac{\text{opposite}}{\tan(\theta)}$$

Knowing which formula to use comes down to what information the problem presents. See the table below.

Unknown Side	Equation to Use
Hypotenuse	$\text{hypotenuse} = \dfrac{\text{opposite}}{\sin(\theta)}$ or $\text{hypotenuse} = \dfrac{\text{adjacent}}{\cos(\theta)}$
Opposite	$\text{opposite} = \sin(\theta) \times \text{hypotenuse}$ or $\text{opposite} = \tan(\theta) \times \text{adjacent}$
Adjacent	$\text{adjacent} = \cos(\theta) \times \text{hypotenuse}$ or $\text{adjacent} = \dfrac{\text{opposite}}{\tan(\theta)}$

Goodheart-Willcox Publisher

Example 36-2

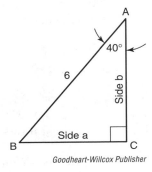

Goodheart-Willcox Publisher

What is angle B and the length of side b?

The total of all of the angles in any triangle is 180°. Subtract the known angles from 180° to find the remaining angle:

$$180° - 90° - 40° = 50°; \angle B = 50°$$

To find the side adjacent to ∠A, side b, first determine what is known in the triangle. In this case, the hypotenuse is known (6), and the adjacent side needs to be found. Therefore, cosine needs to be used to solve the adjacent.

$$\text{adjacent} = \cos(\theta) \times \text{hypotenuse}$$
$$\text{adjacent} = \cos 40° \times 6$$

Use a scientific calculator.

Goodheart-Willcox Publisher

The display shows:

Goodheart-Willcox Publisher

Side b = 4.5963 to the nearest tenth in shop terms (or nearest ten-thousandth in mathematical terms).

Example 36-3

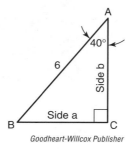

Goodheart-Willcox Publisher

What is the length of side a?

To find the side opposite ∠A, side a, first determine what is known in the triangle. In this case the hypotenuse is known (6), and the opposite needs to be found. Therefore, sine needs to be used to solve the opposite.

$$\text{opposite} = \sin(\theta) \times \text{hypotenuse}$$
$$\text{opposite} = \sin 40° \times 6$$

Use a scientific calculator.

Goodheart-Willcox Publisher

The display shows:

sin 40 × 6 =
3.8567256581192

Goodheart-Willcox Publisher

Side a = 3.8567 to the nearest tenth in shop terms (or nearest ten-thousandth in mathematical terms).

Example 36-4

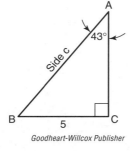

Goodheart-Willcox Publisher

What is angle B and the length of side c?

The total of all angles in any triangle is 180°. Subtract the known angles from 180° to find the remaining angle.

$$180° - 90° - 43° = 47°$$
$$\angle B = 47°$$

(Continued)

To find side c, first determine what is known in the triangle. In this case the opposite is known (5), and the hypotenuse needs to be found. Therefore, sine needs to be used to solve the hypotenuse.

$$\text{hypotenuse} = \frac{\text{opposite}}{\sin(\theta)}$$

$$\text{hypotenuse} = \frac{5}{\sin 43°}$$

Use a scientific calculator.

5 ÷ **SIN** 43° **ENTER =**

Goodheart-Willcox Publisher

The display shows:

> 5 ÷ sin 43 =
> **7.3313959281981**

Goodheart-Willcox Publisher

Side c = 7.331 to the nearest thousandth.

Example 36-5

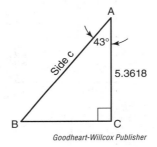

Goodheart-Willcox Publisher

What is the length of side c?

To find side c, first determine what is known in the triangle. In this case the adjacent is known (5.3618), and the hypotenuse needs to be found. Therefore, cosine needs to be used to solve the hypotenuse.

$$\text{hypotenuse} = \frac{\text{adjacent}}{\cos(\theta)}$$

$$\text{hypotenuse} = \frac{5.3618}{\cos 43°}$$

Use a scientific calculator.

5.3618 ÷ **COS** 43° **ENTER =**

Goodheart-Willcox Publisher

The display shows:

> 5.3618 ÷ cos 43 =
> **7.3313363809185**

Goodheart-Willcox Publisher

Side c = 7.331 to the nearest thousandth.

Example 36-6

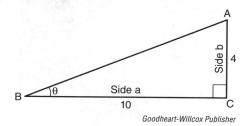

Goodheart-Willcox Publisher

What is angle B to the nearest whole degree?

The opposite side (side b) and the adjacent side (side a) are known, so use the tangent function to solve:

$$\theta = \tan^{-1}\left(\frac{\text{opposite}}{\text{adjacent}}\right)$$

$$\theta = \tan^{-1}\left(\frac{4}{10}\right)$$

Use the tan key on the scientific calculator or a table of tangents.

Goodheart-Willcox Publisher

The display shows:

$$\tan^{-1}(4 \div 10) =$$
$$21.801409486352$$

Goodheart-Willcox Publisher

Pressing the degree, minutes, seconds key ⊙','' will convert this to 21° 48′ 5.07″.
∠B = 21.80140949° or 21° 48′ 5.07″.
∠B = 22° to the nearest whole degree.

Example 36-7

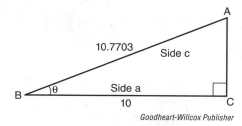

Goodheart-Willcox Publisher

What is angle B to the nearest whole degree?

The adjacent side (side a), and the hypotenuse (side c) are known, so use the cosine function to solve:

$$\theta = \cos^{-1}\left(\frac{\text{adjacent}}{\text{hypotenuse}}\right)$$

$$\theta = \cos^{-1}\left(\frac{10}{10.7703}\right)$$

(Continued)

Use the cos key on the scientific calculator or a table of tangents.

2nd   COS   (10 ÷ 10.7703   )   ENTER =

Goodheart-Willcox Publisher

The display shows:

$$\cos^{-1}(10 \div 10.7703) =$$
$$21.801015628426$$

Goodheart-Willcox Publisher

Pressing the degree, minutes, seconds key ∘, '' will convert this to 21° 48′ 3.66″.
∠B = 21.80101563° or 21° 48′ 3.66″.
∠B = 22° to the nearest whole degree.

Example 36-8

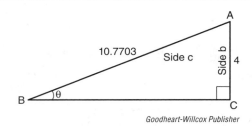

Goodheart-Willcox Publisher

What is angle B to the nearest whole degree?

The opposite side (side b) and the hypotenuse (side c) are known, so use the sine function to solve:

$$\theta = \sin^{-1}\left(\frac{\text{opposite}}{\text{hypotenuse}}\right)$$

$$\theta = \sin^{-1}\left(\frac{4}{10.7703}\right)$$

Use the sin key on the scientific calculator or a table of tangents.

2nd   SIN   (4 ÷ 10.7703   )   ENTER =

Goodheart-Willcox Publisher

The display shows:

$$\sin^{-1}(4 \div 10.7703) =$$
$$21.801472503092$$

Goodheart-Willcox Publisher

Pressing the degree, minutes, seconds key ∘, '' will convert this to 21° 48′ 5.3″.
∠B = 21.8014725° or 21° 48′ 5.3″.
∠B = 22° to the nearest whole degree.

Unit 36 Review

Name _____ Date _____ Class _____

Fill in the blanks in the following review questions.

1. The lengths of the sides of a right triangle are described by the _____ _____, which states that the sum of the squares of the two shortest sides is equal to the square of the _____.

2. The Pythagorean theorem holds true only for triangles with a _____ angle.

3. The unknown side of any right triangle can be found with one of these equations as long as _____ sides are known.

4. _____ is the study of the ratios of sides and angles in triangles.

5. Using three basic trigonometry functions, it is possible to calculate the _____ of an unknown side or the _____ of an unknown corner of a triangle.

6. The side opposite the right angle in a right triangle is always the _____, which is also always the longest side of a right-angle triangle.

7. One of the sides touching a given angle other than the right angle is called the _____ side. The side that is across from the given angle and not touching it is called the _____ side.

Practice

Calculate the length of the unknown side. Round your answers to the nearest tenth (in mathematical terms). Show your work.

1.
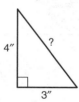
4"   ?
3"

Goodheart-Willcox Publisher

_____

2.
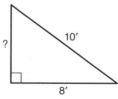
?   10'
8'

Goodheart-Willcox Publisher

_____

3.
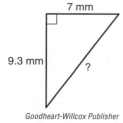
7 mm
9.3 mm   ?

Goodheart-Willcox Publisher

_____

4.

?   7"
5"

Goodheart-Willcox Publisher

_____

5.

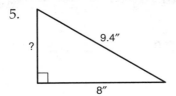

9.4″
?
8″
Goodheart-Willcox Publisher

_____

6.

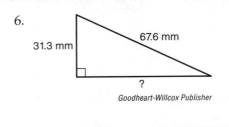

67.6 mm
31.3 mm
?
Goodheart-Willcox Publisher

_____

7.

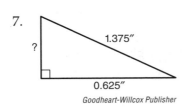

1.375″
?
0.625″
Goodheart-Willcox Publisher

_____

8.

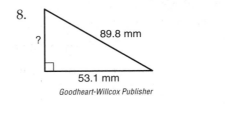

89.8 mm
?
53.1 mm
Goodheart-Willcox Publisher

_____

9.

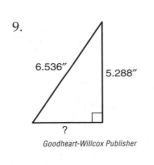

6.536″ 5.288″
?
Goodheart-Willcox Publisher

_____

10.

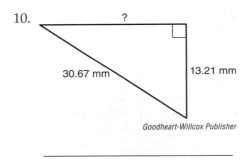

?
30.67 mm 13.21 mm
Goodheart-Willcox Publisher

_____

Name the sides of the following triangles as hypotenuse, opposite, or adjacent.

11.

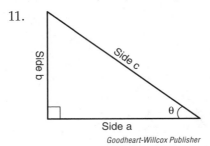

Side b
Side c
θ
Side a
Goodheart-Willcox Publisher

Side a = _____

Side b = _____

Side c = _____

12.

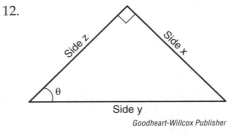

Side z
Side x
θ
Side y
Goodheart-Willcox Publisher

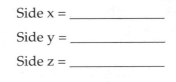

Side x = _____

Side y = _____

Side z = _____

Name _____   **Date** _____   **Class** _____

13.

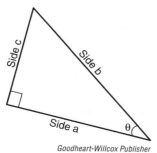

Goodheart-Willcox Publisher

Side a = _____

Side b = _____

Side c = _____

14.

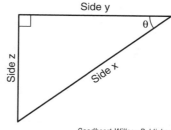

Goodheart-Willcox Publisher

Side x = _____

Side y = _____

Side z = _____

15.

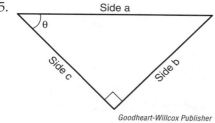

Goodheart-Willcox Publisher

Side a = _____

Side b = _____

Side c = _____

16.

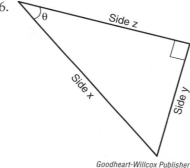

Goodheart-Willcox Publisher

Side x = _____

Side y = _____

Side z = _____

17.

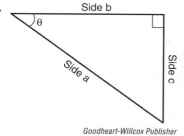

Goodheart-Willcox Publisher

Side a = _____

Side b = _____

Side c = _____

18.

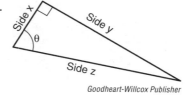

Goodheart-Willcox Publisher

Side x = _____

Side y = _____

Side z = _____

19.

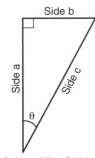

Goodheart-Willcox Publisher

Side a = _____

Side b = _____

Side c = _____

20.

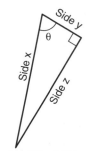

Goodheart-Willcox Publisher

Side x = _____

Side y = _____

Side z = _____

21.

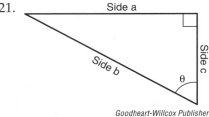

Goodheart-Willcox Publisher

Side a = _____

Side b = _____

Side c = _____

22.

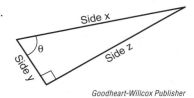

Goodheart-Willcox Publisher

Side x = _____

Side y = _____

Side z = _____

23.

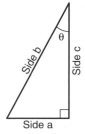

Goodheart-Willcox Publisher

Side a = _____

Side b = _____

Side c = _____

24.

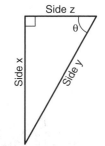

Goodheart-Willcox Publisher

Side x = _____

Side y = _____

Side z = _____

Name _____   **Date** _____   **Class** _____

For the following right triangles, label each side (hypotenuse, opposite, or adjacent), find the measure of each angle indicated (θ) to the nearest second, and provide the equation used to solve the problem.

25.

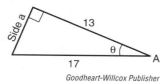

Goodheart-Willcox Publisher

Name of side a: _____

Name of side that measures 13: _____

Name of side that measures 17: _____

Equation to solve for ∠A: _____

∠A (θ) = _____

26.

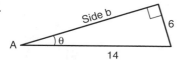

Goodheart-Willcox Publisher

Name of side b: _____

Name of side that measures 6: _____

Name of side that measures 14: _____

Equation to solve for ∠A: _____

∠A (θ) = _____

27.

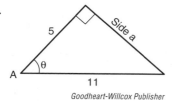

Goodheart-Willcox Publisher

Name of side a: _____

Name of side that measures 5: _____

Name of side that measures 11: _____

Equation to solve for ∠A: _____

∠A (θ) = _____

28.

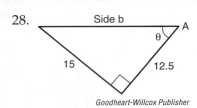

Goodheart-Willcox Publisher

Name of side b: _____

Name of side that measures 12.5: _____

Name of side that measures 15: _____

Equation to solve for ∠A: _____

∠A (θ) = _____

29.

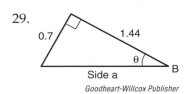

Goodheart-Willcox Publisher

Name of side a: _____

Name of side that measures 0.7: _____

Name of side that measures 1.44: _____

Equation to solve for ∠B: _____

∠B (θ) = _____

30.

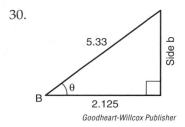

Goodheart-Willcox Publisher

Name of side b: _____

Name of side that measures 2.125: _____

Name of side that measures 5.33: _____

Equation to solve for ∠B: _____

∠B (θ) = _____

Name _____   **Date** _____   **Class** _____

31.

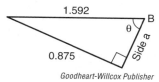

Goodheart-Willcox Publisher

Name of side a: _____

Name of side that measures 0.875: _____

Name of side that measures 1.592: _____

Equation to solve for ∠B: _____

∠B (θ) = _____

32.

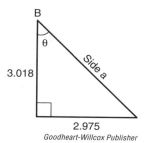

Goodheart-Willcox Publisher

Name of side a: _____

Name of side that measures 2.975: _____

Name of side that measures 3.018: _____

Equation to solve for ∠B: _____

∠B (θ) = _____

For the following right triangles, label each side (hypotenuse, opposite, or adjacent), find the measure of each side indicated to the nearest four decimal places (0.0000), and provide the equation to use to solve the problem.

33.

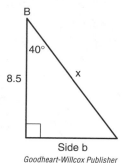

Side b

Goodheart-Willcox Publisher

Name of side b: _____

Name of side x: _____

Name of side that measures 8.5: _____

Equation to solve for side x: _____

Length of side x = _____

34.

Goodheart-Willcox Publisher

Name of side b: _____

Name of side x: _____

Name of side that measures 17.58: _____

Equation to solve for side x: _____

Length of side x = _____

35.

Goodheart-Willcox Publisher

Name of side b: _____

Name of side x: _____

Name of side that measures 6.25: _____

Equation to solve for side x: _____

Length of side x = _____

36.

Goodheart-Willcox Publisher

Name of side b: _____

Name of side x: _____

Name of side that measures 23.762: _____

Equation to solve for side x: _____

Length of side x = _____

Name _____   **Date** _____   **Class** _____

37.

Side b
Goodheart-Willcox Publisher

Name of side b: _____

Name of side x: _____

Name of side that measures 3.778: _____

Equation to solve for side x: _____

Length of side x = _____

38.

Goodheart-Willcox Publisher

Name of side a: _____

Name of side x: _____

Name of side that measures 9.33: _____

Equation to solve for side x: _____

Length of side x = _____

39.

Side a
Goodheart-Willcox Publisher

Name of side a: _____

Name of side x: _____

Name of side that measures 7.125: _____

Equation to solve for side x: _____

Length of side x = _____

40.

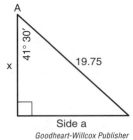

Goodheart-Willcox Publisher

Name of side a: _____

Name of side x: _____

Name of side that measures 19.75: _____

Equation to solve for side x: _____

Length of side x = _____

For the following right triangles, find the unknown angle to the nearest second, label each side (hypotenuse, opposite, or adjacent), find the measure of each unknown side to four decimal places (0.0000), and provide the equations to use to solve each problem.

41.

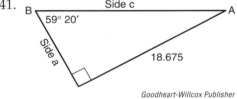

Goodheart-Willcox Publisher

Unknown angle (A) = _____

Name of side a: _____

Name of side c: _____

Name of side that measures 18.675: _____

Equation to solve for side a: _____

Length of side a = _____

Equation to solve for side c: _____

Length of side c = _____

Name _____   **Date** _____   **Class** _____

42.

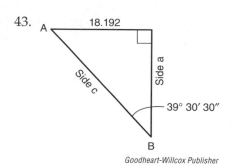

Side b

49° 1′ 30″

A

4.792

Side c

B

Goodheart-Willcox Publisher

Unknown angle (B) = _____

Name of side b: _____

Name of side c: _____

Name of side that measures 4.792: _____

Equation to solve for side b: _____

Length of side b = _____

Equation to solve for side c: _____

Length of side c = _____

43.   A   18.192

Side c   Side a

39° 30′ 30″

B

Goodheart-Willcox Publisher

Unknown angle (A) = _____

Name of side a: _____

Name of side c: _____

Name of side that measures 18.192: _____

Equation to solve for side a: _____

Length of side a = _____

Equation to solve for side c: _____

Length of side c = _____

44.

Goodheart-Willcox Publisher

Unknown angle (A) = _____

Name of side a: _____

Name of side b: _____

Name of side that measures 25.375: _____

Equation to solve for side a: _____

Length of side a = _____

Equation to solve for side b: _____

Length of side b = _____

45.

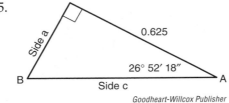

Goodheart-Willcox Publisher

Unknown angle (B) = _____

Name of side a: _____

Name of side c: _____

Name of side that measures 0.625: _____

Equation to solve for side a: _____

Length of side a = _____

Equation to solve for side c: _____

Length of side c = _____

Name _____   **Date** _____   **Class** _____

46.

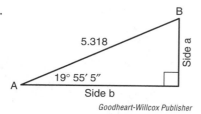

Goodheart-Willcox Publisher

Unknown angle (B) = _____

Name of side a: _____

Name of side b: _____

Name of side that measures 5.318: _____

Equation to solve for side a: _____

Length of side a = _____

Equation to solve for side b: _____

Length of side b = _____

47.

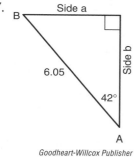

Goodheart-Willcox Publisher

Unknown angle (B) = _____

Name of side a: _____

Name of side b: _____

Name of side that measures 6.05: _____

Equation to solve for side a: _____

Length of side a = _____

Equation to solve for side b: _____

Length of side b = _____

48.

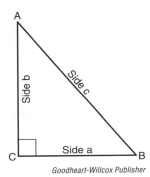

Goodheart-Willcox Publisher

Unknown angle (A) = _____

Name of side b: _____

Name of side c: _____

Name of side that measures 10.829: _____

Equation to solve for side b: _____

Length of side b = _____

Equation to solve for side c: _____

Length of side c = _____

Complete the table. Round all answers to one decimal place. (It may help to draw and label a right triangle as you solve these questions.)

Goodheart-Willcox Publisher

	Angle A	Angle B	Side a (Opposite Angle A)	Side b (Adjacent to Angle A)	Hypotenuse
49.			3″		5″
50.		25°	5″		
51.			22 mm	40 mm	
52.			1.5″	2.75″	
53.	41°			5.4″	
54.		18° 30′			2.75″
55.	33° 25′ 10″		3.172″		
56.				4.652″	7.252″
57.		41° 12′ 18″		3.75″	
58.	52.5°				75 mm

Goodheart-Willcox Publisher

Name _____   **Date** _____   **Class** _____

Applications

Use the figure below to solve problem 1. Show your work.

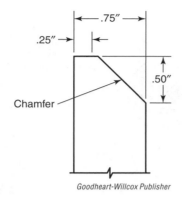

Goodheart-Willcox Publisher

1. What is the length of the chamfer? Round your answer to the nearest hundredth in mathematical terms.

    _____

2. A 0.6″ thick tool is ground to a 48° included angle (the angle between the two ground surfaces) as shown in the figure. How far back from the tip of the tool does the ground surface extend? Show all of your work and round your answer to the nearest 0.001″.

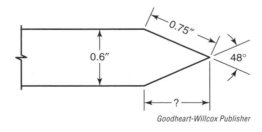

Goodheart-Willcox Publisher

    _____

3. What is dimension X to the nearest .001"?

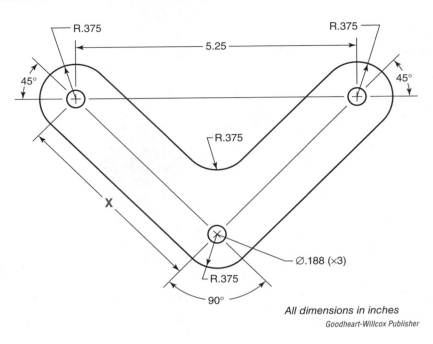

All dimensions in inches

Goodheart-Willcox Publisher

_____

4. What is the angle of the taper shown in the drawing below to the nearest .1°?

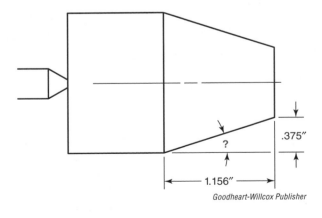

Goodheart-Willcox Publisher

_____

Oblique Triangles

Objectives

Information in this unit will enable you to:

- Describe the customary method of labeling an oblique triangle.
- Discuss the law of sines and how it is used.
- Describe the law of cosines and how it is used.

Oblique Triangles

An oblique triangle is any triangle that does not include a 90° angle. If three parts of the triangle including one or more sides are known, it is possible to find the values of the remaining parts, using the law of sines or the law of cosines. The explanations of these laws are easiest if the parts are named consistently. It is customary to label angles A, B, and C, and to label the side opposite each angle with the lower case of the same letter: the side opposite angle A is side a, the side opposite angle B is side b, etc.

Law of Sines

The law of sines is that each side of a triangle is in proportion to the sine of the opposite angle. In mathematical terms, the law of sines is:

$$\frac{a}{\sin A} = \frac{b}{\sin B} = \frac{c}{\sin C}$$

The example here explains how the law of sines can be used to solve a triangle with two known angles and the side opposite one of them.

Example 37-1

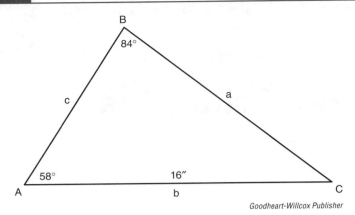

Goodheart-Willcox Publisher

Find the third angle and the lengths of the other two sides to the nearest tenth.

First, find angle C. The sum of all the angles of any triangle equals 180°:

$$\angle C = 180° - 58° - 84°. \ \angle C = 38°.$$

Use the law of sines to calculate the length of side a.

$$\frac{a}{\sin A} = \frac{b}{\sin B}$$

Multiply both sides by sinA.

$$a = \frac{b \times \sin A}{\sin B}$$

Substitute the known values in the equation. Use a sine table or scientific calculator to find the values of sinA and sinB.

$$a = \frac{16 \times .848}{.995} = 13.6''$$

Repeat steps 2 through 4 to calculate side c.

$$\frac{c}{\sin C} = \frac{b}{\sin B}$$

$$c = \frac{b \times \sin C}{\sin B}$$

$$c = \frac{16 \times .616}{.995}$$

$$c = 9.9''$$

Law of Cosines

The law of cosines is based on the Pythagorean theorem that was explained in Unit 36, $c^2 = a^2 + b^2$. The Pythagorean theorem applies only to right triangles, those having a right angle. By adding something to that equation, it can be used with any triangle. In the law of cosines, $2ab(\cos C)$ is added to the equation. The **law of cosines** is $c^2 = a^2 + b^2 - 2ab(\cos C)$. The law of cosines can also be expressed as $b^2 = a^2 + c^2 - 2ac(\cos B)$ or as $a^2 = b^2 + c^2 - 2bc(\cos A)$.

The law of cosines is useful for finding the third side of a triangle when two sides and the angle between them are known; or for finding the angles of a triangle when all three sides are known.

Example 37-2

Find side a of the triangle in the figure to the nearest one-tenth millimeter.

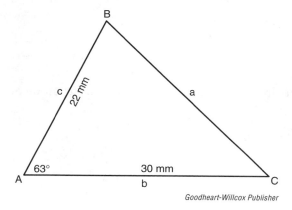

Goodheart-Willcox Publisher

Use the law of cosines in the arrangement that has a^2 alone on the left.

$$a^2 = b^2 + c^2 - 2bc(\cos A)$$

Substitute the known information.

$$a^2 = 30^2 + 22^2 - 2 \times 30 \times 22 \times .454$$
$$a^2 = 900 + 484 - 599.3$$
$$a^2 = 784.7$$

Find the square root.

$$a = \sqrt{784.7} = 28.0 \text{ mm}$$

When the angle between the two sides is greater than 90°, the cosine of that angle is a negative number.

Example 37-3

Find side c of the triangle below to the nearest hundredth inch.

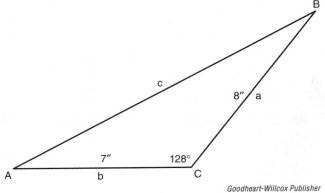

Goodheart-Willcox Publisher

Use the law of cosines in the arrangement that has c^2 alone on the left.

$$c^2 = a^2 + b^2 - 2ab(\cos C)$$

Substitute the known information.

$$c^2 = 8^2 + 7^2 - 2 \times 8 \times 7 \times (-.616)$$

$$c^2 = 113 - 112 \times (-.616)$$

$$c^2 = 113 + 68.992$$

(When a negative number is multiplied by another negative number, the product is a positive number.)

$$c^2 = 181.992$$

Find the square root.

$$c = 13.490''$$

The law of cosines can be rearranged to make it easier to find unknown angles. The **angle version** of the law of cosines can be in any of these forms:

$$\cos(C) = \frac{a^2 + b^2 - c^2}{2ab} \qquad \cos(A) = \frac{b^2 + c^2 - a^2}{2bc} \qquad \cos(B) = \frac{c^2 + a^2 - b^2}{2ca}$$

Example 37-4

Find the angles of the triangle in the figure to the nearest whole degree.

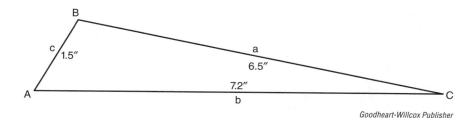

Goodheart-Willcox Publisher

Use the angle version of the law of cosines.

$$\cos C = \frac{a^2 + b^2 - c^2}{2ab}$$

Substitute known information.

$$\cos C = \frac{6.5^2 + 7.2^2 - 1.5^2}{2 \times 6.5 \times 7.2}$$

Reduce the fraction.

$$\cos C = \frac{42.25 + 51.84 - 2.25}{93.6}$$

$$\cos C = \frac{91.84}{93.6}$$

Use a scientific calculator or table of cosines to solve for ∠C.

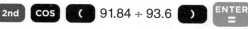

Goodheart-Willcox Publisher

The display shows:

$$\angle C = 11°$$

Goodheart-Willcox Publisher

Repeat this process to find one of the other angles, then subtract from 180° to find the third angle.

Work Space/Notes

Unit 37 Review

Name _____ **Date** _____ **Class** _____

Fill in the blanks in the following review questions.

1. An oblique triangle is any triangle that does not include a _____°
   angle.

2. It is customary to label angles A, B, and C, and to label the side
   _____ each angle with the lower case of the same letter.

3. The law of sines is that each side of a triangle is in _____ to the sine
   of the opposite angle.

4. The law of cosines is based on the _____ theorem.

5. The law of cosines is useful for finding the third side of a triangle when
   _____ sides and the _____ between them are known, or for
   finding the angles of a triangle when all _____ are known.

6. When the exponent applies to both the numerator and denominator, the
   fraction is enclosed in parentheses and the exponent is written _____
   the parentheses.

7. When parentheses are used in any mathematical expression, the operations
   inside the parentheses are performed _____ whatever is indicated
   outside the parentheses.

8. If the fraction inside the parentheses can be reduced, that is done
   _____ raising to the indicated power.

Practice

*Using the law of sines, find the missing information (lengths of sides and angles) for the
triangles. Round your answers to the nearest hundredth.*

Use this triangle for problems 1, 2, and 3.   *Use this triangle for problems 4, 5, and 6.*

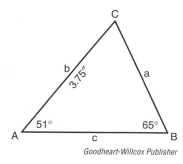

Goodheart-Willcox Publisher

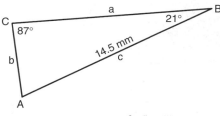

Goodheart-Willcox Publisher

1. ∠C = _____   4. ∠A = _____

2. side a = _____   5. side a = _____

3. side c = _____   6. side b = _____

Use this triangle for problems 7, 8, and 9.   *Use this triangle for problems 10, 11, and 12.*

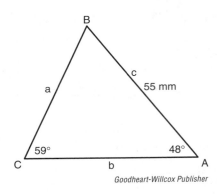

Goodheart-Willcox Publisher

7. ∠B = _____

8. Side a = _____

9. Side b = _____

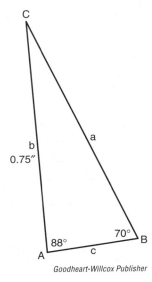

Goodheart-Willcox Publisher

10. ∠C = _____

11. Side a = _____

12. Side c = _____

Using the law of cosines, find each of the angles in the triangles. Round your answers to the nearest tenth of a degree.

Use this triangle for problems 13, 14, and 15.

Use this triangle for problems 16, 17, and 18.

Goodheart-Willcox Publisher

13. ∠A = _____

14. ∠B = _____

15. ∠C = _____

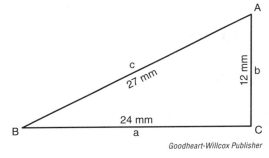

Goodheart-Willcox Publisher

16. ∠A = _____

17. ∠B = _____

18. ∠C = _____

Name _____   **Date** _____   **Class** _____

Using the law of cosines, find the missing side of each triangle.

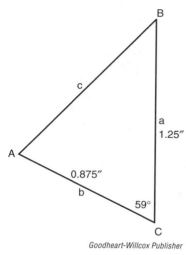

Goodheart-Willcox Publisher

19. Side c = _____

20. Side a = _____

Goodheart-Willcox Publisher

Applications

Use the following drawing to answer questions 1 and 2.

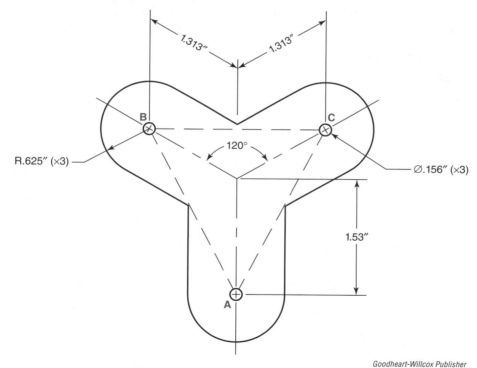

Goodheart-Willcox Publisher

What is the distance between holes (dashed lines) to the nearest thousandth of an inch?

1. Hole A to hole B?

2. Hole B to hole C?

_____   _____

Use the following drawing to answer questions 3 through 5.

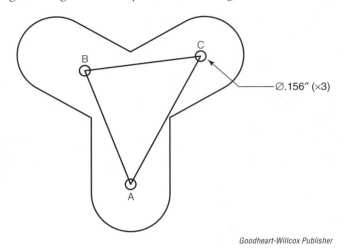

Ø.156″ (×3)

Goodheart-Willcox Publisher

Center-to-center distances between holes:

Hole B to hole C = 50 mm

Hole C to hole A = 76 mm

Hole A to hole B = 67 mm

Calculate the angles formed in the above triangle given the center-to-center distances between the holes. Round answers to the nearest whole degree.

3. Angle A = _____

4. Angle B = _____

5. Angle C = _____

Shop Math

Key Terms

altitude

bisect

congruent

constant

dovetail slide

gage blocks

point angle

point of tangency

sine bar

sine plate

tangent line

tangent to a circle
  theorem

transversal

two-tangent theorem

UNIT 38

Sine Bars and Sine Plates

Objectives

Information in this unit will enable you to:

- Describe a sine bar and how it is used.
- Use a sine bar to determine an unknown angle of a workpiece.
- Describe a sine plate and how it is used.
- Use a sine bar to determine the height of a stack of gage blocks to form a right triangle.

Sine Bar

A **sine bar** is a precisely machined tool consisting of a hardened, precision-ground flat body typically no wider than 1", with one of two precision-ground cylinders of equal diameter (sometimes referred to as *rolls*) attached at each end. The parts that make up the sine bar are often made from tool steel to avoid wear and tear when used. An end rail can also be mounted at either end. The body often has relief holes that help reduce the weight. The two cylinders (rolls) are placed at a precise distance, usually 5", 10", or 20" apart. They are also provided in metric sizes of 100 mm, 200 mm, and 300 mm.

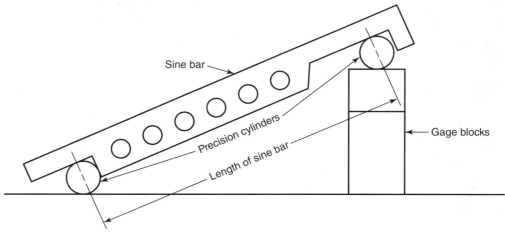

Sine bar

Precision cylinders

Length of sine bar

Gage blocks

Goodheart-Willcox Publisher

A sine bar is frequently used with **gage blocks**, also called *jo blocks* (after inventor Carl Edvard Johansson), which are made from hardened steel. Gage blocks are ground to precise measurements and can be stacked in various combinations of sizes to any desired height. Most gage block sets have nine blocks in increments of ten-thousandths of an inch (0.1001″–0.1009″). In metric gage block sets, the first series consists of increments of micrometers, commonly known as microns (1.001–1.009 mm).

When the sine bar is placed on a flat surface, such as a surface plate, the top surface of the sine bar is parallel to that surface. When one end rests on a stack of gage blocks, a right triangle is formed. The sine bar forms a hypotenuse of a known length (the dimension between the cylinders). The stack of gage blocks forms a side of known length (the total height of the gage blocks). The gage blocks always form the opposite side of the right-angle triangle. Using the sine function, it is possible to compute the angle the sine bar forms with the surface it rests on. The sine bar is useful for determining unknown angles of small parts to a high degree of accuracy.

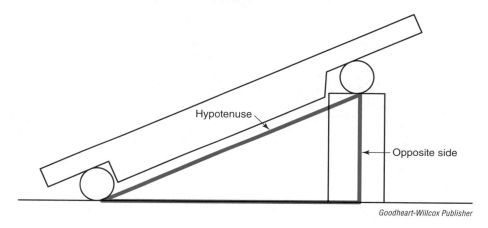

Goodheart-Willcox Publisher

Sine Plates

Sine plates are used in a fashion very similar to sine bars, but sine plates are much wider than sine bars and can sometimes be used to hold workpieces or fixtures at a specific angle to guide accurate machining. Sine plates are sometimes referred to as *sine tables*. A sine plate has a top plate that tilts out from the base plate to set up an exact angle. Gage blocks are used to raise the top plate to a specific height to create the desired angle. The top plate can be locked in place once the angle has been set.

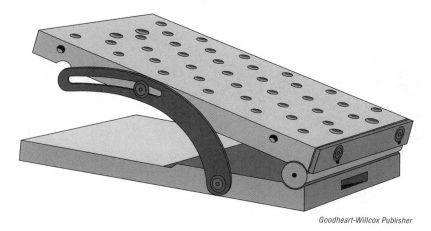

Goodheart-Willcox Publisher

Using a Sine Bar

To measure the angle of small parts or workpieces, set up the sine bar at an angle to the surface plate by placing one of the cylinders of the sine bar on a stack of gage blocks. Place the part to be measured on the sine bar against the end rail if needed. Use a dial indicator to check whether the surface of the workpiece is parallel to the surface plate.

Move the dial indicator over the length of the workpiece surface from one side to the other to detect any variation in the parallelism of the surface of the workpiece to the surface plate. If the part is not parallel, adjust the height of the gage blocks and check the surface again for parallelism. Repeat these steps until the indicator reads zero across the entire surface.

You know the height of the gage blocks simply by adding their individual heights. Now you can use trigonometry to precisely calculate the angle formed with the part. (Note: During angular measurement using a sine bar, the sine bar length should be greater than or equal to the length of the component you are inspecting.)

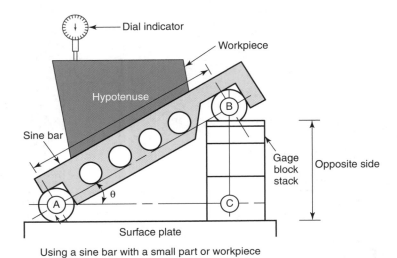

Using a sine bar with a small part or workpiece

Goodheart-Willcox Publisher

The sine bar forms the hypotenuse, and the gage blocks form the opposite side. The angle can be calculated using the following equation:

$$\angle = \sin^{-1}\left(\frac{\text{opposite}}{\text{hypotenuse}}\right)$$

Example 38-1

To the nearest second, what is the angle of the part being checked with a sine bar in the following illustration?

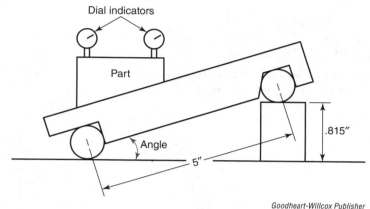

Goodheart-Willcox Publisher

The distance between the cylinders of the sine bar is 5", which is the hypotenuse of the right triangle. The height of the gage block stack is 0.815", which is the opposite side of the right triangle. The angle can be calculated by using the following equation:

$$\angle = \sin^{-1}\left(\frac{\text{opposite}}{\text{hypotenuse}}\right)$$

$$\angle = \sin^{-1}\left(\frac{0.815}{5}\right)$$

Use a scientific calculator.

Goodheart-Willcox Publisher

The display shows:

$$\sin^{-1}(0.815 \div 5) =$$
9.3810700615213

Goodheart-Willcox Publisher

Pressing the Degree, Minutes, Seconds key  converts this to 9° 22′ 51.85″.

$$\angle = 9.381070062° \text{ or } 9° \ 22′ \ 51.85″$$
$$\angle = 9° \ 22′ \ 52″ \text{ rounded to the nearest second}$$

To calculate the required height of the gage block stack (the opposite side of the right triangle), use trigonometry. The sine bar forms the hypotenuse, and the necessary angle is given, so use the following formula to calculate the height for the stack of gage blocks:

$$\text{opposite} = \sin(\theta) \times \text{hypotenuse}$$

Example 38-2

What is the height of the stack of gage blocks in the following illustration, rounded to the nearest ten-thousandth of an inch?

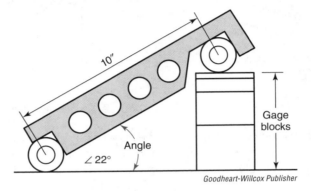

Goodheart-Willcox Publisher

The distance between the cylinders of the sine bar is the hypotenuse, which is 10″. The required angle of the sine bar is 22°.

The height of the gage block stack can be calculated using the following equation:

$$\text{opposite} = \sin(\theta) \times \text{hypotenuse}$$
$$\text{opposite} = \sin \angle 22° \times 10$$

Use a scientific calculator.

Goodheart-Willcox Publisher

The display shows:

Goodheart-Willcox Publisher

The height of the gage block stack is 3.7461″ to the nearest ten-thousandth of an inch.

Unit 38 Review

Name _____ **Date** _____ **Class** _____

Fill in the blanks in the following review questions.

1. A sine bar is a precisely machined _____ consisting of a hardened, precision-ground flat body typically no wider than 1″.

2. The two cylinders (rolls) are placed at a precise distance, usually _____, _____, or _____ inches apart.

3. A sine bar is frequently used with _____ blocks, also called *jo blocks*.

4. When the sine bar is placed on a flat surface, such as a surface plate, the top surface of the sine bar is _____ to that surface.

5. The sine bar forms a _____ of a known length, and the gage blocks always form the _____ side of the right-angle triangle.

6. Sine plates are used in a fashion very similar to sine bars, but sine plates are much _____ than sine bars.

7. During angular measurement using a sine bar, the sine bar length should be _____ than or equal to the length of the component you are inspecting.

Practice

1. To the nearest second, what is the angle of the part being checked with a sine bar in the following illustration? _____

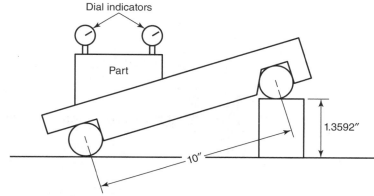

Goodheart-Willcox Publisher

2. To the nearest three decimal places, what is the angle of the part being checked with a sine bar in the following illustration? _____

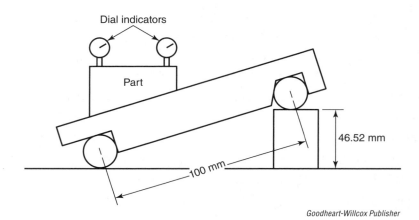

Goodheart-Willcox Publisher

3. To the nearest second, what is the angle of the sine bar in the following illustration? _____

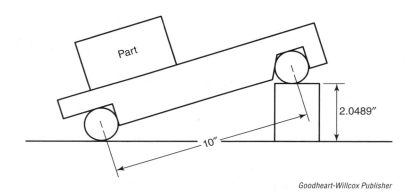

Goodheart-Willcox Publisher

4. To the nearest three decimal places, what is the angle of the sine bar in the following illustration? _____

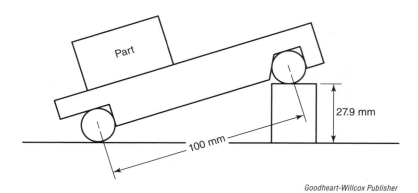

Goodheart-Willcox Publisher

Name _____   **Date** _____   **Class** _____

5. To the nearest three decimal places, what is the angle of the sine bar in the following illustration? _____

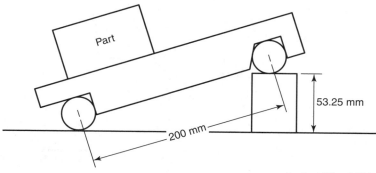

Goodheart-Willcox Publisher

6. If a 5″ sine bar needs to be set to an angle of 15° 30′, what would the height of the gage block stack need to measure? Round the answer to the nearest ten-thousandth of an inch. _____

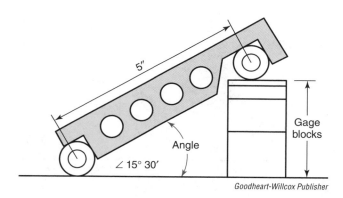

Goodheart-Willcox Publisher

7. If a 10″ sine plate needs to be set to an angle of 41° 15′ 30″, what would the height of the raised end of the plate need to measure? Round the answer to the nearest ten-thousandth of an inch. _____

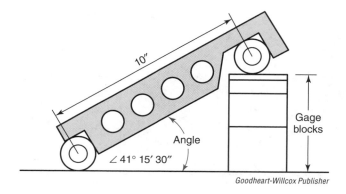

Goodheart-Willcox Publisher

8. If a 20″ sine bar needs to be set to an angle of 8° 30′ 45″, what would the height of the gage block stack need to measure? Round the answer to the nearest ten-thousandth of an inch. _____

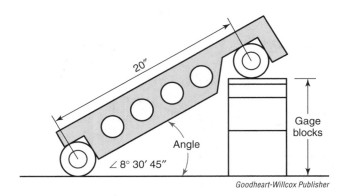

Goodheart-Willcox Publisher

9. If a 100 mm sine bar needs to be set to an angle of 29.33°, what would the height of the gage block stack need to measure? Round the answer to three decimal places. _____

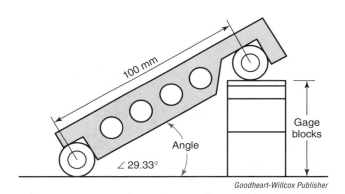

Goodheart-Willcox Publisher

10. If a 200 mm sine bar needs to be set to an angle of 38.75°, what would the height of the gage block stack need to measure? Round the answer to three decimal places. _____

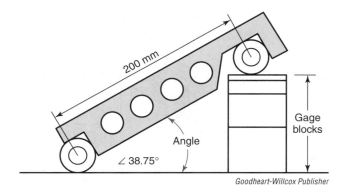

Goodheart-Willcox Publisher

Name _____ **Date** _____ **Class** _____

Applications

1. To what angle is a 10″ sine bar set if it is sitting on a gage block stack measuring 3.4591″ high? Round the answer to the nearest second. Sketch a drawing here.

   _____

2. To what angle is a 100 mm sine bar set if it is sitting on a gage block stack measuring 26.721 mm high? Round the answer to three decimal places. Sketch a drawing.

   _____

3. A quality control inspector requires an angle of 18° 28′ 41″. Using a 10″ sine bar, what height of gage block stack is required? Round the answer to the nearest ten-thousandth of an inch. Sketch a drawing.

   _____

4. Considering question 3, what would the height of the gage blocks be if a 5″ sine bar were used instead? Round the answer to the nearest ten-thousandth of an inch. Sketch a drawing.

   _____

5. The gage block stack height for a 200 mm sine bar used in measuring an angle is 83.701 mm. What is the resulting angle measurement? Round the answer to three decimal places. Sketch a drawing.

    _____

6. A machinist wants to indicate the head on a conventional mill to an angle of 20° 30′. Using a 10″ sine bar, what height of gage block stack is required? Round the answer to the nearest ten-thousandth of an inch. Sketch a drawing.

    _____

7. Determine the height of gage block stack required to set the following angles on a 10″ sine bar. Round answers to the nearest ten-thousandth.

    A. 35° _____

    B. 13° 10′ _____

    C. 36° 50′ _____

    D. 9° 44′ _____

    E. 28° 32′ _____

8. Determine the height of stacked gage blocks required to set the following angles with a 200 mm sine bar. Round answers to three decimal places.

    A. 40.25° _____

    B. 7.333° _____

    C. 12.75° _____

    D. 34.459° _____

    E. 21.785° _____

UNIT 39

Drill Point Angles

Objectives

Information in this unit will enable you to:

- Describe the point angle of a drill.
- Calculate drill point size using right-angle trigonometry.
- Calculate a drill point constant.
- Use a drill point constant to find drill point size and total hole depth.

Point Angle of a Drill

The **point angle** of a drill (the included angle) is the angle formed by the cutting edges (lips) at the drill point. Drills are available in a variety of point angles. The 118° angle has proven to work well on most materials over time and is the most commonly used point angle. However, different materials in industry require different cutting parameters, such as specific speeds, feeds, depths of cut, coolant flow, tool coatings, clearance angles, etc. A point angle other than 118° is sometimes recommended for increased efficiency with certain types of materials.

In general, a high point angle (flatter point or blunter angle) is recommended for harder and tougher materials because the drill will have stronger cutting edges, meaning cutting force is distributed across a greater area on a steep point. A lower point angle (sharper point or more acute angle), sharper than 118°, is generally used for soft nonferrous materials with good machinability. A drill with a sharper point enters soft materials more quickly and cuts through them more efficiently.

Calculating Drill Point Size

As the angle of the drill point changes, so does its size. On a print, a drilled hole's depth does not include the drill point unless specified. If a machinist drills to the depth indicated on the print, the hole may not actually be deep enough, depending on tolerances. To ensure the correct drilled depth, the size of the drill point should be added to the depth dimension indicated on the print.

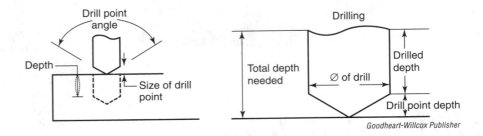

Goodheart-Willcox Publisher

The size of the drill point can be calculated using right-angle trigonometry. Drawing a line from one cutting edge to the other creates an isosceles triangle. The two cutting edges are of equal size (the legs), and the line created forms the base. The base is the diameter of the drill. The included angle or the angle of the drill point is the vertex angle. Drawing an altitude from the vertex to the base of an isosceles triangle creates two congruent (equal) triangles. An **altitude** is a line drawn from the vertex of an angle in a triangle to the opposite side to form a right angle. The altitude bisects (divides into two equal parts) the vertex angle and the base and creates the needed right triangles for trigonometry.

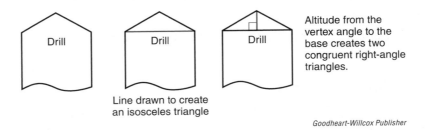

Goodheart-Willcox Publisher

Now that a right-angle triangle is formed, the size of the drill point can be calculated. The base (diameter of the drill) is cut in half, as is the vertex angle (the drill point angle). Those two pieces of information can be applied to the right-angle triangle.

Goodheart-Willcox Publisher

The formula to find the drill point size for any tool angle is:

$$\text{Size of drill point} = \left(\frac{\varnothing \text{ of drill}}{2}\right) \div \tan\left(\frac{\text{drill point angle}}{2}\right)$$

This could be rearranged to:

$$\text{Size of drill point} = \frac{\varnothing \text{ of drill}}{2\tan\left(\dfrac{\text{drill point angle}}{2}\right)}$$

Example 39-1

Calculate the size of the drill point on a ∅1/2″ drill with a drill point angle of 118°.

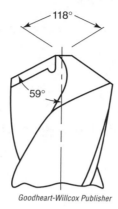

Goodheart-Willcox Publisher

$$\text{Size of drill point} = \frac{\varnothing \text{ of drill}}{2 \tan\left(\dfrac{\text{drill point angle}}{2}\right)}$$

Substitute known information.

$$\text{Size of drill point} = \frac{0.5}{2 \tan\left(\dfrac{118}{2}\right)}$$

$$\text{Size of drill point} = \frac{0.5}{2 \tan(59)}$$

$$\text{Size of drill point} = \frac{0.5}{3.3286}$$

The size of the drill point is 0.1502″ rounded to the nearest ten-thousandth of an inch.

Calculating and Using Drill Point Constants

Performing the drill point size calculation for every drill in the shop would not be an efficient use of time. Using percent, it is possible to calculate a constant value for any given angle. A **constant** is a number with a fixed value in a given situation.

In percent applications (see Unit 20), 100% = 1.00 (moving the decimal point two places to the left). If 1 is substituted for the diameter of the tool, the size of the drill point can be calculated as a percent of its diameter.

Example 39-2

To calculate for the constant (percent) of a 118° drill, use the following:

$$\text{Size of drill point} = \frac{1}{2 \tan \left(\dfrac{\text{drill point angle}}{2} \right)}$$

$$\text{Size of drill point} = \frac{1}{2 \tan \left(\dfrac{118}{2} \right)}$$

$$\text{Size of drill point} = \frac{1}{2 \tan(59)}$$

$$\text{Size of drill point} = \frac{1}{3.3286}$$

The size of the drill point is 0.3004, or 30.04% of the tool diameter. Some reference books may refer to this number as a constant. For any 118° tool, the drill point will be 30.04% of its diameter.

$$\text{Size of drill point} = \text{Percent (constant)} \times \text{Tool diameter}$$

Looking back at the previous example, calculate the size of the drill point on a Ø1/2″ (0.5″) drill with a drill point angle of 118°:

$$\text{Size of drill point} = \text{Percent (constant)} \times \text{Tool diameter}$$

$$\text{Size of drill point} = 0.3004 \times 0.5$$

The size of the drill point is 0.1502″ rounded to the nearest ten-thousandth of an inch.

Using the formula from Example 39-2, a table of constants for common included angles found in the machine shop has been calculated.

Shortcut Formulas for Drill Point Size Using Common Constants	
Drill Point Angle (DPA)	**Shortcut Formula for Size of Drill Point**
60°	0.866 × ∅ of tool
82°	0.5752 × ∅ of tool
90°	0.5 × ∅ of tool
100°	0.4195 × ∅ of tool
110°	0.3501 × ∅ of tool
118°	0.3004 × ∅ of tool
120°	0.2887 × ∅ of tool
130°	0.2332 × ∅ of tool
135°	0.2071 × ∅ of tool
140°	0.1820 × ∅ of tool
150°	0.134 × ∅ of tool

Goodheart-Willcox Publisher

Example 39-3

Calculate the size of the drill point on a ∅29/64″ (0.4531″) drill with a drill point angle of 135°.

Goodheart-Willcox Publisher

Looking at the provided table, a constant of 0.2071 is given for a 135° drill point angle (DPA).

$$\text{Size of drill point} = \text{Percent (constant)} \times \text{Tool diameter}$$

$$\text{Size of drill point} = 0.2071 \times 0.4531''$$

The size of the drill point is 0.0938″ rounded to the nearest ten-thousandth of an inch.

Example 39-4

Calculate the total depth needed if the print calls for a drilled hole of ∅11/16″ (0.6875″) with a 1.0″ depth, and the material requires a drill point angle of 150°.

$$\text{Size of drill point} = \text{Percent (constant)} \times \text{Tool diameter}$$

Looking at the provided table of constants, for a 150° DPA, the formula is $0.134 \times \varnothing$ of tool to find drill point size:

$$\text{Size of drill point} = 0.134 \times 0.6875″$$
$$\text{Size of drill point} = 0.0921″$$

To find the total depth:

$$\text{Total depth} = \text{Size of drill point} + \text{Depth of hole}$$
$$\text{Total depth} = 0.0921″ + 1.0″$$

The total depth is 1.0921″ rounded to the nearest ten-thousandth of an inch.

Unit 39 Review

Name _____ **Date** _____ **Class** _____

Fill in the blanks in the following review questions.

1. The _____ _____ of a drill (the included angle) is the angle formed by the cutting edges (lips) at the drill point.

2. The size of a drill point can be calculated using right-angle _____.

3. Drawing a line from one cutting edge to the other creates a(n) _____ triangle. The base is the _____ of the drill. The included angle or angle of the drill point is the _____ angle.

4. Drawing a(n) _____ from the vertex to the base of an isosceles triangle creates two congruent (equal) triangles. The _____ bisects (divides into two equal parts) the vertex angle and the base and creates the needed right triangles for trigonometry.

5. Once a right-angle triangle is formed, the size of the _____ _____ can be calculated.

6. Using percent, it is possible to calculate a _____ value for any given angle. The _____ multiplied by the diameter of the tool provides the size of the drill point. Total hole _____ can then be calculated.

Practice

Calculate the total depth needed for the blind holes when the size of the drill point is also taken into consideration. Round answers to the nearest ten-thousandth of an inch. For metric questions, round to three decimal places.

	Depth Dimension on Print	Diameter of Drill	Angle of Drill Point	Formula to Solve Size of Drill Point	Size of Drill Point	Total Depth
1.	0.5″	0.3125″	118°			
2.	0.44″	0.156″	118°			
3.	0.275″	0.125″	135°			
4.	0.625″	0.188″	150°			
5.	1.5″	0.257″	90°			
6.	0.94″	0.312″	118°			
7.	0.75″	0.201″	60°			
8.	2.0″	0.25″	135°			
9.	0.7″	0.213″	150°			
10.	0.475″	0.136″	120°			
11.	1.0″	0.312″	135°			

	Depth Dimension on Print	Diameter of Drill	Angle of Drill Point	Formula to Solve Size of Drill Point	Size of Drill Point	Total Depth
12.	0.3"	0.246"	118°			
13.	19 mm	8.5 mm	118°			
14.	10 mm	5 mm	150°			
15.	28.5 mm	12 mm	135°			
16.	44.5 mm	9.5 mm	118°			
17.	16 mm	6 mm	90°			
18.	25 mm	7 mm	118°			

Goodheart-Willcox Publisher

Center-to-Center Distances

Objectives

Information in this unit will enable you to:

- Use trigonometry to find the center-to-center distances between holes on a bolt circle.
- Determine the measurement between and over adjacent pins in holes in a bolt circle.
- Calculate distance measurements between nonadjacent holes in a bolt circle.
- Use triangles and trigonometry to calculate the diameter of a bolt circle.

Distance between Holes

When checking the distance between holes on a bolt circle, the center-to-center distance (chord) between holes cannot be directly measured. However, by using gage pins placed in the holes of a bolt circle, it is possible to take a measurement over the pins with an outside micrometer, or alternatively between pins with an inside micrometer. The length of the chord can also be calculated using right-angle trigonometry. A right-angle triangle needs to be constructed to solve the problem.

Center-to-Center Distance

Using trigonometry, the distance between hole centers in a bolt circle in both equally spaced holes and unequally spaced holes can be checked.

Example 40-1

Eight holes of ∅0.25″ are equally spaced on a 4″ diameter bolt circle. Calculate the distance between the centers of hole 1 and hole 2 in the figure below.

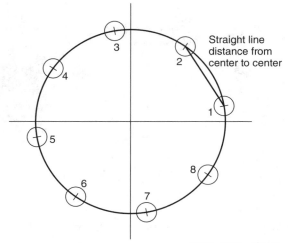

Drawing lines from the center of the bolt circle to the centers of both hole 1 and hole 2 creates an isosceles triangle. The distance between the holes becomes the base, and the lines created are radii. A radius can be calculated by dividing the diameter by two. Bolt circles are often dimensioned with a bolt circle diameter, so:

$$\text{radius} = \frac{\text{∅ of bolt circle}}{2}$$

$$r = \frac{4}{2}$$

$$r = 2″$$

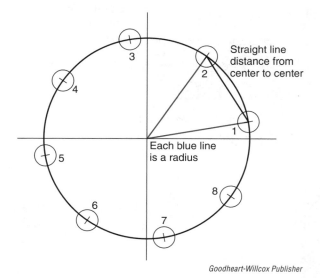

(Continued)

The angle between holes or between the radius lines can be calculated the same way as in bolt circle applications. When holes are equally spaced on a bolt circle, the angular spacing is calculated by dividing 360° by the number of holes:

$$\angle \text{ between holes} = \frac{360°}{\text{\# of holes}}$$

$$\angle = \frac{360°}{8 \text{ holes}}$$

$$\angle = 45°$$

Drawing an altitude from the vertex to the base of an isosceles triangle creates two congruent (equal) triangles. An altitude is a line drawn from the vertex of one angle to the opposite side forming a right angle. The altitude bisects (divides into two equal parts) the vertex angle and the base and creates the needed right triangles for trigonometry.

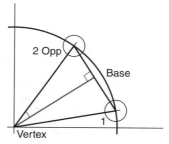

Goodheart-Willcox Publisher

The radius forms the hypotenuse of the right-angle triangle. The angle of the right-angle triangle is half of the angle between the holes. In this example, the radius is 2″ and the angle of the right-angle triangle is 22° 30′, or 22.5° (45° ÷ 2).

The base (chord) is equal to the opposite side of the right-angle triangle times two. Two equal right-angle triangles were formed.

To calculate the opposite using the hypotenuse and the angle:

$$\text{opposite} = \sin(\theta) \times \text{hypotenuse}$$

$$\text{opposite} = \sin(22.5°) \times 2$$

Use a scientific calculator.

Goodheart-Willcox Publisher

The display shows:

Goodheart-Willcox Publisher

The center-to-center distance (base) = opposite side × 2
The center-to-center distance (base) = 0.76536686473018 × 2

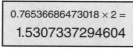

Goodheart-Willcox Publisher

The center-to-center distance between hole 1 and hole 2 = 1.5307 rounded to the nearest ten-thousandth of an inch.

Pin-to-Pin Distance

To determine the measurement between adjacent pins in a bolt circle, calculate the center-to-center distance and subtract the radius of each pin.

Example 40-2

Eight holes of ⌀0.25″ are equally spaced on a 3″ diameter bolt circle. Calculate the measurement between pins inserted in holes 1 and hole 2.

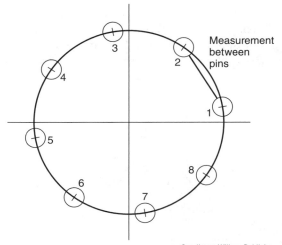

Goodheart-Willcox Publisher

$$\text{radius} = \frac{\varnothing \text{ of bolt circle}}{2}$$

$$r = \frac{3}{2}$$

$$r = 1.5″$$

$$\angle \text{ between holes} = \frac{360°}{\text{\# of holes}}$$

$$\angle = \frac{360°}{8 \text{ holes}}$$

$$\angle = 45°$$

Create an isosceles triangle between holes 1 and 2 by drawing lines from the center of the bolt circle to the center of each hole.

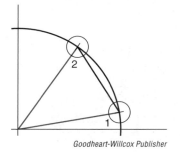

Goodheart-Willcox Publisher

(Continued)

Bisect the isosceles triangle to form equal right-angle triangles.

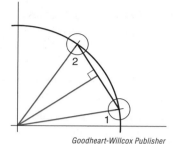

Goodheart-Willcox Publisher

The hypotenuse (radius) is 1.5″, and the angle within the right-angle triangle is 22.5°. To calculate the opposite using the hypotenuse and the angle:

$$\text{opposite} = \sin(\theta) \times \text{hypotenuse}$$

$$\text{opposite} = \sin(22.5°) \times 1.5$$

Use a scientific calculator.

Goodheart-Willcox Publisher

The display shows:

sin 22.5 × 1.5 =

0.57402514854764

Goodheart-Willcox Publisher

The center-to-center distance (base) = opposite side × 2
The center-to-center distance (base) = 0.57402514854764 × 2

0.57402514854764 × 2 =

1.1480502970953

Goodheart-Willcox Publisher

The center-to-center distance between hole 1 and hole 2 = 1.1480502970953.

The measurement between pins equals the center-to-center distance minus the radius of each pin.

Center to center distance = 1.1480502970953; the radius of each pin = 0.125″ (Ø0.25 ÷ 2).

Measurement between pins = 1.1480502970953 − 0.125 − 0.125.

Measurement between pins = 0.8981″ rounded to the nearest ten-thousandth of an inch.

Measurement over Pins

To determine the measurement over pins, the center-to-center distance is calculated and then the radius of each pin is added.

Example 40-3

Eight holes of ⌀0.25″ are equally spaced on a 5″ diameter bolt circle. Calculate the measurement over pins inserted in hole 1 and hole 2.

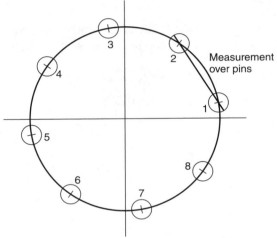

Goodheart-Willcox Publisher

$$\text{radius} = \frac{\varnothing \text{ of bolt circle}}{2}$$

$$r = \frac{5}{2}$$

$$r = 2.5''$$

$$\angle \text{ between holes} = \frac{360°}{\# \text{ of holes}}$$

$$\angle = \frac{360°}{8 \text{ holes}}$$

$$\angle = 45°$$

Create an isosceles triangle between holes 1 and 2 by drawing lines from the center of the bolt circle to the center of each hole.

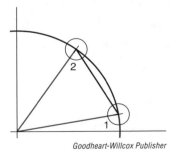

Goodheart-Willcox Publisher

(Continued)

Bisect the isosceles triangle to form equal right-angle triangles.

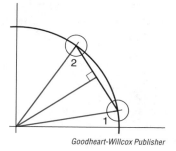

Goodheart-Willcox Publisher

The hypotenuse (radius) is 2.5″, and the angle within the right-angle triangle is 22.5°. To calculate the opposite using the hypotenuse and the angle:

$$opposite = \sin(\theta) \times hypotenuse$$
$$opposite = \sin(22.5°) \times 2.5$$

Use a scientific calculator.

Goodheart-Willcox Publisher

The display shows:

sin 22.5 × 2.5 =
0.95670858091272

Goodheart-Willcox Publisher

The center-to-center distance (base) = opposite side × 2
The center-to-center distance (base) = 0.95670858091272 × 2

0.95670858091272 × 2 =
1.9134171618255

Goodheart-Willcox Publisher

The center-to-center distance between hole 1 and hole 2 = 1.9134171618255.

The measurement over pins equals the center-to-center distance plus the radius of each pin.

Center-to-center distance = 1.9134171618255; the radius of each pin = 0.125″ (∅0.25 ÷ 2).

Measurement over pins = 1.9134171618255 + 0.125 + 0.125.

Measurement over pins = 2.1634″ rounded to the nearest ten-thousandth of an inch.

Measuring between Nonadjacent Holes

You can use trigonometry to measure the distances between nonadjacent holes in a bolt circle, including the center-to-center distance, distance between pins, and distance over pins.

Example 40-4

Eight holes of ⌀0.25″ are equally spaced on a 4″ diameter bolt circle. Calculate the distance between the centers of hole 1 and hole 4 in the figure below.

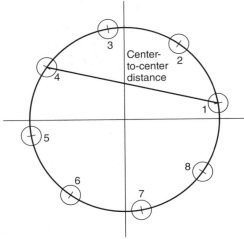

Goodheart-Willcox Publisher

Drawing lines from the center of the bolt circle to the centers of both hole 1 and hole 4 creates an isosceles triangle, with the distance between holes being the base. The lines created are radii. A radius can be calculated by dividing the diameter by two. Bolt circles are often dimensioned with a bolt circle diameter, so:

$$r = \frac{\varnothing \text{ of bolt circle}}{2}$$

$$r = \frac{4}{2}$$

$$r = 2''$$

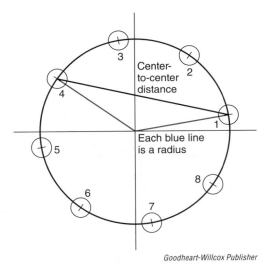

Goodheart-Willcox Publisher

(Continued)

The angle between holes or between the radii can be calculated the same way as in bolt circle applications. When holes are equally spaced on a bolt circle, the angular spacing is calculated by dividing 360° by the number of holes.

$$\angle = \frac{360°}{\text{\# of holes}}$$

$$\angle \text{ between holes} = \frac{360°}{\text{\# of holes}}$$

$$\angle = \frac{360°}{8 \text{ holes}}$$

$$\angle = 45°$$

The angle from hole 1 to hole 2 is 45°, the angle from hole 2 to hole 3 is 45°, and the angle from hole 3 to hole 4 is 45°. So, the total angular distance between hole 1 and hole 4 is 135° (45° + 45° + 45°).

Drawing an altitude from the vertex to the base of an isosceles triangle creates two congruent (equal) triangles. The altitude bisects (divides into two equal parts) the vertex angle and the base and creates the needed right triangles for trigonometry.

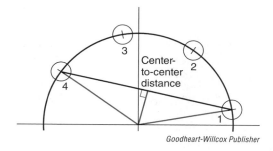

Goodheart-Willcox Publisher

The radius forms the hypotenuse of the right-angle triangle. The angle of the right-angle triangle is half of the angle between the holes. In this example, the radius is 2″, and the angle of the right-angle triangle is 67° 30′, or 67.5° (135° ÷ 2).

The base (chord) is equal to the opposite side of the right-angle triangle times two. Two equal right-angle triangles were formed.

To calculate the opposite using the hypotenuse and the angle:

$$\text{opposite} = \sin(\theta) \times \text{hypotenuse}$$

$$\text{opposite} = \sin(67.5°) \times 2$$

Use a scientific calculator.

Goodheart-Willcox Publisher

The display shows:

sin 67.5 × 2 =
1.8477590650226

Goodheart-Willcox Publisher

The center-to-center distance (base) = opposite side × 2
The center-to-center distance (base) = 1.8477590650226 × 2

(Continued)

$$1.8477590650226 \times 2 =$$
$$3.6955181300452$$

The center-to-center distance between hole 1 and hole 4 = 3.6955" rounded to the nearest ten-thousandth of an inch.

Measurement between Pins

The measurement between pins equals the center-to-center distance minus the radius of each pin.

Center to center distance = 3.6955181300452; the radius of each pin = 0.125" (∅0.25 ÷ 2).

Measurement between pins = 3.6955181300452 − 0.125 − 0.125.

Measurement between pins = 3.4455" rounded to the nearest ten-thousandth of an inch.

Measurement over Pins

The measurement over pins equals the center-to-center distance plus the radius of each pin.

Center to center distance = 3.6955181300452; the radius of each pin = 0.125" (∅0.25 ÷ 2).

Measurement over pins = 3.6955181300452 + 0.125 + 0.125.

Measurement over pins = 3.9455" rounded to the nearest ten-thousandth of an inch.

Bolt Circle Diameter

You can use triangles and trigonometry to calculate the diameter of a bolt circle.

Example 40-5

Eight holes of ∅0.25" are equally spaced on a bolt circle. Calculate the diameter of the bolt circle when the distance between the centers of hole 1 and hole 4 equals 5.5433".

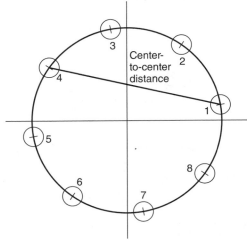

(Continued)

The angle between holes or between the radii can be calculated the same way as in bolt circle applications. When holes are equally spaced on a bolt circle, the angular spacing is calculated by dividing 360° by the number of holes.

$$\angle = \frac{360°}{\# \text{ of holes}}$$

$$\angle \text{ between holes} = \frac{360°}{\# \text{ of holes}}$$

$$\angle = \frac{360°}{8 \text{ holes}}$$

$$\angle = 45°$$

The angle from hole 1 to hole 2 is 45°, the angle from hole 2 to hole 3 is 45°, and the angle from hole 3 to hole 4 is 45°. So, the total angular distance between hole 1 and hole 4 is 135° (45° + 45° + 45°).

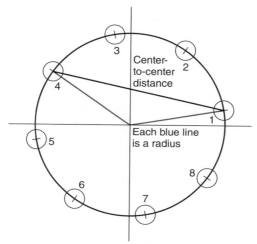

Goodheart-Willcox Publisher

Drawing an altitude from the vertex to the base of an isosceles triangle creates two congruent (equal) triangles. The altitude bisects the vertex angle and the base and creates the needed right triangles for trigonometry.

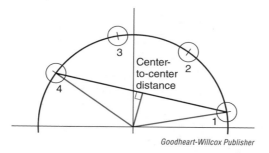

Goodheart-Willcox Publisher

The radius forms the hypotenuse of the right-angle triangle. The angle of the right-angle triangle is half of the angle between the holes and amounts to 67° 30′, or 67.5° (135° ÷ 2).

The base (chord) is 5.5433″.

The opposite side in the right-angle triangle is half of the base, which comes to 2.77165″ (5.5433″ ÷ 2).

(Continued)

The bolt circle diameter is equal to two times the radius.

To calculate the hypotenuse (radius) using the opposite and the angle:

$$\text{hypotenuse} = \frac{\text{opposite}}{\sin(\theta)}$$

$$\text{hypotenuse} = \frac{2.77165}{\sin(67.5)}$$

Use a scientific calculator.

2.77165 ÷ **SIN** 67.5° **ENTER =**

Goodheart-Willcox Publisher

The display shows:

```
2.77165 ÷ sin 67.5 =
3.0000123419404
```

Goodheart-Willcox Publisher

The bolt circle diameter = hypotenuse × 2

The bolt circle diameter = 3.0000123419404 × 2

```
3.0000123419404 × 2 =
6.0000246838808
```

Goodheart-Willcox Publisher

The bolt circle diameter = 6.0000″ rounded to the nearest ten-thousandth of an inch.

Name _____ **Date** _____ **Class** _____

Fill in the blanks in the following review questions.

1. Using _____ _____ placed in the holes of a bolt circle, it is possible to take a measurement with a micrometer.

2. The length of the chord can also be calculated by constructing a right-angle triangle and using _____.

3. Using trigonometry, the distance can be found between both _____ and _____ holes on a bolt circle.

4. Center-to-center distance between holes on a bolt circle can be found using pins and either subtracting or adding the pins' _____.

5. You can use triangles and trigonometry to calculate the _____ of a bolt circle.

Practice

1. Three holes are equally spaced around a 3-inch diameter bolt circle. Find the center-to-center distance between any two holes.

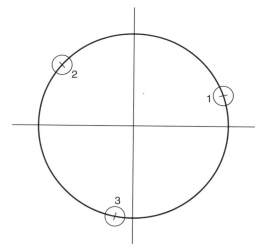

Goodheart-Willcox Publisher

Center-to-center distance = _____

2. If the holes in problem 1 are machined to a diameter of 0.375″, and gage pins are placed inside the holes for inspection, calculate the following:

Measurement over the pins = _____

Measurement between the pins = _____

3. Five holes are equally spaced around a 100-mm diameter bolt circle. Find the center-to-center distance between the holes listed below.

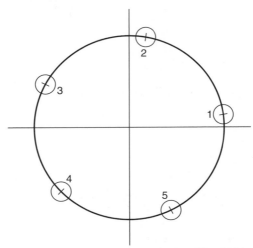

Goodheart-Willcox Publisher

Hole 1 to hole 2 = _____

Hole 1 to hole 3 = _____

4. If the holes in problem 3 are machined to a diameter of 12 mm and gage pins are placed inside of the holes for inspection, calculate the following:

Measurement over the pins for hole 1 to hole 2 = _____

Measurement between the pins for hole 1 to hole 3 = _____

5. Six holes each 10 mm in diameter are equally spaced around a 150-mm diameter bolt circle. Find the center-to-center distance between the following holes:

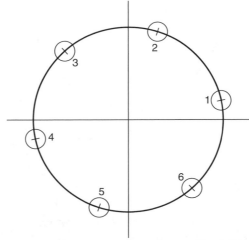

Goodheart-Willcox Publisher

Hole 1 to hole 2 = _____

Hole 1 to hole 3 = _____

Hole 1 to hole 4 = _____

Name _____   **Date** _____   **Class** _____

6. Nine holes each 0.5″ in diameter are equally spaced around a 5″ diameter bolt circle. Find the center-to-center distance between the following holes:

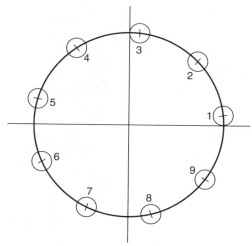

Goodheart-Willcox Publisher

     Hole 1 to hole 2 = _____

     Hole 1 to hole 3 = _____

     Hole 1 to hole 4 = _____

     Hole 1 to hole 5 = _____

7. Ten holes each 16 mm in diameter are equally spaced around a 225-mm diameter bolt circle. Find the center-to-center distance between the following holes:

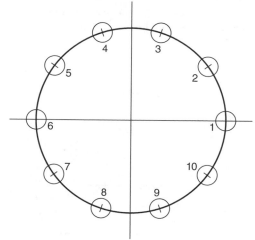

Goodheart-Willcox Publisher

     Hole 1 to hole 2 = _____

     Hole 1 to hole 3 = _____

     Hole 1 to hole 4 = _____

     Hole 1 to hole 5 = _____

     Hole 1 to hole 6 = _____

8. Twelve holes with ⌀0.4375″ are equally spaced around a 12″ diameter bolt circle. Find the center-to-center distance between the following holes:

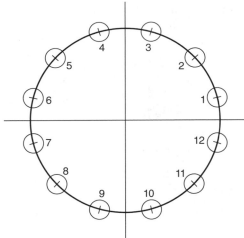

Goodheart-Willcox Publisher

Hole 1 to hole 2 = _____

Hole 1 to hole 3 = _____

Hole 1 to hole 4 = _____

Hole 1 to hole 5 = _____

Hole 1 to hole 6 = _____

Hole 1 to hole 7 = _____

9. Eight holes of ⌀0.375″ are equally spaced on a bolt circle. Calculate the radius and the diameter of the bolt circle when the distance between the centers of hole 1 and hole 3 is 5.4801″. Round answers to the nearest ten-thousandth of an inch.

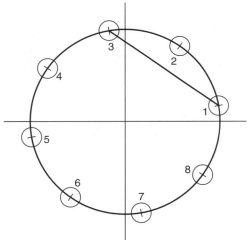

Goodheart-Willcox Publisher

Radius of bolt circle = _____

Diameter of bolt circle = _____

Name _____   **Date** _____   **Class** _____

Applications

In the table below, for a bolt circle with equally spaced holes and radius of 1, calculate the angle between adjacent holes, the angle that would be formed bisecting the isosceles triangle, and the center-to-center distance between adjacent holes (chord). Do not round answers for the chord.

	# of Holes on Bolt Circle	Angle between Adj. Holes	Angle formed in Right Triangle	Length of Chord
1.	3			
2.	4			
3.	5			
4.	6			
5.	7			
6.	8			
7.	9			
8.	10			
9.	11			
10.	12			
11.	13			
12.	14			
13.	15			
14.	16			
15.	17			
16.	18			
17.	19			
18.	20			
19.	21			
20.	22			
21.	23			
22.	24			
23.	25			

Goodheart-Willcox Publisher

Using the table from problems 1 through 23, the center-to-center distance of adjacent holes can be quickly calculated by multiplying the diameter of the bolt circle by the length of the chord. What is the center-to-center distance between adjacent holes for each of the 23 bolt circles in the table?

24. Bolt circle 1 = _____

25. Bolt circle 2 = _____

26. Bolt circle 3 = _____

27. Bolt circle 4 = _____

28. Bolt circle 5 = _____

29. Bolt circle 6 = _____

30. Bolt circle 7 = _____

31. Bolt circle 8 = _____

32. Bolt circle 9 = _____

33. Bolt circle 10 = _____

34. Bolt circle 11 = _____

35. Bolt circle 12 = _____

36. Bolt circle 13 = _____

37. Bolt circle 14 = _____

38. Bolt circle 15 = _____

39. Bolt circle 16 = _____

40. Bolt circle 17 = _____

41. Bolt circle 18 = _____

42. Bolt circle 19 = _____

43. Bolt circle 20 = _____

44. Bolt circle 21 = _____

45. Bolt circle 22 = _____

46. Bolt circle 23 = _____

UNIT 41

Dovetails

Objectives

Information in this unit will enable you to:

- Understand and explain what a dovetail slide is.
- Understand tangent lines and tangent points and theorems that use them.
- Know how to use gage pins with dovetail slides.
- Use trigonometry to calculate unknown dimensions with dovetail slides.

Dovetail Slides

A **dovetail slide** is a type of plain linear bearing that relies on direct contact between sliding surfaces to support and move a heavier load. The full-contact design of the sliding surfaces, or ways, makes dovetail slides well suited for absorbing and damping the shocks and vibrations of the machine tool. The base and saddle of the slides are generally constructed from close-grained cast iron and are normalized (a process of heating and slow cooling) to prevent distortion.

When used in machining or inspection applications, dovetail slides are commonly driven by a handle or dial connected to a lead screw. Dovetail slides usually have large contact areas, which helps ensure accurate, smooth, and reliable movement along an axis, as well as precise positioning of the machine tool. Dovetail slides require very little maintenance because of their durability. The compound rest located on top of the cross slide on a conventional lathe is a great example of a dovetail commonly used in machining applications.

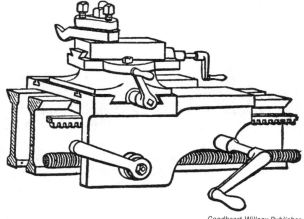

Goodheart-Willcox Publisher

A dovetail slide consists of two main parts: the external dovetail, often referred to as the male dovetail; and the internal dovetail, or female dovetail. The angle of the dovetail comes in many different sizes, 45°, 50°, 55°, and 60° being the most common. Dovetails are usually symmetrical, which helps when trying to calculate missing dimensions. The dovetails in this unit will all be considered as symmetrical.

Goodheart-Willcox Publisher

When a dovetail is to be machined, the depth/height and the angle are given, with the width of the slide either at the small or large end of the dovetail. Using trigonometry, we can calculate for the missing dimension.

Example 41-1

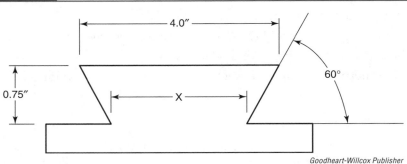

Goodheart-Willcox Publisher

The height of the above dovetail is 0.75", the angle is 60°, and the distance across the large end is 4". To calculate dimension X, a right triangle must first be constructed. Drawing a vertical line down to the horizontal line of the slide creates a 90° angle.

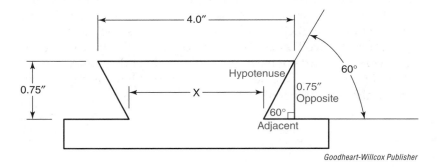

Goodheart-Willcox Publisher

In the right-angle triangle, the angle is given as 60°, and the height (the side opposite) is 0.75". Dimension X can be calculated by taking the width and subtracting the adjacent side two times. The adjacent is subtracted two times because the part is symmetrical, and the triangle formed would be identical on the other side.

(Continued)

To calculate the side adjacent to ∠60°, first determine what is known in the triangle. In this case, the opposite is known (0.75"), and the adjacent needs to be found. Therefore, tangent needs to be used to solve the adjacent.

$$\text{adjacent} = \frac{\text{opposite}}{\tan(\theta)}$$

$$\text{adjacent} = \frac{0.75}{\tan(60°)}$$

Use a scientific calculator.

0.75 ÷ **TAN** 60° **ENTER**
                      =

Goodheart-Willcox Publisher

The display shows:

```
              0.75 ÷ tan 60 =
   0.43301270189222
```

Goodheart-Willcox Publisher

Dimension X = the width minus the adjacent two times
Dimension X = 4.0" − (0.433012701 × 2)
Dimension X = 4.0" − (0.866025403)
Dimension X = 3.133974596
Dimension X = 3.1340" rounded to the nearest ten-thousandth of an inch.

Example 41-2

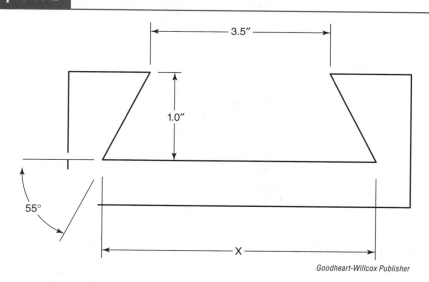

Goodheart-Willcox Publisher

(Continued)

The depth of the dovetail is 1.0″, the angle is 55°, and the distance across the small end is 3.5″. To calculate dimension X, a right triangle must first be constructed. Drawing a vertical line down to the horizontal line of the slide creates a 90° angle.

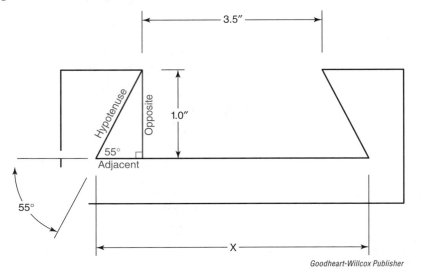

Goodheart-Willcox Publisher

In the right-angle triangle, the angle is 55°, and the opposite side (the depth) is 1.0″. Dimension X can be calculated by taking the width and adding the adjacent side measurement two times. The adjacent is added two times because the part is symmetrical, and the triangle formed would be identical on the other side.

To calculate the side adjacent to ∠55°, first determine what is known in the triangle. In this case, the opposite is known (1.0″), and the adjacent needs to be found. Therefore, tangent needs to be used to solve the adjacent.

$$\text{adjacent} = \frac{\text{opposite}}{\tan(\theta)}$$

$$\text{adjacent} = \frac{1.0}{\tan(55°)}$$

Use a scientific calculator.

1.0 ÷ **TAN** 55° **ENTER =**

Goodheart-Willcox Publisher

The display shows:

1 ÷ tan 55 =
0.70020753820971

Goodheart-Willcox Publisher

Dimension X = the width plus the adjacent two times

Dimension X = 3.5″ + (0.700207538 × 2)

Dimension X = 3.5″ + (1.400415076)

Dimension X = 4.900415076

Dimension X = 4.9004″ rounded to the nearest ten-thousandth of an inch.

Tangent Lines

Dovetail slides that need to be machined accurately to a given width (dimension) are usually measured across gage pins or dowels with a micrometer. Both male dovetails and female dovetails can be gaged in this way.

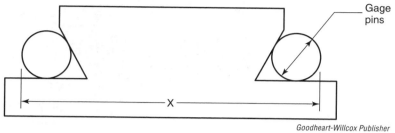

Goodheart-Willcox Publisher

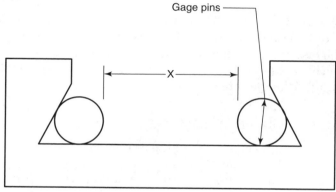

Goodheart-Willcox Publisher

Some important concepts must first be understood before the problem can be solved. In plane geometry, a **tangent line** to a circle is a line that touches the circle at exactly one point or position and never enters the circle's interior. The word tangent is derived from the word *tangere*, meaning "to touch."

Tangent lines to circles form the subject of several geometry theorems, two of which will be described in this unit.

Tangent to a Circle Theorem

As stated previously, a tangent to a circle is a straight line that touches the circle at one point only. That point of contact is called the **point of tangency** or the *tangent point*. The **tangent to a circle theorem** states that at the point of tangency, the tangent of the circle is perpendicular to the radius.

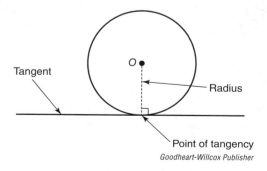

Goodheart-Willcox Publisher

Two-Tangent Theorem

The second geometric theorem is called the **two-tangent theorem**. It states that if two lines are drawn from the same point outside the circle such that both lines are tangent to the circle, the two lines are congruent. The word *equal* is often used in place of congruent.

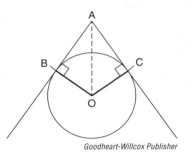

In the above drawing, lines AB and AC are tangent to the circle and share a common point outside the circle; therefore, AB and AC are equal in length. The angle at A can be bisected by drawing a line from the center of the circle (O) to position A. **Bisect** means to divide into two congruent, or equal, parts. In this figure, two identical right-angle triangles are formed.

Using Tangent Lines with Gage Pins

In problems involving dovetails, both of the aforementioned tangent theorems are used. Another important consideration is that the tangent point of the gage pin and dovetail slide must be on the angulur surface. In other words, if the gage pins are too large, the tangent to a circle theorem will not apply and an incorrect answer will be calculated.

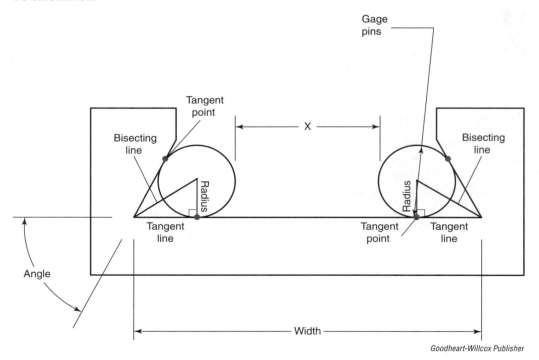

In the above figure, notice the points of tangency, where the gage pins touch the surface of the dovetail. There are four tangent points total, and because the dovetails are symmetrical, either side can be used.

The lines drawn down to form right angles are the radius of each gage pin, which is perpendicular to the tangent point.

The lines tangent to the gage pins meet in the same position outside the circle, in each corner of the dovetail.

The lines drawn from the centers of the gage pins to each corner of the dovetail, where the two lines tangent to each gage pin meet, bisect those angles to create right-angle triangles, which will allow the problem to be solved.

Example 41-3

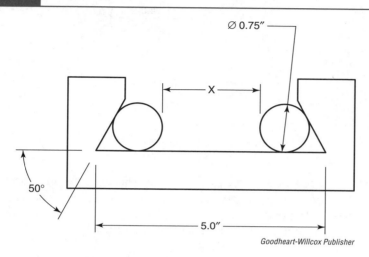

Goodheart-Willcox Publisher

Determine dimension X for the figure above.

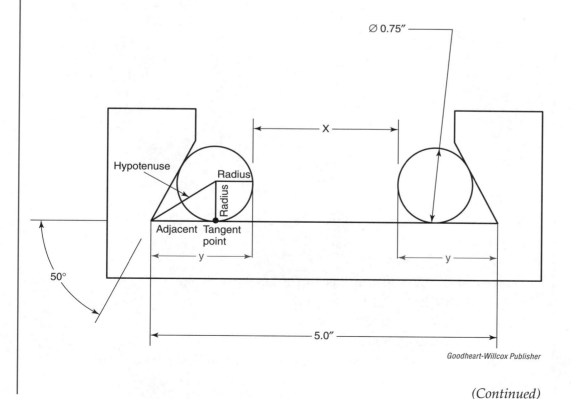

Goodheart-Willcox Publisher

(Continued)

As shown in the figure, drawing a line from the tangent point to the center of the gage pin creates a radius, the side opposite the angle (0.375″), and a 90° angle. Drawing a line from the center of the gage pin to the corner of the dovetail creates a right triangle and the hypotenuse. This line also bisects the dovetail corner angle.

Dimension X can be calculated by taking the width and subtracting dimension y twice.

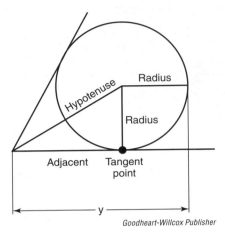

Goodheart-Willcox Publisher

Dimension y can be calculated by adding the adjacent side and the radius together. To calculate the side adjacent, first determine what is known in the triangle. In this case, the opposite is known (0.375″) and the adjacent needs to be found. Therefore, tangent needs to be used to solve the adjacent.

$$\text{adjacent} = \frac{\text{opposite}}{\tan(\theta)}$$

$$\text{adjacent} = \frac{0.375}{\tan(25°)}$$

Use a scientific calculator.

0.375 ÷ **TAN** 25° **ENTER =**

Goodheart-Willcox Publisher

The display shows:

0.375 ÷ tan 25 =
0.80419009519108

Goodheart-Willcox Publisher

Dimension y = adjacent + radius

Dimension y = 0.804190095″ + 0.375″

Dimension y = 1.179190095″

Dimension X = the width minus dimension y two times

Dimension X = 5.0″ − (1.179190095 × 2)

Dimension X = 5.0″ − (2.35838019)

Dimension X = 2.64161981″

Dimension X = 2.6416″ rounded to the nearest ten-thousandth of an inch.

Example 41-4

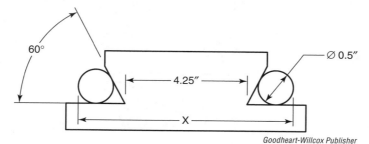

Goodheart-Willcox Publisher

Determine dimension X in the above drawing.

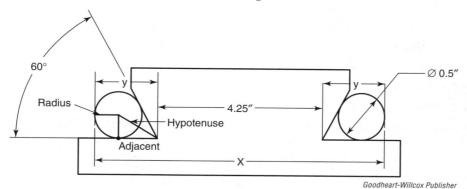

Goodheart-Willcox Publisher

As shown above, drawing a line from the tangent point to the center of the gage pin creates a radius, the side opposite the angle (0.25″), and a 90° angle. Drawing a line from the center of the gage pin to the corner of the dovetail creates a right triangle and the hypotenuse. This line also bisects the angle.

Dimension X can be calculated by taking the width and adding dimension y twice.

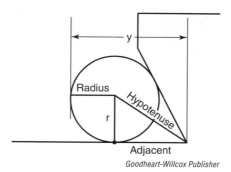

Goodheart-Willcox Publisher

Dimension y can be calculated by adding the adjacent side and the radius together. To calculate the adjacent side, first determine what is known in the triangle. In this case, the opposite is known (0.25″) and the adjacent needs to be found. Therefore, tangent needs to be used to solve the adjacent.

$$\text{adjacent} = \frac{\text{opposite}}{\tan(\theta)}$$

$$\text{adjacent} = \frac{0.25}{\tan(30°)}$$

(Continued)

Use a scientific calculator.

$$0.25 \div \boxed{\textbf{TAN}} \ 30° \ \boxed{\begin{array}{c}\textbf{ENTER}\\ =\end{array}}$$

The display shows:

$$
\boxed{\begin{array}{r}0.25 \div \tan 30 = \\[4pt] 0.43301270189222\end{array}}
$$

Dimension y = adjacent + radius

Dimension y = 0.433012701″ + 0.25″

Dimension y = 0.683012701″

Dimension X = the width plus dimension y two times

Dimension X = 4.25″ + (0.683012701 × 2)

Dimension X = 4.25″ + (1.366025404)

Dimension X = 5.616025404″

Dimension X = 5.6160″ rounded to the nearest ten-thousandth of an inch.

Unit 41 Review

Name _____ **Date** _____ **Class** _____

Fill in the blanks in the following review questions.

1. A _____ slide is a type of plain linear bearing that relies on direct contact between sliding surfaces to support and move a heavier load.

2. A dovetail slide consists of two main parts: the external dovetail, often referred to as the _____ dovetail; and the internal dovetail, or _____ dovetail.

3. A _____ to a circle is a straight line that touches the circle at one point only.

4. That point of contact is called the point of _____ or the _____ point.

5. The _____ to a _____ theorem states that at the point of tangency, the tangent of the circle is perpendicular to the radius.

6. The _____-_____ theorem states that if two lines are drawn from the same point outside the circle such that both lines are tangent to the circle, the two lines are congruent.

7. _____ means to divide into two congruent, or equal, parts.

8. Using gage pins and known dovetail dimensions, unknown dovetail dimensions can be found using _____.

Applications

Calculate dimension X in the following problems. In inch problems, round to the nearest ten-thousandth of an inch (four decimal places). In metric problems, round to three decimal places.

1.

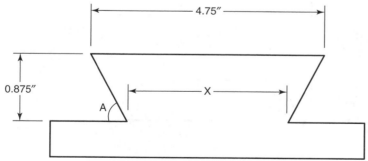

Goodheart-Willcox Publisher

	∠A	Adjacent	Dimension X	Rounded Answer
A.	45°			
B.	50°			
C.	55°			
D.	60°			

Goodheart-Willcox Publisher

2.

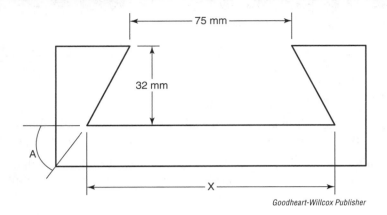

Goodheart-Willcox Publisher

	∠A	Adjacent	Dimension X	Rounded Answer
A.	45°			
B.	50°			
C.	55°			
D.	60°			

Goodheart-Willcox Publisher

3.

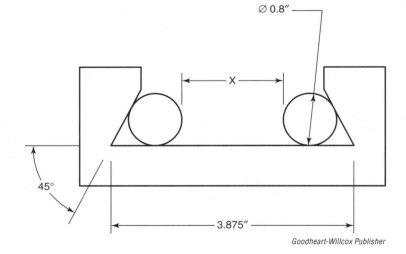

Goodheart-Willcox Publisher

_____

Name _____   **Date** _____   **Class** _____

4.

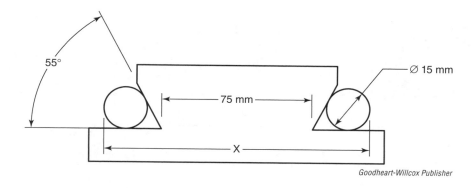

Goodheart-Willcox Publisher

_____

5.

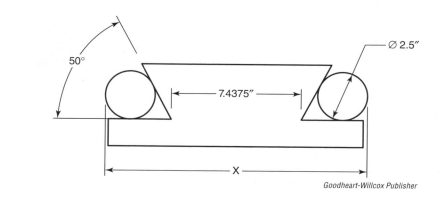

Goodheart-Willcox Publisher

_____

In the following problems, calculate for both dimensions X and W. In inch problems, round to the nearest ten-thousandth of an inch (four decimal places). In metric problems, round to three decimal places.

6.

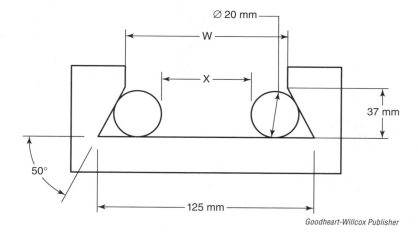

Goodheart-Willcox Publisher

W = _____

X = _____

7.

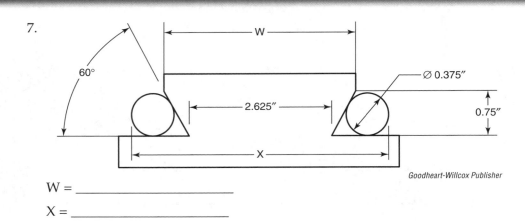

Goodheart-Willcox Publisher

W = _____

X = _____

8.

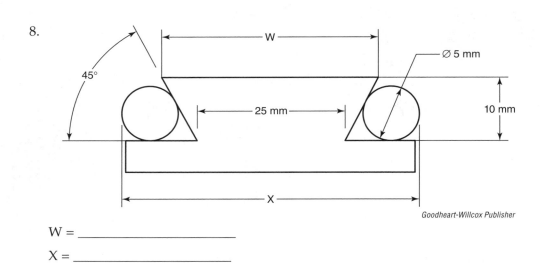

Goodheart-Willcox Publisher

W = _____

X = _____

UNIT 42

Tapers

Objectives

Information in this unit will enable you to:

- Know the elements of tapers.
- Explain how to construct right-angle triangles within tapers.
- Use trigonometry to solve for unknown elements of tapers.
- Understand the use and effects of a transversal intersecting parallel lines.

Elements of Tapers

Another practical use for trigonometry is with tapers, previously discussed in Unit 28. Almost all machine tool spindles have a taper as the primary method of attachment for tools. A male member of conical form (that is, with a taper) fits into the female socket, which has a matching taper of equal angle. Standard tapers typically designate the taper by number, such as Morse taper #3 (MT #3).

Special tapers and precision work require the taper (workpiece/part) to be dimensioned differently. When that is the case, one of the diameters (large or small diameter), the length (axial distance between the two diameters), and the angle are specified. The angle may be given as an included angle or an angle to the centerline (1/2 the included angle). The included angle in this context represents the angle between the two tapered sides. Sometimes the length may be omitted, and both diameters may be given instead. It depends on what features the engineer is trying to control.

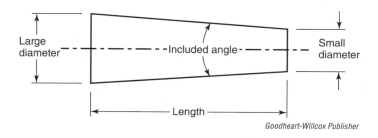

Goodheart-Willcox Publisher

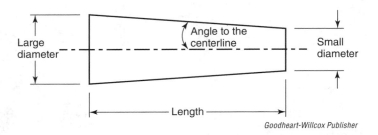

Tapers and Trigonometry

To solve taper problems, a right-angle triangle needs to be constructed within the taper. The right-angle triangle that is created needs to have two known dimensions to solve the subsequent problem. Typically, conical tapers are symmetrical on the horizontal centerline, and all the problems in this unit will use symmetrical tapers.

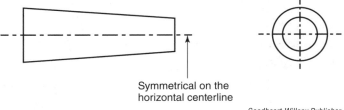

Symmetrical on the horizontal centerline

   If the lines of the tapered sides are extended past the small diameter, eventually they will intersect, creating a vertex. The centerline bisects the included angle, meaning it divides it into two congruent, or equal, parts. So the angle from the tapered side to the centerline is half of the included angle.

$$\text{Angle from tapered side to centerline} = \left(\frac{\text{Included angle}}{2}\right)$$

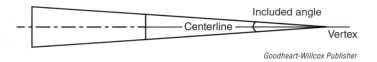

   To solve taper problems, a right-angle triangle needs to be constructed within the taper. Creating or drawing lines parallel to the centerline creates two congruent (equal) right triangles.

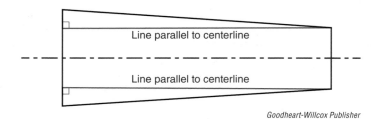

When two parallel lines are intersected by a third line, called the **transversal**, the corresponding angles are congruent (equal).

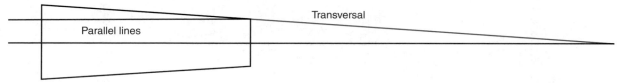

In creating the lines parallel to the centerline, equal angles were created. Both angles are equal to 1/2 the included angle.

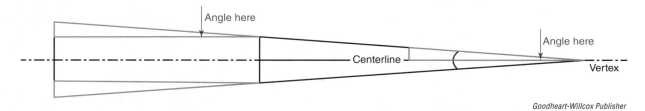

Example 42-1

Calculate the small diameter in the taper. Round the answer to the nearest ten-thousandth of an inch (four decimal places).

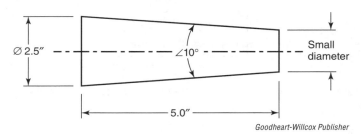

Starting from the corners at the small diameter, create lines parallel to the centerline.

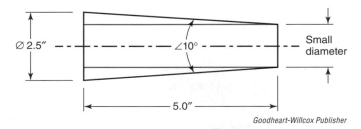

Use the new parallel lines to create congruent right-angle triangles.

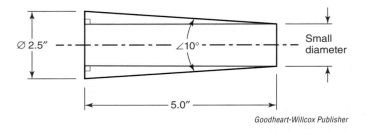

(Continued)

In the right-angle triangle that is constructed, the angle is $5°\left(\dfrac{\text{Included angle}}{2}\right)$, and the length of the adjacent side is 5.0″.

Goodheart-Willcox Publisher

The small diameter of the taper can be calculated by subtracting the opposite side from the large diameter two times. Subtract two times because the part is symmetrical, so the triangles formed are congruent (equal). Two equal triangles means two equal opposites.

The adjacent side (the length) and the angle are known, and the opposite is to be calculated, so use the tangent function to solve:

$$\text{opposite} = \tan(\theta) \times \text{adjacent}$$
$$\text{opposite} = \tan(5°) \times 5$$

Use a scientific calculator.

Goodheart-Willcox Publisher

The display shows:

> tan 5 × 5 =
> **0.43744331762962**

Goodheart-Willcox Publisher

Small diameter = Large diameter minus the opposite two times

Small diameter = 2.5″ − (0.437443317 × 2)

Small diameter = 2.5″ − (0.874886635)

Small diameter = 1.625113365″

Small diameter = 1.6251″ rounded to the nearest ten-thousandth of an inch.

Example 42-2

Calculate the large diameter in the taper. Round the answer to three decimal places.

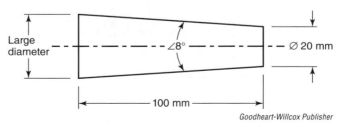

Goodheart-Willcox Publisher

(Continued)

From the corners of the small diameter, draw lines parallel to the centerline.

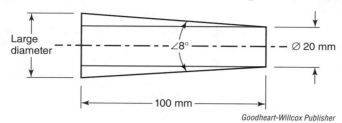

Goodheart-Willcox Publisher

Use the new parallel lines to create congruent right-angle triangles.

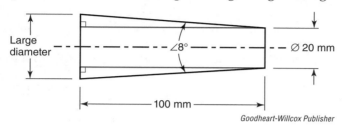

Goodheart-Willcox Publisher

In the right-angle triangle that is constructed, the angle is $4° \left(\dfrac{\text{Included angle}}{2} \right)$, and the length of the adjacent side is 100 mm.

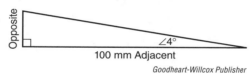

Goodheart-Willcox Publisher

The large diameter can be calculated by adding the opposite side to the small diameter two times. The part is symmetrical, so the triangles formed are congruent (equal). Two equal triangles means two equal opposites.

The adjacent side (the length) and the angle are known, and the opposite is to be calculated, so use the tangent function to solve:

$$\text{opposite} = \tan(\theta) \times \text{adjacent}$$
$$\text{opposite} = \tan(4°) \times 100$$

Use a scientific calculator.

TAN 4° × 100 **ENTER =**

Goodheart-Willcox Publisher

The display shows:

tan 4 × 100 =

6.992681194351

Goodheart-Willcox Publisher

Large diameter = Small diameter plus the opposite two times

Large diameter = 20 mm + (6.992681194 × 2)

Large diameter = 20 mm + (13.98536239)

Large diameter = 33.98536239

Large diameter = 33.985 mm rounded to three decimal places.

Example 42-3

Calculate the length of the taper. Round the answer to the nearest ten-thousandth of an inch (four decimal places).

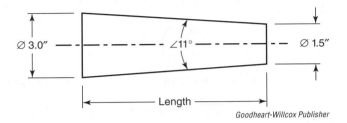

Goodheart-Willcox Publisher

From the corners of the small diameter, draw lines parallel to the centerline.

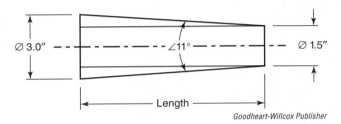

Goodheart-Willcox Publisher

Use the new parallel lines to create congruent right-angle triangles.

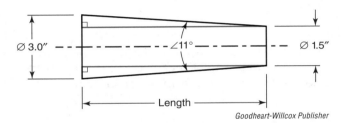

Goodheart-Willcox Publisher

In the right-angle triangle that is constructed, the angle is 5° 30′ $\left(\dfrac{\text{Included angle}}{2}\right)$. As both the small diameter and large diameter are given, the opposite side can be calculated.

$$\text{opposite} = \frac{(\text{Large diameter} - \text{Small diameter})}{2}$$

$$\text{opposite} = \frac{(3 - 1.5)}{2}$$

$$\text{opposite} = \frac{1.5}{2}$$

$$\text{opposite} = 0.75''$$

Goodheart-Willcox Publisher

(Continued)

The adjacent side (length) can be calculated using trigonometry. The opposite side (0.75″) and the angle (5° 30′) are known, and the adjacent side (length) is to be calculated, so use the tangent function to solve:

$$\text{adjacent} = \frac{\text{opposite}}{\tan(\theta)}$$

$$\text{adjacent} = \frac{0.75}{\tan(5° \, 30′)}$$

Use a scientific calculator.

0.75 ÷ **TAN** 5° 30′ **ENTER =**

The display shows:

0.75 ÷ tan 5° 30′ =
7.7890478101036

Length of the taper = Opposite divided by tan ∠
Length of the taper = 0.75″ ÷ tan 5° 30′
Length of the taper = 0.75″ ÷ (0.096289048)
Length of the taper = 7.78904781″
Length of the taper = 7.7890″ rounded to the nearest ten-thousandth of an inch.

Example 42-4

Calculate the included angle in the taper. Round the answer to the nearest second.

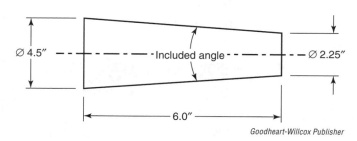

Starting from the corners at the small diameter, create lines parallel to the centerline.

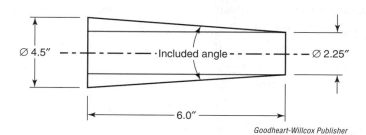

(Continued)

Use the new parallel lines to create congruent right-angle triangles.

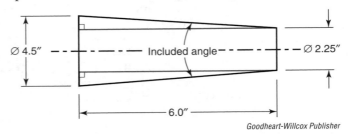

Goodheart-Willcox Publisher

In the right-angle triangle that is constructed, the length (the adjacent side) is 6.0″. As both the small diameter and large diameter are given, the opposite side can be calculated.

$$\text{opposite} = \frac{(\text{Large diameter} - \text{Small diameter})}{2}$$

$$\text{opposite} = \frac{(4.5 - 2.25)}{2}$$

$$\text{opposite} = \frac{2.25}{2}$$

$$\text{opposite} = 1.125″$$

Goodheart-Willcox Publisher

The angle can now be calculated using trigonometry. The opposite side (1.125″) and the adjacent side (6.0″) are known, so use the tangent function to solve for the angle. The included angle is two times the angle in the triangle.

$$\theta = \tan^{-1}\left(\frac{\text{opposite}}{\text{adjacent}}\right)$$

$$\theta = \tan^{-1}\left(\frac{1.125}{6}\right)$$

Use a scientific calculator.

2nd TAN (1.125 ÷ 6) ENTER =

Goodheart-Willcox Publisher

The display shows:

$$\tan^{-1}(1.125 \div 6) =$$
10.619655276155

Goodheart-Willcox Publisher

Included angle = Two times the angle in the triangle

Included angle = $2 \times [\tan^{-1}(1.125 \div 6)]$

Included angle = $2 \times [\tan^{-1}(0.1875)]$

Included angle = $2 \times (10.61965528)$

Included angle = $21.23931055°$

Pressing the Degree, Minutes, Seconds key **° , ,,** will convert this to 21° 14′ 21.52″.

Included angle = 21.23931055°, or 21° 14′ 21.52″

Included angle = 21° 14′ 22″ rounded to the nearest second.

Unit 42 Review

Name _____ **Date** _____ **Class** _____

Fill in the blanks in the following review questions.

1. Almost all machine tool spindles have a _____ as the primary method of attachment for tools.

2. Special tapers and precision work require the taper (workpiece/part) to be _____ differently.

3. When that is the case, _____ of the diameters (large or small diameter), the _____ (axial distance between the two diameters) and the _____ are specified.

4. To solve taper problems, a _____-_____ triangle needs to be constructed within the taper.

5. The centerline _____ the included angle, meaning it divides it into _____ congruent, or equal, parts.

6. Creating or drawing lines _____ to the centerline creates two _____ (equal) right triangles.

7. When two parallel lines are intersected by a third line, called the _____, the corresponding angles are congruent.

8. In creating the lines parallel to the centerline, equal angles were created. Both angles are equal to _____ the included angle.

Practice

Calculate for the missing dimension in the problems below. When calculating for a diameter or length, round the answer to the nearest ten-thousandth of an inch (four decimal places). When calculating for the included angle, round to the nearest second.

1. Find the small diameter of the taper. _____

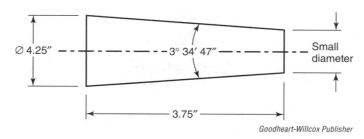

Ø 4.25"   3° 34′ 47″   Small diameter

3.75"

Goodheart-Willcox Publisher

2. Find the large diameter of the taper. _____

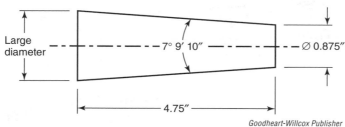

Large diameter   7° 9′ 10″   Ø 0.875"

4.75"

Goodheart-Willcox Publisher

3. Find the length of the taper. _____

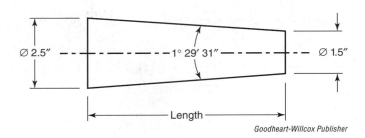

Goodheart-Willcox Publisher

4. Find the included angle of the taper. _____

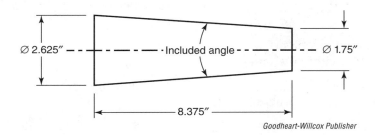

Goodheart-Willcox Publisher

5. Find the small diameter of the taper. _____

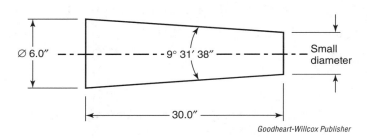

Goodheart-Willcox Publisher

6. Find the large diameter of the taper. _____

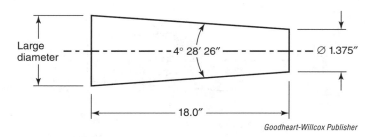

Goodheart-Willcox Publisher

7. Find the length of the taper. _____

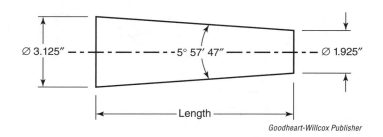

Goodheart-Willcox Publisher

Name _____ **Date** _____ **Class** _____

8. Find the included angle of the taper. _____

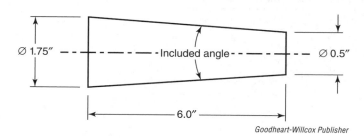

Goodheart-Willcox Publisher

Applications

1. What is the large diameter of the tapered reamer after 14 inches?
    The reamer has an included angle of 1° 11′ 42″.

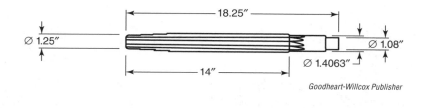

Goodheart-Willcox Publisher

_____

2. What is the large diameter of the tapered three-fluted end mill after 50 mm?

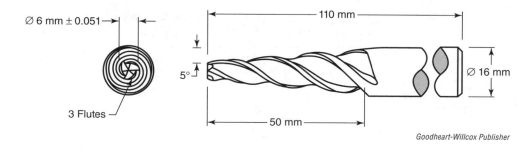

Goodheart-Willcox Publisher

_____

3. What is the included angle of the high-temperature tapered plug?

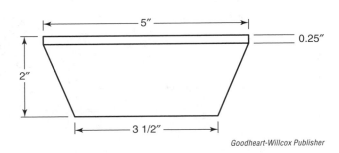

Goodheart-Willcox Publisher

_____

A CAT toolholder is shown below. Given an included angle of 16° 35′ 40″, calculate the small diameter of the taper in each problem. Round answers to the nearest ten-thousandth of an inch (four decimal places).

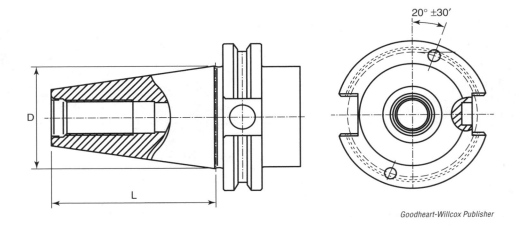

Goodheart-Willcox Publisher

	Toolholder Size	Large Diameter (D)	Length (L)	Small Diameter of Taper
4.	CAT 30	1.25″	1.875″	
5.	CAT 40	1.75″	2.687″	
6.	CAT 50	2.75″	4.0″	
7.	CAT 60	4.25″	6.375″	

Goodheart-Willcox Publisher

Area and Volume

Key Terms

chord	irregular polygon	prism
circular segment	non-polyhedron	radius
cone	octagon	regular polygon
cube	parallelogram	sector
cuboid	pentagon	sphere
cylinder	π (pi)	θ (theta)
diameter	polygon	trapezoid
hexagon	polyhedron	

UNIT 43

Area of Polygons

Objectives

Information in this unit will enable you to:

- Explain what a polygon is and identify examples of polygons.
- Describe how to determine the area of squares and rectangles, parallelograms, triangles, trapezoids, pentagons, octagons, and hexagons.

Polygons

A **polygon** is a two-dimensional shape with straight sides. A polygon may have as few as three sides (a triangle) or many more sides. A few common polygons are triangles, squares, rectangles, trapezoids, parallelograms, pentagons, hexagons, and octagons. A polygon is considered a **regular polygon** if all sides are equal in length and all angles are equal. Equilateral triangles, squares, and hexagons are all regular polygons. A shape with straight sides of different lengths or corners of different angles is an **irregular polygon**. Rectangles, parallelograms, and trapezoids are examples of irregular polygons.

Area of Rectangles and Squares

As was discussed in Unit 31, the area of a square is found by squaring the length of a side. This can be stated mathematically by the formula $A = s^2$, where A is area and s is the length of one side. If a square is 7 inches on a side, its area is 7^2 or 49 square inches. A rectangle is similar to a square except that its width and length are not equal. The area of a rectangle is found by multiplying its length by its width. The formula for the area of a rectangle is $A = lw$, where A is area, l is the length of a long side, and w is the length of a short side.

Example 43-1

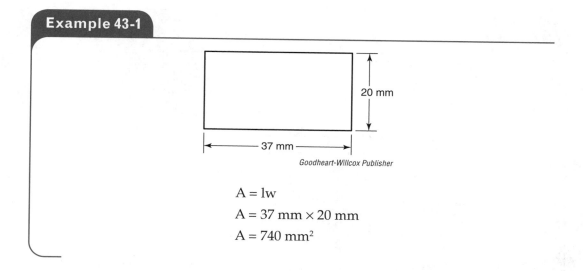

Goodheart-Willcox Publisher

A = lw

A = 37 mm × 20 mm

A = 740 mm²

Area of a Parallelogram

A **parallelogram** is a four-sided shape in which each side is parallel to the opposite side, but not necessarily at right angles to the adjacent sides. A rectangle is a parallelogram in which the short sides are perpendicular to the long sides; all corners are 90°. Generally, when a shape is referred to as a parallelogram, the corners are not 90°.

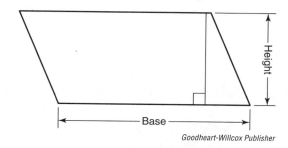

Goodheart-Willcox Publisher

Name one of the long sides the base. (It could be a short side, but a long side is most often called the base.) The distance from the base to the opposite side, measured along a perpendicular line, is the height of the parallelogram. To find the area, multiply the base times the height. The mathematical formula is A = bh, where A is the area, b is the length of the base, and h is the height.

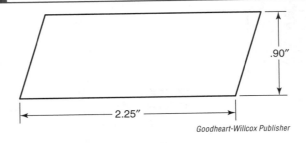

Goodheart-Willcox Publisher

$$A = bh$$
$$A = 2.25'' \times .90''$$
$$A = 2.025 \text{ in}^2$$

The highest degree of precision given in the problem is two decimal places, so round the answer to $A = 2.03 \text{ in}^2$.

Area of a Triangle

Two dimensions are needed to find the area of a triangle. The base is the length of any one side. The height is the dimension from the base to the opposite vertex in a line that is perpendicular to the base, as shown in the following figure. The height may or may not be a side of the triangle.

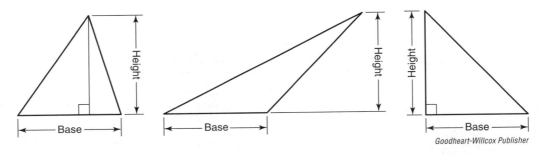

Goodheart-Willcox Publisher

The area of a triangle is found by multiplying 1/2 times the base times the height. The formula for the area of a triangle is $A = 1/2bh$, where A is the area, b is the base, and h is the height.

Example 43-3

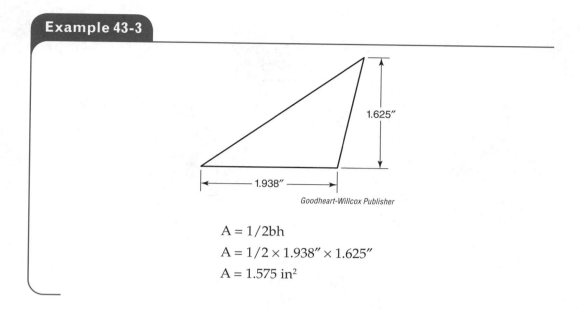

Goodheart-Willcox Publisher

$A = 1/2bh$

$A = 1/2 \times 1.938'' \times 1.625''$

$A = 1.575 \text{ in}^2$

Area of a Trapezoid

A **trapezoid** is a four-sided shape in which two sides are parallel, but the other two sides are not parallel. As with other shapes, it is necessary to establish names for the parts of the trapezoid, so the area can be computed. The trapezoid has two bases, so one is named base 1 and the other is base 2.

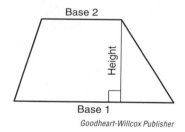

Goodheart-Willcox Publisher

The area of a trapezoid is calculated by finding the average of the two bases and multiplying that average by the height of the trapezoid. The average of two numbers is found by adding them together and then dividing by 2. The formula for the area of a trapezoid is $A = 1/2(b_1 + b_2)h$, where A is the area, b_1 and b_2 are the lengths of the two bases, and h is the height of the trapezoid.

Example 43-4

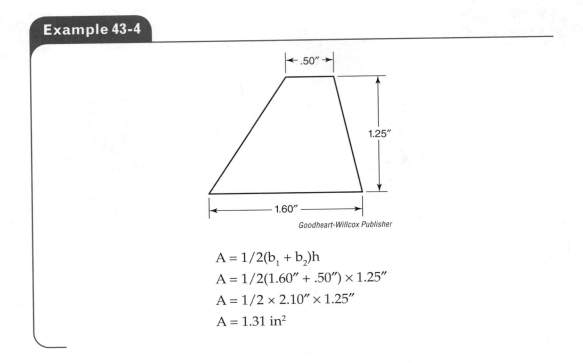

Goodheart-Willcox Publisher

$A = 1/2(b_1 + b_2)h$

$A = 1/2(1.60'' + .50'') \times 1.25''$

$A = 1/2 \times 2.10'' \times 1.25''$

$A = 1.31 \text{ in}^2$

Areas of Pentagons, Hexagons, and Octagons

A **pentagon** is a five-sided shape, a **hexagon** is a six-sided shape, and an **octagon** is an eight-sided shape. Regular pentagons, hexagons, and octagons are those shapes with all sides being equal in length.

Regular polygons can be divided into triangles.

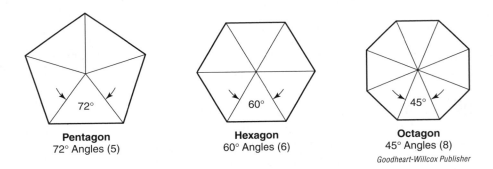

Pentagon
72° Angles (5)

Hexagon
60° Angles (6)

Octagon
45° Angles (8)

Goodheart-Willcox Publisher

There are multiple ways to find the areas of these polygons, but if the lengths of the sides are known, all their areas can be computed by solving the triangles that make them up. Use the tangent function of the known angle of the polygon to find the height of the triangles, then use the formula for the area of a triangle.

Example 43-5

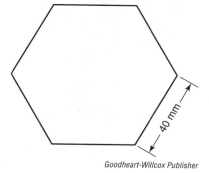

Goodheart-Willcox Publisher

Divide the hexagon into six triangles with an angle of 60° at the center.

Draw a line from the center of the hexagon to the center of one side of the hexagon. This line forms a 90° angle with the base of the triangle and forms a new, smaller right triangle. The base of the right triangle is 20 mm and the angle at the top of the new triangle is 30°.

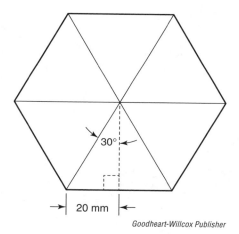

Goodheart-Willcox Publisher

Use the tangent function to find the height of the triangle.

$$\tan = \text{opp}/\text{adj}$$

$$\tan 30° = 20/\text{height of triangle}$$

$$.5774 = 20/\text{height of triangle}$$

$$.5774 \times \text{height} = 20$$

$$\text{height} = 20/.5774 = 34.64 \text{ mm}$$

Use the formula for the area of a triangle to find the area of the larger triangle.

$$A = 1/2bh$$

$$A = 1/2 \times 40 \text{ mm} \times 34.64 \text{ mm}$$

$$\text{Area of larger triangles} = 692.8 \text{ mm}^2$$

Multiply by 6 triangles.

$$\text{Area of hexagon} = 692.8 \text{ mm}^2 \times 6 = 4156.8 \text{ mm}^2$$

Work Space/Notes

Unit 43 Review

Name _____ Date _____ Class _____

Fill in the blanks in the following review questions.

1. A _____ is a two-dimensional shape with straight sides. It may have as few as three sides (a triangle) or many more sides.

2. A few common polygons are triangles, squares, rectangles, trapezoids, parallelograms, _____, _____, and _____.

3. A polygon is considered a _____ polygon if all sides are equal in length and all angles are equal. Equilateral triangles, squares, and hexagons are all _____ polygons.

4. A shape with straight sides of different lengths or corners of different angles is a(n) _____ polygon. Rectangles, parallelograms, and trapezoids are examples of _____ polygons.

5. The area of a square is found by _____ the length of a side. This can be stated mathematically by the formula _____.

6. The formula for the area of a rectangle is A = lw, where *A* is _____, *l* is the length of a _____ side, and *w* is the length of a _____ side.

7. A _____ is a four-sided shape in which each side is parallel to the opposite side but not necessarily at right angles to the adjacent sides. Generally, when a shape is referred to as a _____, the corners are not 90°.

8. To find the area of a parallelogram, the mathematical formula is _____, where _____ is area, _____ is the length of the base, and *h* is the _____.

9. The formula for the area of a triangle is _____, where *A* is the _____, *b* is the _____, and _____ is the height.

10. A _____ is a four-sided shape in which two sides are parallel, but the other two sides are not parallel. A _____ has two bases, named base 1 and base 2.

11. The area of a trapezoid is calculated by finding the _____ of the two bases and multiplying it by the height of the trapezoid. The formula for the area of a trapezoid is _____, where b_1 and b_2 are the lengths of the two _____, and _____ is the height of the trapezoid.

12. A _____ is a five-sided shape, a _____ is a six-sided shape, and an octagon is an _____-sided shape.

13. Regular polygons can be divided into triangles to find their area. Use the tangent function of the known angle of a regular polygon to find the _____ of the triangles, then use the formula for the area of a triangle.

Practice

Compute the areas of the polygons to the nearest hundredth.

1.
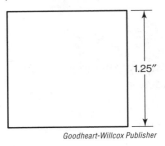
1.25"
Goodheart-Willcox Publisher

32.5 mm
Goodheart-Willcox Publisher

A. _____

B. _____

2.
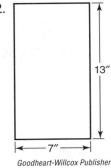
13"
7"
Goodheart-Willcox Publisher

8.5 mm
18 mm
Goodheart-Willcox Publisher

A. _____

B. _____

3.
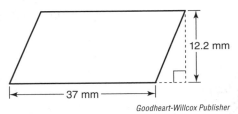
12.2 mm
37 mm
Goodheart-Willcox Publisher

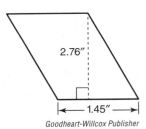
2.76"
1.45"
Goodheart-Willcox Publisher

A. _____

B. _____

4.

18 mm
22 mm
Goodheart-Willcox Publisher

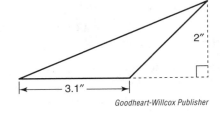
2"
3.1"
Goodheart-Willcox Publisher

A. _____

B. _____

Name _____   **Date** _____   **Class** _____

5.

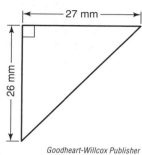

27 mm

26 mm

Goodheart-Willcox Publisher

1.4″

4.25″

Goodheart-Willcox Publisher

A. _____

B. _____

6.

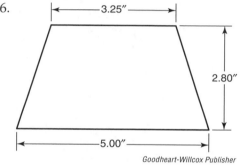

3.25″

2.80″

5.00″

Goodheart-Willcox Publisher

17 mm

11 mm

28 mm

Goodheart-Willcox Publisher

A. _____

B. _____

7.

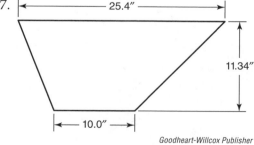

25.4″

11.34″

10.0″

Goodheart-Willcox Publisher

25 mm

37.5 mm

75 mm

Goodheart-Willcox Publisher

A. _____

B. _____

8.

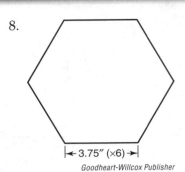

← 3.75″ (×6) →

Goodheart-Willcox Publisher

A. _____

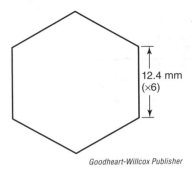

12.4 mm
(×6)

Goodheart-Willcox Publisher

B. _____

9.

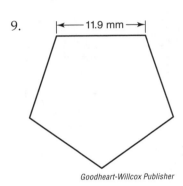

← 11.9 mm →

Goodheart-Willcox Publisher

A. _____

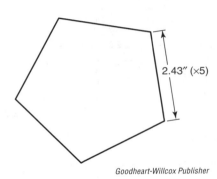

2.43″ (×5)

Goodheart-Willcox Publisher

B. _____

10.

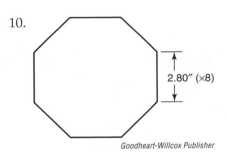

2.80″ (×8)

Goodheart-Willcox Publisher

A. _____

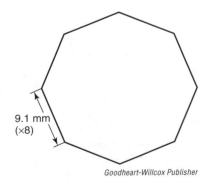

9.1 mm
(×8)

Goodheart-Willcox Publisher

B. _____

Name _____   **Date** _____   **Class** _____

Applications

Find the shaded areas of the following shapes. Round answers to the nearest hundredth.

1.

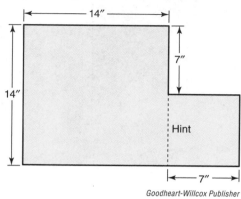

Goodheart-Willcox Publisher

_____

2.

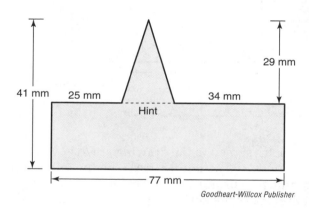

Goodheart-Willcox Publisher

_____

3.

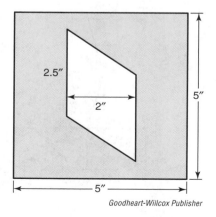

Goodheart-Willcox Publisher

_____

4.

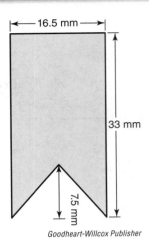

16.5 mm

33 mm

7.5 mm

Goodheart-Willcox Publisher

_____

5.

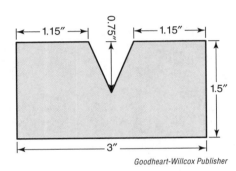

1.15″   0.75″   1.15″

1.5″

3″

Goodheart-Willcox Publisher

_____

6. Seventy of the plates shown in the drawing are to be acid-washed on both sides. The amount of acid solution is based on the total surface area to be washed. Compute the surface area of 70 of these plates to the nearest square inch. Deduct the area of the keyhole shape that is cut out of the plates. (Hint: The keyhole shape is made up of two trapezoids.)

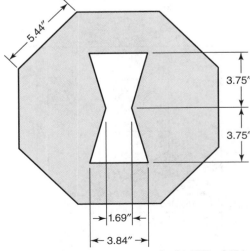

5.44″

3.75″

3.75″

1.69″

3.84″

Goodheart-Willcox Publisher

_____

Area of Circles and Circular Segments

Objectives

Information in this unit will enable you to:

- Explain what pi and theta are and what they represent in geometry.
- Discuss how to determine the area of circles, circular sectors, and circular segments.

Greek Letters

Pi was discussed in Unit 23. The circumference of any circle is approximately 3.14159 times its diameter (alternatively, $2\pi r$). The actual number is an unending decimal, and it is typically rounded off to 3.14159 or even 3.1416. Since the mid-eighteenth century, mathematicians have used the Greek letter π (**pi**, pronounced "pie") to represent this number. Computations having to do with circles use π extensively. Scientific calculators have a button for use with formulas.

The Greek letter θ (**theta**) is commonly used as a symbol for the angle at the center of the circle when discussing sectors and segments of circles.

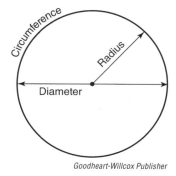

Goodheart-Willcox Publisher

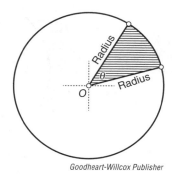

Goodheart-Willcox Publisher

Area of a Circle

The area of a circle can be computed by multiplying π by the square of the radius. The mathematical formula for the area of a circle is $A = \pi r^2$, where A is area and r is the **radius** (1/2 the diameter) of the circle.

Example 44-1

Goodheart-Willcox Publisher

Calculate the area of the circle.

$$A = \pi r^2$$

radius is 1/2 the diameter

$$A = 3.14159 \times 1.5^2$$

$$A = 7.06858 \text{ in}^2$$

Area of a Circlular Sector

A **sector** is a slice of a circle. It is formed by drawing two straight lines from the center to the circumference.

Sector

Goodheart-Willcox Publisher

If the radius and the central angle of the sector are known, the sector's area can be found by comparing it to the area of the complete circle.

Example 44-2

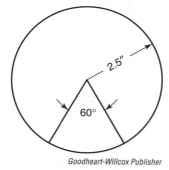

Goodheart-Willcox Publisher

Calculate the area of the circle.

$$A = \pi r^2$$
$$A = \pi \times 2.5^2$$
$$A = 19.635 \text{ in}^2$$

The ratio of the sector area to the circle area is in proportion to the ratio of the sector's inner angle to the angle of the circle (360°).

$$\frac{60}{360} = \frac{\text{sector area}}{19.635}$$

Multiply both sides by 19.635.

$$\frac{60 \times 19.635}{360} = \text{sector area}$$
$$3.2725 \text{ in}^2 = \text{sector area}$$

These steps can be combined to make the formula

$$\text{sector} = \frac{\theta}{360}(\pi r^2)$$

Area of a Circular Segment

A **circular segment** is a part of a circle that is cut off from the rest of the circle by a **chord**, which is a straight line intersecting the circumference of a circle in two places. A chord that passes through the center of the circle is the **diameter**.

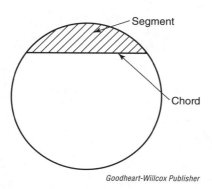

Goodheart-Willcox Publisher

The sides of a sector are straight lines drawn from the ends of the chord to the center of the circle. The portion of the sector that is not the segment is a triangle. The area of a segment is the area of the sector minus the area of the triangle formed by the chord and the sides of the sector. If the inner angle and the radius of the circle are known, the length of the chord can be computed with the law of cosines, explained in Unit 37. The formula for the area of a segment when the radius and the inner angle are known is:

$$A = \frac{r^2}{2}[(\pi/180) \times \theta - \sin\theta]$$

Example 44-3

Use the formula for area of a segment:

$$A = \frac{r^2}{2}[(\pi/180) \times \theta - \sin\theta]$$

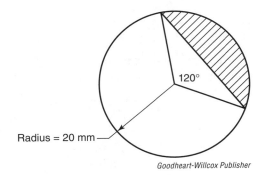

120°

Radius = 20 mm

Goodheart-Willcox Publisher

$$A = \frac{20^2}{2}[(\pi/180) \times 120 - 0.866]$$

$$A = \frac{20^2}{2}(2.094 - 0.866)$$

$$A = 200 \times 1.228$$

$$A = 245.6 \text{ mm}^2 \text{ rounded}$$

Unit 44 Review

Name _____ **Date** _____ **Class** _____

Fill in the blanks in the following review questions.

1. The _____ of any circle is approximately 3.14159 times its diameter. Mathematicians use the Greek letter _____ (pi, pronounced "pie") to represent this number.

2. The Greek letter _____ (theta) is commonly used as a symbol for the angle at the center of the circle when discussing circle sectors and segments.

3. The area of a circle is $A = \pi r^2$, where A is _____, and r is the _____ (1/2 the diameter) of the circle.

4. A _____ is a slice of a circle formed by drawing two straight lines from the center to the circumference.

5. If the _____ and central _____ of the sector are known, the sector's area can be found by comparing it to the area of the complete circle. The steps can be combined to make the formula _____.

6. A circular _____ is a part of a circle that is cut off from the rest of the circle by a _____, which is a straight line intersecting the circumference of a circle in two places. A _____ that passes through the center of the circle is the diameter.

7. The sides of a _____ are straight lines drawn from the ends of the chord that forms the segment to the center of the circle. The portion of the sector that is not the segment is a _____.

8. The area of a _____ is the area of the sector minus the area of the triangle formed by the chord and the sides of the sector.

9. If the inner _____ and the _____ of the circle are known, the length of the chord can be computed with the law of cosines.

10. The formula for the area of a _____ when the radius and the inner angle are known is $A = r^2/2[(\pi/180) \times \theta - \sin \theta]$.

Practice

Identify the indicated features.

Goodheart-Willcox Publisher   Goodheart-Willcox Publisher

1. _____   2. _____

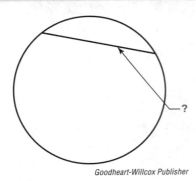

Goodheart-Willcox Publisher

3. _____

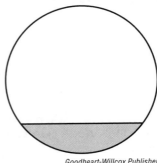

Goodheart-Willcox Publisher

4. _____

Goodheart-Willcox Publisher

5. _____

Goodheart-Willcox Publisher

6. _____

Goodheart-Willcox Publisher

7. _____

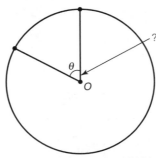

Goodheart-Willcox Publisher

8. _____

Name _____ **Date** _____ **Class** _____

Compute the areas of the shapes to the nearest thousandth.

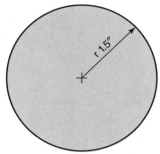

Goodheart-Willcox Publisher

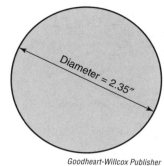

Goodheart-Willcox Publisher

9. _____

10. _____

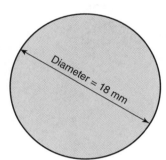

Goodheart-Willcox Publisher

Goodheart-Willcox Publisher

11. _____

12. _____

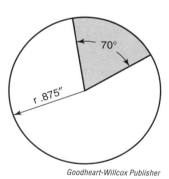

Goodheart-Willcox Publisher

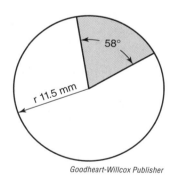

Goodheart-Willcox Publisher

13. _____

14. _____

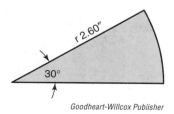

Goodheart-Willcox Publisher

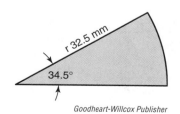

Goodheart-Willcox Publisher

15. _____

16. _____

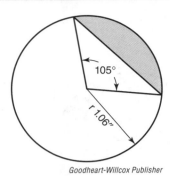

Goodheart-Willcox Publisher

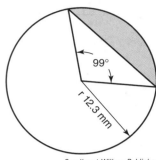

Goodheart-Willcox Publisher

17. _____

18. _____

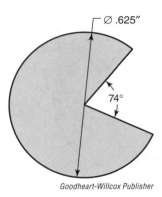

Goodheart-Willcox Publisher

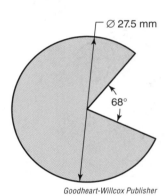

Goodheart-Willcox Publisher

19. _____

20. _____

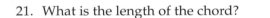

21. What is the length of the chord?

22. What is the length of the chord?

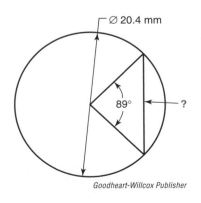

Goodheart-Willcox Publisher

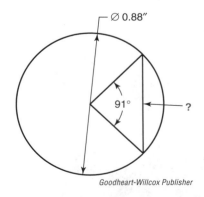

Goodheart-Willcox Publisher

_____

_____

Name _____ **Date** _____ **Class** _____

Applications

Find the shaded areas of the following shapes. Round answers to the nearest hundredth.

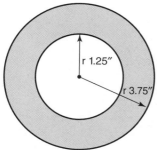

Goodheart-Willcox Publisher

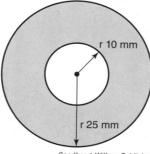

Goodheart-Willcox Publisher

1. _____

2. _____

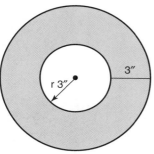

Goodheart-Willcox Publisher

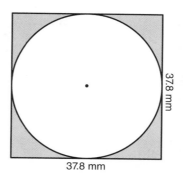

Goodheart-Willcox Publisher

3. _____

4. _____

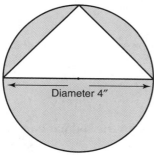

Goodheart-Willcox Publisher

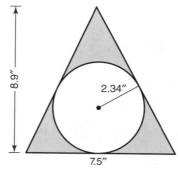

Goodheart-Willcox Publisher

5. _____

6. _____

Fifteen of these plates must be sent out for electroplating. What is the total area of the 15 plates to be plated? The plates are thin, so it is not necessary to include the edges in your calculations. However, this exercise requires several steps, so write the results of your calculations for each of the steps listed. Include several decimal places in your intermediate steps, but round the final result to the nearest whole square millimeter.

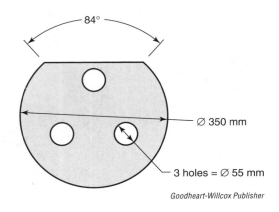

Goodheart-Willcox Publisher

7. Area of base circle:

    _____

8. Area of segment to be removed:

    _____

9. Area of each hole:

    _____

10. Total area of the three holes:

    _____

11. Area of one side of one plate after the deductions for segment and holes:

    _____

12. Total area of two sides of one plate after deductions for the segment and holes:

    _____

13. Total area of two sides of 15 plates after deductions for segment and holes:

    _____

UNIT 45

Volume of Cubes, Rectangular Solids, and Prisms

Objectives

Information in this unit will enable you to:

- Explain what a polyhedron is.
- Identify and describe different types of polyhedrons.
- Discuss how to determine the volume of a polyhedron.

Polyhedrons

There are two main types of solid shapes: polyhedrons and non-polyhedrons. This unit explains how to determine the volume of polyhedrons. A **polyhedron** is a three-dimensional shape that has flat faces. A **cube** is a six-sided polyhedron in which all sides are identical squares. A **cuboid** is a polyhedron in which all of the faces are squares or rectangles and all the angles are right angles. A shoebox is a cuboid. Cuboids are sometimes referred to as rectangular solids. A **prism** is a solid shape with identically shaped ends and sides that are rectangles. The ends of a prism may be triangles, pentagons, or any other polygon. The following figure shows several polyhedrons.

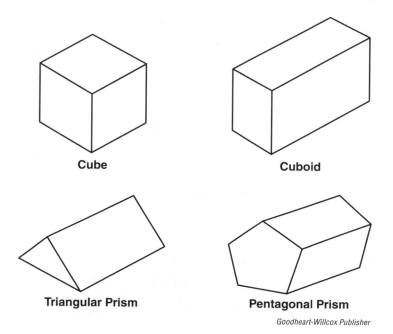

Cube   **Cuboid**

Triangular Prism   **Pentagonal Prism**

Goodheart-Willcox Publisher

Volume of Polyhedrons

The volume of a cube, cuboid, or prism can be found by multiplying the area of one face by the third dimension. The formula for the area of a square, explained in Unit 43, is the length of one side to the second power or the length of one side squared. The formula for the volume of a cube is $V = s^3$, where V is volume and s is the length of one side. That is the same as length × width × depth. The volume of a cuboid or rectangular solid is found by multiplying the length × width × depth. The formula for the volume of a cuboid is $V = l \times w \times d$, or simply $V = lwd$.

Example 45-1

Find the volume of the cuboid shown in the figure.

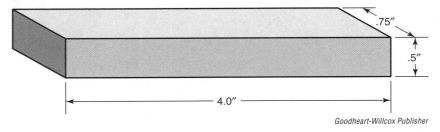

.75"

.5"

4.0"

Goodheart-Willcox Publisher

Use the formula for the volume of a cuboid.

$$V = l \times w \times d$$
$$V = 4.0" \times .5" \times .75"$$
$$V = 1.50 \text{ in}^3$$

When the ends are a shape other than a square or rectangle, as in a prism, find the area of one end first, then multiply that area by the length of the prism.

Example 45-2

Find the volume of the hexagonal prism in the figure.
Find the area of the hexagon, as explained in Unit 43.

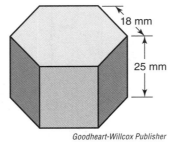

18 mm

25 mm

Goodheart-Willcox Publisher

$$A = 842 \text{ mm}^2$$

Multiply by the height or length of the prism.

$$V = 842 \text{ mm}^2 \times 25 \text{ mm}$$
$$V = 21,050 \text{ mm}^3$$

Unit 45 Review

Name _____ **Date** _____ **Class** _____

Fill in the blanks in the following review questions.

1. There are two main types of solid shapes: _____ and _____.
   A _____ is a three-dimensional shape that has flat faces.

2. A _____ is a six-sided polyhedron in which all sides are identical
   squares. A _____ is a polyhedron in which all faces are squares or
   _____ and all angles are right angles. _____ are sometimes
   referred to as rectangular solids.

3. A _____ is a solid shape with identically shaped ends and sides that
   are rectangles. The ends of a _____ may be triangles, pentagons, or
   any other _____.

4. The _____ of a cube, cuboid, or prism can be found by
   _____ the area of one face by the third dimension.

5. The formula for the volume of a cube is $V = s^3$, where V is _____,
   and s is the _____ of one side.

6. The volume of a cuboid or rectangular solid is found by multiplying the
   _____ × _____ × _____. The formula for the
   volume of a cuboid is _____.

7. To find volume when the ends are a shape other than a square or rectangle, as
   in a prism, find the _____ of one end first, then multiply that by the
   length of the prism.

Practice

Compute the volumes of the polyhedrons. Round answers to the nearest thousandth.

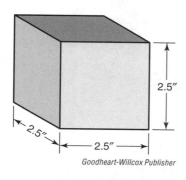

2.5"

2.5"

2.5"

Goodheart-Willcox Publisher

50 mm

50 mm

50 mm

Goodheart-Willcox Publisher

1. _____

2. _____

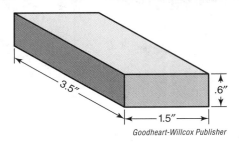

3.5"

1.5"

.6"

Goodheart-Willcox Publisher

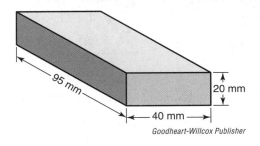

95 mm

40 mm

20 mm

Goodheart-Willcox Publisher

3. _____

4. _____

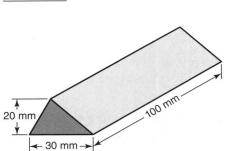

20 mm

30 mm

100 mm

Goodheart-Willcox Publisher

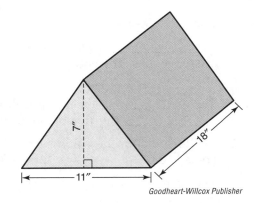

7"

11"

18"

Goodheart-Willcox Publisher

5. _____

6. _____

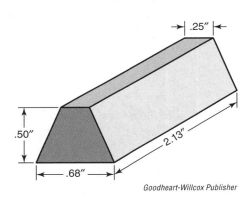

.25"

.50"

2.13"

.68"

Goodheart-Willcox Publisher

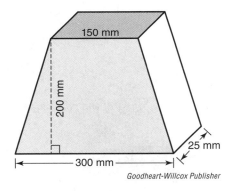

150 mm

200 mm

300 mm

25 mm

Goodheart-Willcox Publisher

7. _____

8. _____

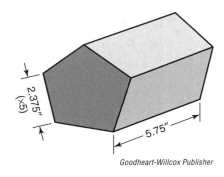

2.375" (×5)

5.75"

Goodheart-Willcox Publisher

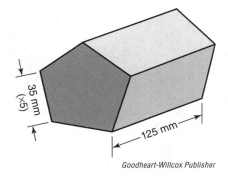

35 mm (×5)

125 mm

Goodheart-Willcox Publisher

9. _____

10. _____

Name _____   **Date** _____   **Class** _____

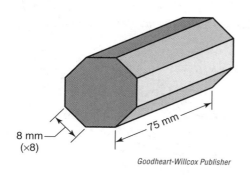

8 mm
(×8)

75 mm

Goodheart-Willcox Publisher

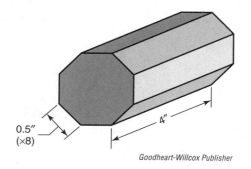

0.5"
(×8)

4"

Goodheart-Willcox Publisher

11. _____   12. _____

Applications

1. A spindle lubrication reservoir on a CNC machining center measures 8″ × 6″ × 5″. Find the volume of the reservoir.

    Volume in in³: _____

2. How many liters of spindle oil would be required to completely fill the reservoir in question 1 from empty? Round to the nearest thousandth. (To convert cubic inches to liters, divide the volume value by 61.024 for an approximate result.)

    Liters: _____

3. The spindle oil comes in a rectangular tin can measuring 10″ × 6.5″ × 4″. What volume is the tin can?

    Volume in in³: _____

4. How many liters of lubrication oil can the tin can from question 3 hold? (To convert cubic inches to liters, divide the volume value by 61.024 for an approximate result.)

    Liters: _____

5. The coolant tank on a CNC turning center needs refilling with coolant from empty. The tank measures 48″ × 24″ × 11″. What is the volume of the tank?

    Volume in in³: _____

6. How many gallons of coolant will be needed to fill the tank? Round to the nearest whole number. (To convert cubic inches to US liquid gallons, divide the volume value by 231.)

    Gallons: _____

7. Fourteen parts are rectangular solids measuring 1.5″ × 3.25″ × 5.45″. They are to be packed in cartons. Which carton size has enough volume to hold the parts with the least amount of extra space? Three sizes of cartons are available:

    Small: 6 1/2″ × 6 1/2″ × 6 1/2″

    Medium: 7 1/4″ × 7 1/4″ × 7 1/4″

    Large: 8 3/4″ × 8 3/4″ × 8 3/4″

    _____

Use these instructions for problems 8 through 11. The hexagonal bar shown here is made of brass that weighs .31 lb per cubic inch. To find the weight of the 36" bar, do the calculations for the steps listed below. Round each step to the nearest thousandth.

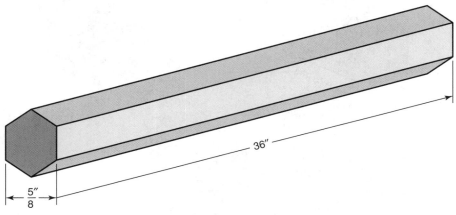

$\frac{5''}{8}$

36"

Goodheart-Willcox Publisher

8. Use the sine function to determine the width of each flat in the drawing.

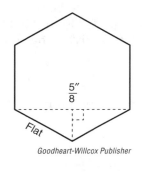

$\frac{5''}{8}$

Flat

Goodheart-Willcox Publisher

_____

9. Use the tangent function to determine the area of the end of the bar.

_____

10. Multiply the end area by the length of the bar to find the volume of the bar.

_____

11. Multiply the volume of the bar by the weight per cubic inch to find the weight of the bar.

_____

UNIT 46

Volume of Cylinders, Cones, and Spheres

Objectives

Information in this unit will enable you to:

- Explain what non-polyhedrons are and provide examples of them.
- Discuss how to determine the volume of cylinders, cones, and spheres.

Non-Polyhedrons

A **non-polyhedron** is a solid shape that has at least one surface that is not flat. Cylinders, cones, and spheres are examples of non-polyhedrons. There are other non-polyhedrons, but these three are the most frequently encountered, so they are the ones explained here.

Non-Polyhedrons

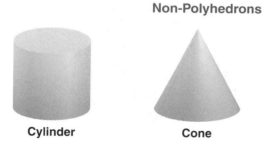 

Cylinder	**Cone**	**Sphere**

Goodheart-Willcox Publisher

Volume of a Cylinder

A **cylinder** is similar to a prism. It has flat ends or bases that are circles, and it has length. The flat ends are parallel to one another. The volume of a cylinder is found by multiplying the area of the circular end by the length. The formula for the area of a circle, $A = \pi r^2$, was discussed in Unit 44. The formula for the volume of a cylinder adds the length to the area formula: $V = \pi r^2 l$.

Example 46-1

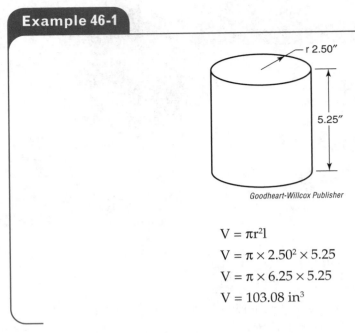

Goodheart-Willcox Publisher

$$V = \pi r^2 l$$
$$V = \pi \times 2.50^2 \times 5.25$$
$$V = \pi \times 6.25 \times 5.25$$
$$V = 103.08 \text{ in}^3$$

Volume of a Cone

A cone is a solid shape with a flat circular base and sides tapering to a point or vertex. The formula for finding the volume of a cone is very similar to the formula for the volume of a cylinder. The cone tapers to a point at one end, so its volume is 1/3 that of a cylinder with the same radius. The volume formula for a cone is $V = 1/3 \, \pi r^2 l$.

Example 46-2

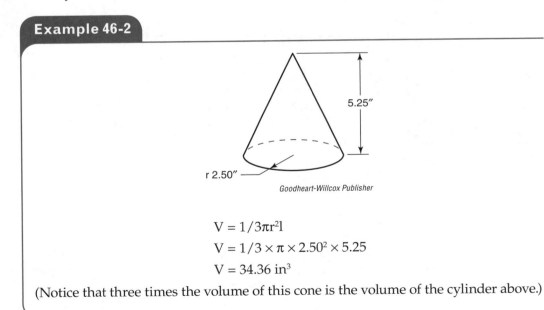

Goodheart-Willcox Publisher

$$V = 1/3 \pi r^2 l$$
$$V = 1/3 \times \pi \times 2.50^2 \times 5.25$$
$$V = 34.36 \text{ in}^3$$

(Notice that three times the volume of this cone is the volume of the cylinder above.)

Volume of a Sphere

A **sphere** is a solid shape with a curved surface, and every point on its surface is equal distance from its center. The volume of a sphere is $V = 4/3\pi r^3$.

Example 46-3

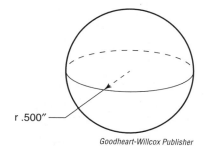

r .500″

Goodheart-Willcox Publisher

$$V = 4/3\pi r^3$$
$$V = 4/3 \times \pi \times .500^3$$
$$V = .524 \text{ in}^3$$

Work Space/Notes

Unit 46 Review

Name _____ **Date** _____ **Class** _____

Fill in the blanks in the following review questions.

1. A _____ is a solid shape that has at least one surface that is not flat. Cylinders, cones, and spheres are examples of _____.

2. A _____ is similar to a prism. It has flat ends or bases that are circles, and it has length.

3. The _____ of a cylinder is found by multiplying the _____ of the circular end by the length. The formula for _____ of a cylinder is V = _____.

4. A _____ is a solid shape with a flat circular base and sides tapering to a point or vertex. The formula for finding the volume of a _____ is very similar to the formula for the volume of a cylinder.

5. The _____ tapers to a point at one end, so its volume is _____ that of a cylinder with the same radius. The volume formula for a cone is _____.

6. A _____ is a solid shape with a curved surface, and every point on its surface is _____ distance from its center. The volume of a _____ is V = $4/3\pi r^3$.

Practice

Compute the volumes of the non-polyhedrons shown to the nearest thousandth inch or hundredth millimeter.

⌀14.5″

16.0″

Goodheart-Willcox Publisher

r 25 mm

127 mm

Goodheart-Willcox Publisher

1. _____

2. _____

Radius = .458″
Length = 3.250″

Goodheart-Willcox Publisher

Radius = 6.35 mm
Length = 75 mm

Goodheart-Willcox Publisher

3. _____

4. _____

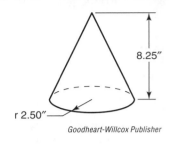

8.25″

r 2.50″

Goodheart-Willcox Publisher

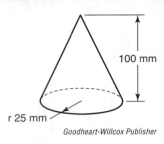

100 mm

r 25 mm

Goodheart-Willcox Publisher

5. _____

6. _____

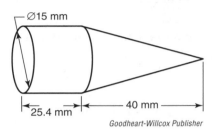

⌀15 mm

25.4 mm

40 mm

Goodheart-Willcox Publisher

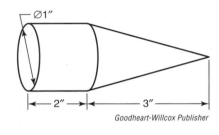

⌀1″

2″

3″

Goodheart-Willcox Publisher

7. _____

8. _____

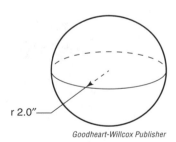

r 2.0″

Goodheart-Willcox Publisher

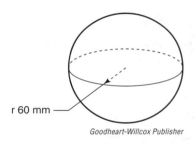

r 60 mm

Goodheart-Willcox Publisher

9. _____

10. _____

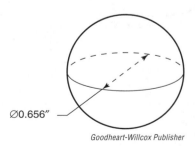

⌀0.656″

Goodheart-Willcox Publisher

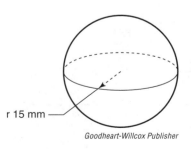

r 15 mm

Goodheart-Willcox Publisher

11. _____

12. _____

Name _____   **Date** _____   **Class** _____

Applications

1. A water-soluble coolant is delivered to the shop in a steel drum. The drum has a diameter of 22″ and is 34″ in height. What is the volume of the steel drum?

    Volume in in³: _____

2. To convert cubic inches to US liquid gallons, divide the volume value by 231. How many gallons of coolant concentrate can the steel drum from question 1 hold? Round to the nearest whole number.

    Gallons: _____

3. What is the volume of material in the punch pin below?

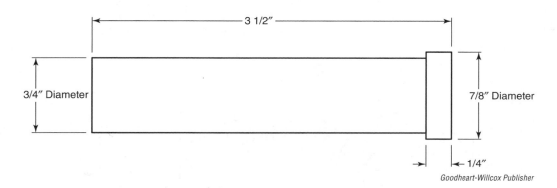

Goodheart-Willcox Publisher

    Volume in in³: _____

4. In the washer below, the O.D. is 4.5″, the I.D. is 2.625″, and the thickness is 0.3″.

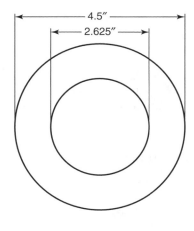

Goodheart-Willcox Publisher

    What is the volume of material in one washer? _____

    What is the volume of material in 50 washers? _____

5. In the pipe below, the O.D. is 2.5″, the I.D. is 2.25″, and the length is 6″. Calculate the volume.

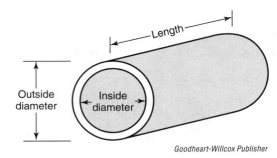

Goodheart-Willcox Publisher

Volume in in³: _____

6. Using the following drawing of a sleeve washer, find the volume of material in one sleeve washer.

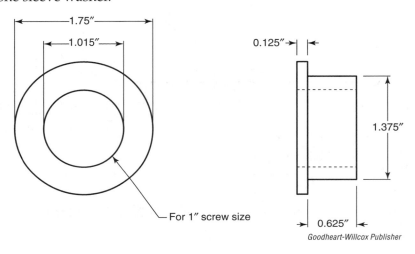

For 1″ screw size

Goodheart-Willcox Publisher

Volume in in³: _____

7. Three thousand of the nuts shown in the figure below must be shipped to a customer. If the steel used to make the nuts weighs .280 lb per cubic inch, what is the weight of the shipment to the nearest ounce? (There are 16 ounces in a pound.)

Ø.563″

Goodheart-Willcox Publisher

_____

Name _____   **Date** _____   **Class** _____

8. One thousand six hundred precisely ground steel ball bearings that have a radius of .3750" are to be stored in a canister of oil. The canister contains one gallon (231 cubic inches) of oil. Which is the smallest canister that would hold the bearings and the oil?

   Small: 6" diameter × 10" deep

   Medium: 8" diameter × 12" deep

   Large: 10" diameter × 16" deep

   _____

9. A 2 1/2" thick cast iron plate has six cones bored into each side to reduce the weight of the plate. The cones are 5" in diameter and 1 1/2" deep. To the nearest hundredth, how many cubic inches of material have been removed?

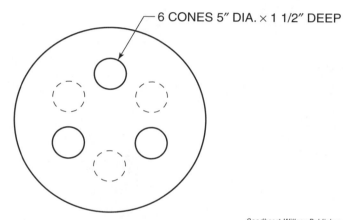

6 CONES 5" DIA. × 1 1/2" DEEP

Goodheart-Willcox Publisher

   _____

Work Space/Notes

Coordinate Systems

Key Terms

absolute positioning

Cartesian coordinate
  system

incremental positioning

plane

UNIT 47

CNC Milling

Objectives

Information in this unit will enable you to:

- Know what the Cartesian coordinate system is and identify the common four quadrants.
- Understand a three-dimensional coordinate system.
- Describe the difference between absolute and incremental positioning when using a CNC machine tool.

Cartesian Coordinate System

The **Cartesian coordinate system**, an idea published in 1637 by French mathematician René Descartes, is a system made of planes or lines that run horizontally and vertically to one another with perpendicular (90°) intersections. It is often referred to as the rectangular coordinate system. A **plane** is a two-dimensional flat surface that theoretically extends indefinitely.

The following figure shows the intersection of the XY, XZ, and YZ planes in a three-dimensional coordinate system. The three planes intersect each other at 90° and form the X axis, Y axis, and Z axis.

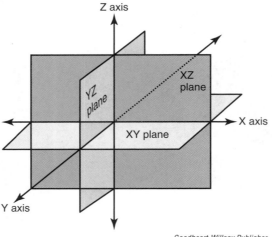

Goodheart-Willcox Publisher

In machining applications, the range of travel is limited by the model or size of the machine. The top of the worktable of a mill, where the vice is attached, could be thought of as a plane. (Although theoretically a plane has no thickness.)

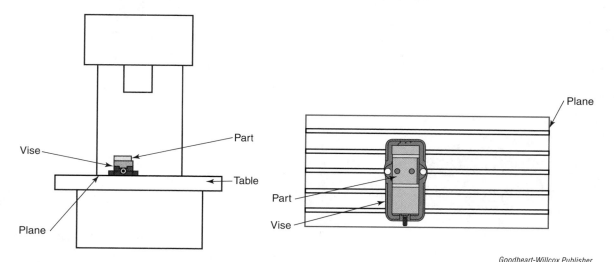

Goodheart-Willcox Publisher

Coordinates with CNC Machining Centers (Mills)

In computer numerical control (CNC) programming, movements of the machine, whether rapid movements, linear movements, or circular interpolation movements, can all be stated within the program using the Cartesian coordinate system. The ability to establish these positions or points within three-dimensional space is crucial for creating and understanding CNC programs.

Each axis can be thought of as a number line. The horizontal number line—movement going left or right—represents the X axis on a mill and has an origin or zero-reference position, often called an absolute zero. The absolute zero may be placed at any point along the line. Either side of the absolute zero are numbered increments. On a CNC machine, those increments are often 0.0001″ (one ten-thousandth of an inch). To the left of absolute zero, the increments are represented by negative increments or negative numbers. To the right of absolute zero, the increments are positive. When machining, the numbers within the program usually represent inches or millimeters, depending on what mode the machine is set up for or what G-code has been activated within the program.

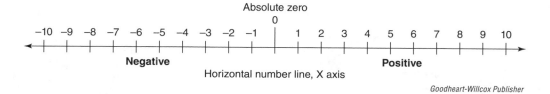

Goodheart-Willcox Publisher

Example 47-1

In this example, the position has a value of –2.

Goodheart-Willcox Publisher

Example 47-2

In this example, the position has a value of +3.

Goodheart-Willcox Publisher

When working within a two-axis coordinate system (plane), there are two number lines perpendicular to one another. The number lines are used to determine the distance and direction of a point or position from the absolute zero (origin). The position where the number lines (axes) intersect is taken as the origin for both.

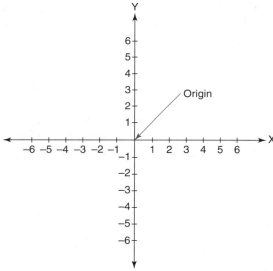

Goodheart-Willcox Publisher

The coordinates are usually written as two numbers in parentheses, separated by a comma, as in (2, –2). In mathematics, this is defined as an ordered pair. The horizontal number line, the X axis, is always given first, followed by the vertical number line, in this case the Y axis, listed second.

Example 47-3

In the following example, position A has a location of (0, 0), position B (2, 1), position C (–1, 2), position D (–3, –1), and position E (1, –2).

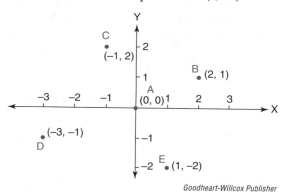

Goodheart-Willcox Publisher

The two axes divide the plane into four parts, or regions, called quadrants. The quadrants are often numbered from first to fourth and denoted by Roman numerals. The first quadrant is in the upper right, with the numbering going counterclockwise. By using positive and negative numbers or increments for positions, descriptions such as "left of zero" or "below zero" are avoided. The two-axis Cartesian coordinate system is a straightforward and precise way of telling the machine where to move.

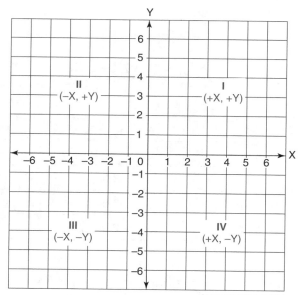

Goodheart-Willcox Publisher

I. Any point or position that falls in quadrant I will have positive values on the X axis and Y axis.

II. Any point or position that falls in quadrant II will have a negative value on the X axis and positive value on the Y axis.

III. Any point or position that falls in quadrant III will have negative values on both the X axis and Y axis.

IV. Any point or position that falls in quadrant IV will have a positive value on the X axis and negative value on the Y axis.

In CNC programs, an address character needs to be specified for the appropriate axis to move, so ordered pairs cannot be used. The X address character is used to indicate motion for the X axis, specifying a position or distance along the X axis. The Y address character is used to indicate motion for the Y axis, specifying a position or distance along the Y axis. The address character is followed by a positive or negative (+/−) number, specified in inches or millimeters. If no decimal point is entered, the last digit is assumed to be 1/10,000 inch or 1/1,000 mm (1 micron or μm).

Example 47-4

In the following example, we have a two-axis coordinate system with four positions plotted. The table shows the position of each point location from the origin point. If each square represents one inch, the values are as follows:

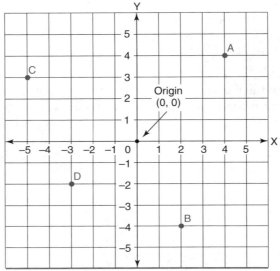

Goodheart-Willcox Publisher

Position	Ordered Pair	X Coordinate	Y Coordinate
A	(4, 4)	4.0	4.0
B	(2, −4)	2.0	−4.0
C	(−5, 3)	−5.0	3.0
D	(−3, −2)	−3.0	−2.0

Goodheart-Willcox Publisher

Omitting the decimal point for position A and writing X4 Y4 in a CNC program would have the machine move to a position of (X0.0004, Y0.0004), a movement too small for the naked eye to distinguish. It would look like the tool was still at the origin point!

When we combine the three axes (X, Y, and Z) or three number lines together, we form the three-dimensional Cartesian coordinate system. The intersection point of the lines is the origin (absolute zero).

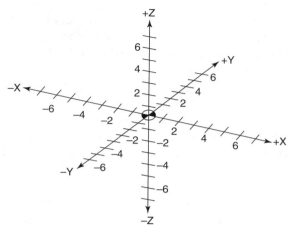

Goodheart-Willcox Publisher

You will see this symbol ⏚ on the examples and images in this unit to indicate where the PART ZERO or ORIGIN has been established.

A typical CNC machining center (mill) has three axes of movement, programmed as X, Y, and Z. The X movements consist of the cutting tool moving either left or right. The X movement has the greatest amount of travel on the machine. The Y axis is the in and out, or forward and back, movement of the cutting tool. The Z axis moves the cutting tool up or down.

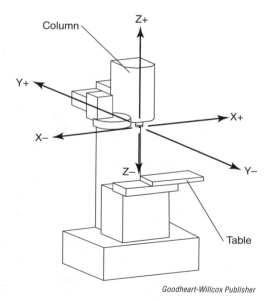

Goodheart-Willcox Publisher

There are two methods of specifying a location or position to the machine control unit (MCU) of a CNC machine: absolute positioning and incremental positioning. Most machines can use both systems, but absolute positioning is the most common positioning system. Remember, when moving the machine, we are concerned with positioning the center of the spindle (cutting tool) in relation to X, Y, and Z zero. Although the machine table is the moving part for the X and Y axes, we must keep in mind our coordinates are based off the theoretical spindle movement.

Absolute Positioning

When using **absolute positioning**, all positions or points are located from the origin, or absolute zero, often referred to as a fixed reference point. Most CNC machines allow the origin point to be located at any convenient position the programmer determines. This could be the corner of a part, the center of a part, or even the center of a hole or boss on a CNC machining center. Once the origin point is determined, all points or positions are measured from this fixed reference point.

Example 47-5

In the following example, absolute positioning is used. The plate has 12 holes, with the origin in the bottom-left corner of the part. Given the origin point, we know all other positions on the part land in quadrant I, which means they will have positive numbers as coordinates.

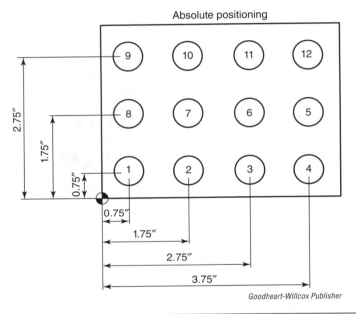

Absolute positioning

Goodheart-Willcox Publisher

Hole #	X Coordinate	Y Coordinate
1	0.75	0.75
2	1.75	0.75
3	2.75	0.75
4	3.75	0.75
5	3.75	1.75
6	2.75	1.75
7	1.75	1.75
8	0.75	1.75
9	0.75	2.75
10	1.75	2.75
11	2.75	2.75
12	3.75	2.75

Goodheart-Willcox Publisher

Incremental Positioning

In **incremental positioning**, the reference point constantly changes. That is, there is no fixed reference point. The machine tool moves to the next position relative to its current position, which is why incremental positioning is often referred to as *relative positioning*. The machine will read movement coordinates as if its current position is the origin, and the coordinate system shifts with each new position. This can also be considered a floating reference point.

When working in incremental positioning, the four quadrants discussed earlier do not necessarily determine whether the move is positive or negative. The specified direction along the axis (or number line) from the machine's current position determines if it's a plus (+) or minus (–) movement. To recap, after the machine has moved to a new location, the position where the tool is now is the new zero point from which the next move is to be made.

The same part used in Example 47-5 for absolute positioning is used in the example that follows, except with incremental positioning used in this case.

Example 47-6

The plate has 12 holes, with the origin in the bottom-left corner of the part. Incremental positioning is used to indicate the coordinates for each of the 12 successive holes.

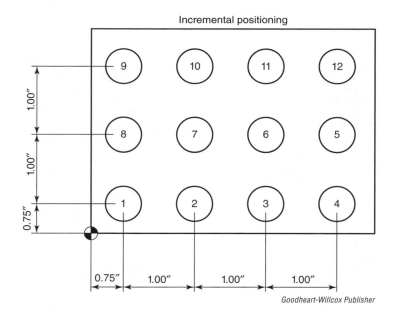

Incremental positioning

Goodheart-Willcox Publisher

(Continued)

Hole #	X Coordinate	Y Coordinate
Distance from origin to hole 1	0.75	0.75
Distance from hole 1 to hole 2	1.0	0
Distance from hole 2 to hole 3	1.0	0
Distance from hole 3 to hole 4	1.0	0
Distance from hole 4 to hole 5	0	1.0
Distance from hole 5 to hole 6	−1.0	0
Distance from hole 6 to hole 7	−1.0	0
Distance from hole 7 to hole 8	−1.0	0
Distance from hole 8 to hole 9	0	1.0
Distance from hole 9 to hole 10	1.0	0
Distance from hole 10 to hole 11	1.0	0
Distance from hole 11 to hole 12	1.0	0

As stated previously, even though the whole plate (part) is in quadrant I, some of the X values are still negative because of the use of incremental positioning.

Unit 47 Review

Name _____ **Date** _____ **Class** _____

Fill in the blanks in the following review questions.

1. The _____ _____ system is a system made of planes or lines that run horizontally and vertically to one another with perpendicular (90°) intersections.

2. In _____ _____ _____ (CNC) programming, movements of the machine can be stated within the program using the Cartesian coordinate system.

3. The ability to establish positions or points within three-dimensional space is crucial for creating and understanding _____ programs.

4. Two axes divide a plane into four parts, or regions, called _____, which are often numbered from first to fourth and denoted by Roman numerals. The first _____ is in the upper right, with the numbering going counterclockwise.

5. When using _____ positioning, all positions or points are located from the origin, or absolute zero, often referred to as a fixed reference point.

6. In _____ positioning, the reference point constantly changes. The machine tool moves to the next position relative to its current position.

Practice

Plot the points using a two-axis Cartesian coordinate system (XY plane).

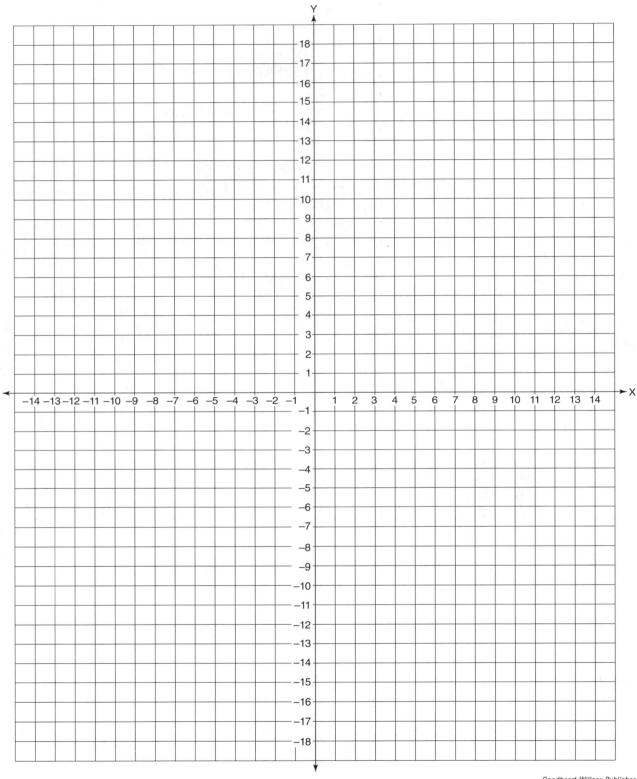

Goodheart-Willcox Publisher

Name _____   **Date** _____   **Class** _____

1. Plot the following coordinates using absolute coordinates:

A = X3.0 Y2.0   B = X–5.0 Y–4.0   C = X7.0 Y–8.0

D = X–12.0 Y–17.0   E = X13.0 Y1.0   F = X1.0 Y7.0

G = X–4.0 Y16.0   H = X6.0 Y–15.0   I = X2.0 Y–3.0

J = X8.0 Y9.0   K = X–9.0 Y–6.0   L = X–11.0 Y–6.0

M = X5.0 Y8.0   N = X–10.0 Y–2.0   O = X14.0 Y3.0

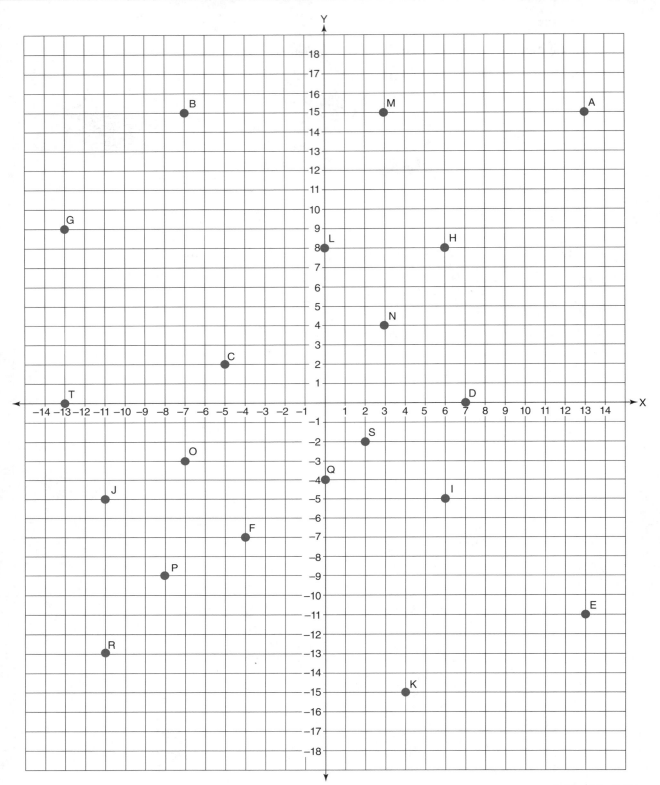

Name _____   **Date** _____   **Class** _____

2. What is the X value and Y value for each position using absolute positioning?

A = X____ Y____   B = X____ Y____   C = X____ Y____

D = X____ Y____   E = X____ Y____   F = X____ Y____

G = X____ Y____   H = X____ Y____   I = X____ Y____

J = X____ Y____   K = X____ Y____   L = X____ Y____

M = X____ Y____   N = X____ Y____   O = X____ Y____

P = X____ Y____   Q = X____ Y____   R = X____ Y____

S = X____ Y____   T = X____ Y____

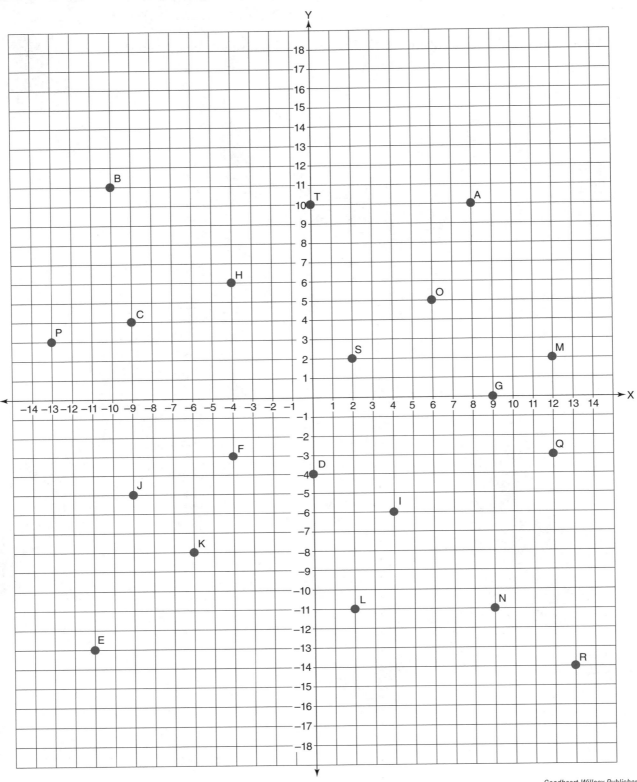

Name _____ **Date** _____ **Class** _____

3. What is the X value and Y value for each move using incremental positioning?

A → B = X____ Y____   B → C = X____ Y____   C → D = X____ Y____

D → E = X____ Y____   E → F = X____ Y____   F → G = X____ Y____

G → H = X____ Y____   H → I = X____ Y____   I → J = X____ Y____

J → K = X____ Y____   K → L = X____ Y____   L → M = X____ Y____

M → N = X____ Y____   N → O = X____ Y____   O → P = X____ Y____

P → Q = X____ Y____   Q → R = X____ Y____   R → S = X____ Y____

S → T = X____ Y____   T → A = X____ Y____

Applications

Use the following part drawing to answer questions 1 and 2.

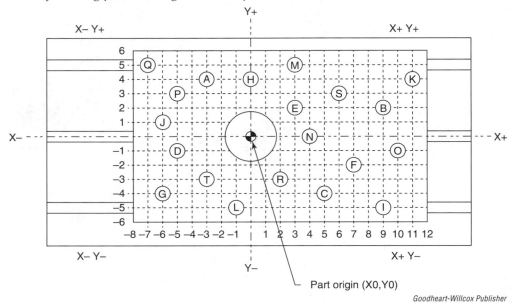

Part origin (X0,Y0)

1. What is the X value and Y value for each hole position using absolute positioning?

Hole Position	X Coordinate	Y Coordinate
A		
B		
C		
D		
E		
F		
G		
H		
I		
J		
K		
L		
M		
N		
O		
P		
Q		
R		
S		
T		

Name _____ **Date** _____ **Class** _____

2. What is the value of each move on the X and Y planes for each hole position using incremental positioning?

Hole Position	X Coordinate	Y Coordinate
Origin → A		
A → B		
B → C		
C → D		
D → E		
E → F		
F → G		
G → H		
H → I		
I → J		
J → K		
K → L		
L → M		
M → N		
N → O		
O → P		
P → Q		
Q → R		
R → S		
S → T		

Goodheart-Willcox Publisher

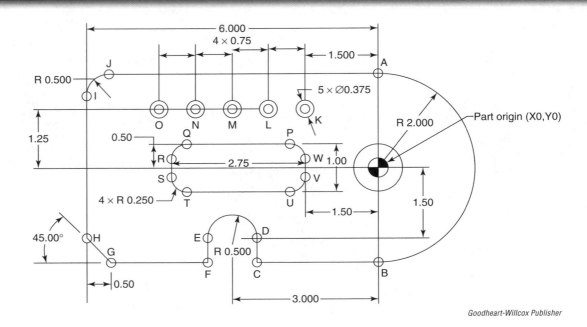

Goodheart-Willcox Publisher

3. What is the X value and Y value for each position using absolute positioning?

Hole Position	X Coordinate	Y Coordinate
A		
B		
C		
D		
E		
F		
G		
H		
I		
J		
K		
L		
M		
N		
O		
P		
Q		
R		
S		
T		
U		
V		
W		

Goodheart-Willcox Publisher

UNIT 48

CNC Turning

Objectives

Information in this unit will enable you to:

- Know what axis and plane are used when programming a CNC turning center.
- Understand the two-dimensional coordinate system typically used on a CNC turning center.
- Describe the difference between absolute and incremental positioning when using a CNC turning center.

CNC Turning Coordinate System

The two-dimensional Cartesian coordinate system for the CNC turning center uses the X and Z axes, or the XZ plane. A typical CNC turning center has two axes, whereas a CNC machining center has three axes (X, Y, and Z).

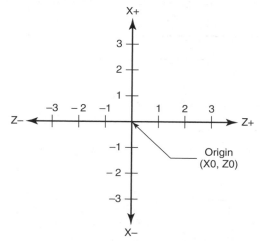

Goodheart-Willcox Publisher

The axis that moves left-to-right, or the horizontal axis, is the Z axis on a turning center. The Z axis controls lengths and/or depths. Movements to the left are negative increments or numbers, with movements to the right being positive increments or numbers. If we place a similar axis front-to-back, or the vertical axis on the machine, this is the X axis. Movement toward the operator is negative increments or numbers, and movement away from the operator is positive increments or numbers. The X axis controls the diameter of features, such as outside diameters (ODs) and inside diameters (IDs).

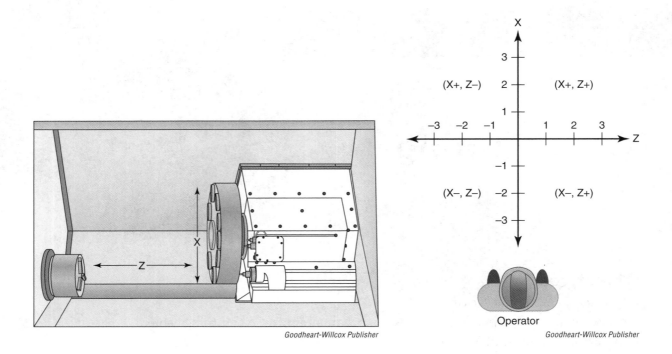

Goodheart-Willcox Publisher

Goodheart-Willcox Publisher

Due to the part rotating on a CNC turning center, the tool only needs to cut on one side. The Z origin is usually set on the front face of the part, and the X origin is set on the centerline of the spindle. Alternatively, some programmers/machines set the origin on the back of the part or the face of the lathe chuck.

When the origin is on the front face of the part, all X-axis cutting will be in a positive range of travel, whereas the Z-axis cutting would be in the negative range of travel (quadrant II). When the origin is on the back face of the part or the face of the chuck, all X- and Z-axis cutting will be in the positive range of travel (quadrant I).

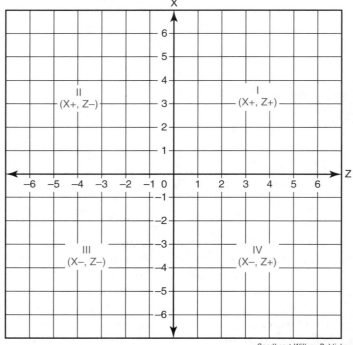

Goodheart-Willcox Publisher

Although both ways of setting the origin point have their advantages, the front face is usually used, as it is typically easier to set the tool length offsets. It is exceedingly rare to ever be working in quadrants III and IV.

Example 48-1

The following drawing shows an example of the origin or part zero being set on the front face of the part.

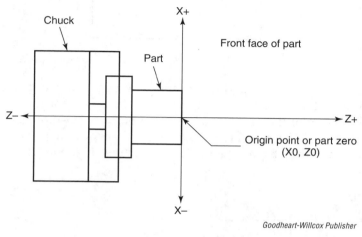

Goodheart-Willcox Publisher

Absolute Positioning

As with the CNC machining center (mill), when using absolute positioning, all positions or points are located from the origin (absolute zero), often referred to as a fixed reference point. In absolute positioning, all coordinate points or positions are given in X-axis and Z-axis commands in relationship to the fixed reference point, part zero or origin point. This is the most common type of positioning.

In the following images, we have a typical lathe setup and a close-up dimensioned drawing of the part.

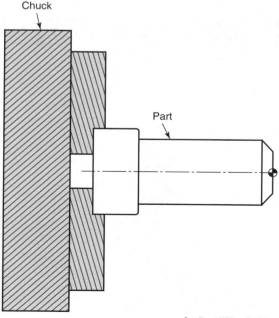

Goodheart-Willcox Publisher

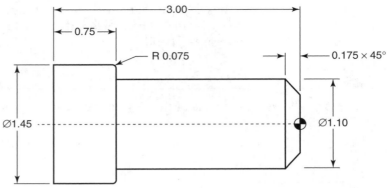

We can now take our part and record the geometry points or positions needed to program the part. As the center of the part is the origin for the X axis, the distance from the center of the part (origin) to each point or position is a radial value. As stated earlier, the X axis controls diameters, and that is what needs to be programmed. So X-axis moves are defined as the actual diameter value and not the radius from center.

Example 48-2

Here is an example of absolute positioning for turning a part.

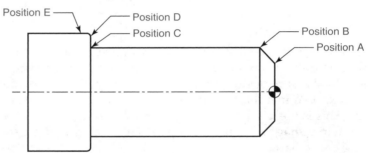

Position	X Coordinate (Diameter)	Z Coordinate
A	⌀0.75	0
B	⌀1.1	−0.175
C	⌀1.1	−2.25
D	⌀1.3	−2.25
E	⌀1.45	−2.325

Incremental Positioning

When working with incremental positioning on a turning center, the reference point constantly changes, just as it does on a machining center (mill). That is, there is no fixed reference point. The machine tool moves to the next position relative to its current position. The machine will read movement coordinates as if its current position is the origin, and the coordinate system shifts with each new position. Incremental positioning on a turning center is, however, a bit different.

The difference with a turning center compared to a machining center is that the characters U and W represent incremental moves. The U character is used to specify incremental motion in the X axis, and the W character is used to specify incremental motion in the Z axis. Another important consideration is that any incremental move on the X axis, using the character U, is a radial move. When programming for diameters, the radial distance or length needs to be doubled to account for the diametric size!

Example 48-3

Here is an example of incremental positioning for turning the part we featured earlier.

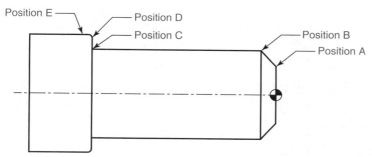

Goodheart-Willcox Publisher

Position	U Value	W Value
Origin → A	0.75	0
A → B	0.35	−0.175
B → C	0	−2.075
C → D	0.2	0
D → E	0.15	−0.075

Goodheart-Willcox Publisher

Work Space/Notes

Unit 48 Review

Name _____ **Date** _____ **Class** _____

Fill in the blanks in the following review questions.

1. The two-dimensional Cartesian coordinate system for the CNC turning center uses the _____ and _____ axes, or the _____ plane.

2. The axis that moves left-to-right, or the horizontal axis, is the _____ axis on a turning center. The _____ axis controls lengths and/or depths.

3. The _____ axis controls the diameter of features, such as outside diameters (ODs) and inside diameters (IDs).

4. In _____ positioning, all coordinate points or positions are given in X-axis and Z-axis commands in relationship to the fixed reference point, part zero or origin point.

5. With a turning center, the characters _____ and _____ represent incremental moves. The _____ character is used to specify incremental motion in the X axis, and the _____ character is used to specify incremental motion in the Z axis.

6. An important consideration is that any incremental move on the _____ axis using the character U is a radial move. When programming for diameters, the radial distance needs to be doubled.

Practice

Identify the plotted points using the two-axis Cartesian coordinate system (XZ plane).

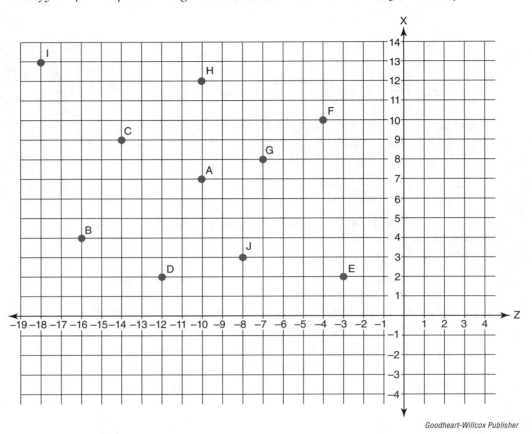

1. What is the X value and Z value for each position using absolute positioning?

   A = X_____ Z_____   B = X_____ Z_____   C = X_____ Z_____

   D = X_____ Z_____   E = X_____ Z_____   F = X_____ Z_____

   G = X_____ Z_____   H = X_____ Z_____   I = X_____ Z_____

   J = X_____ Z_____

Name _____   **Date** _____   **Class** _____

Use the following part drawing to determine the numbered points using absolute and incremental positioning.

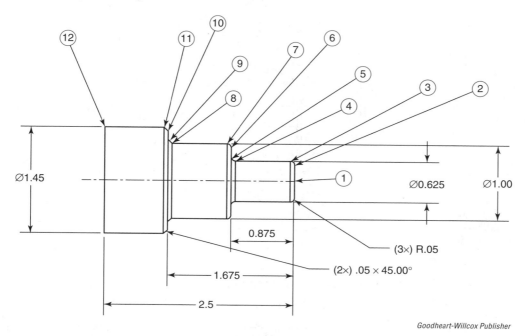

Goodheart-Willcox Publisher

2. What is the X value and Z value for each point or position using absolute positioning?

	Point or Position	X Coordinate (Diameter)	Z Coordinate
A.	1		
B.	2		
C.	3		
D.	4		
E.	5		
F.	6		
G.	7		
H.	8		
I.	9		
J.	10		
K.	11		
L.	12		

Goodheart-Willcox Publisher

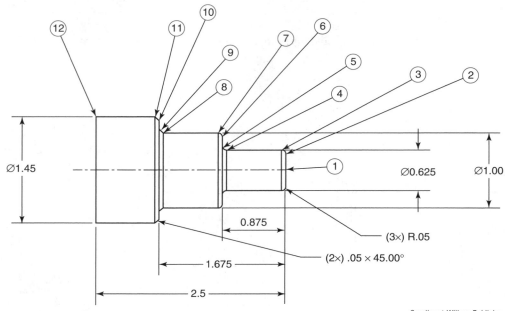

Goodheart-Willcox Publisher

3. What is the U value and W value for each point or position using incremental positioning?

	Point or Position	U Value	W Value
A.	1		
B.	2		
C.	3		
D.	4		
E.	5		
F.	6		
G.	7		
H.	8		
I.	9		
J.	10		
K.	11		
L.	12		

Goodheart-Willcox Publisher

Name _____   **Date** _____   **Class** _____

Applications

Use the following drawing of a toolpath to answer questions 1 and 2, providing coordinates for the lettered points.

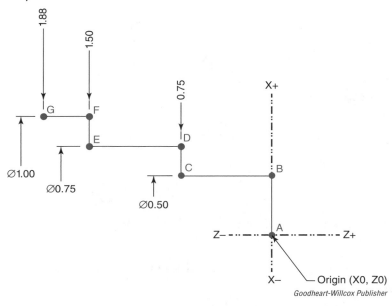

Goodheart-Willcox Publisher

1. What is the X value and Z value for each point or position using absolute positioning?

	Point or Position	X Coordinate (Diameter)	Z Coordinate
A.	A		
B.	B		
C.	C		
D.	D		
E.	E		
F.	F		
G.	G		

Goodheart-Willcox Publisher

2. What is the U value and W value for each point or position using incremental positioning?

	Point or Position	U Value	W Value
A.	A		
B.	B		
C.	C		
D.	D		
E.	E		
F.	F		
G.	G		

Goodheart-Willcox Publisher

Use the following drawing of a toolpath to answer questions 3 and 4, providing coordinates for the lettered points.

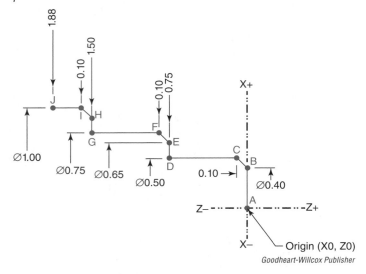

Goodheart-Willcox Publisher

3. What is the X value and Z value for each point or position using absolute positioning?

	Point or Position	X Coordinate (Diameter)	Z Coordinate
A.	A		
B.	B		
C.	C		
D.	D		
E.	E		
F.	F		
G.	G		
H.	H		
I.	I		
J.	J		

Goodheart-Willcox Publisher

4. What is the U value and W value for each point or position using incremental positioning?

	Point or Position	U Value	W Value
A.	A		
B.	B		
C.	C		
D.	D		
E.	E		
F.	F		
G.	G		
H.	H		
I.	I		
J.	J		

Goodheart-Willcox Publisher

Name _____   **Date** _____   **Class** _____

Use the following drawing of a toolpath to answer question 5, providing coordinates for the lettered points.

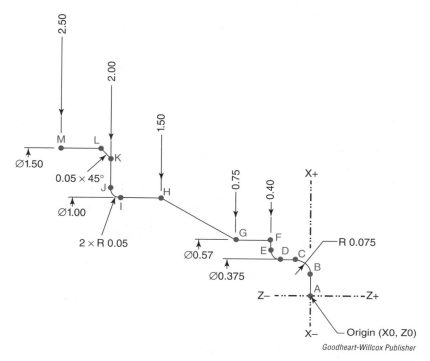

Goodheart-Willcox Publisher

5. What is the X value and Z value for each point or position using absolute positioning?

	Point or Position	X Coordinate (Diameter)	Z Coordinate
A.	A		
B.	B		
C.	C		
D.	D		
E.	E		
F.	F		
G.	G		
H.	H		
I.	I		
J.	J		
K.	K		
L.	L		
M.	M		

Goodheart-Willcox Publisher

Use the following drawing of a toolpath to answer question 6, providing coordinates for the lettered points.

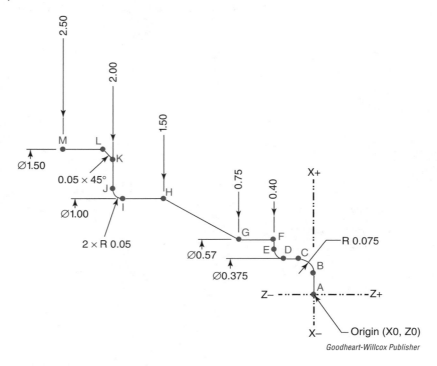

Goodheart-Willcox Publisher

6. What is the U value and W value for each point or position using incremental positioning?

	Point or Position	U Value	W Value
A.	A		
B.	B		
C.	C		
D.	D		
E.	E		
F.	F		
G.	G		
H.	H		
I.	I		
J.	J		
K.	K		
L.	L		
M.	M		

Goodheart-Willcox Publisher

UNIT 49

Bolt Circles

Objectives

Information in this unit will enable you to:

- Know how to create right triangles and use trigonometry with bolt circles.
- Understand how bolt circle positions can be described using coordinates.
- Describe how quadrants in the coordinate system affect the coordinates of holes in a bolt circle.

Trigonometry and Coordinates with Bolt Circles

Another common application of right-angle trigonometry is in work with bolt circles. A bolt circle is a theoretical circle (a circular centerline) on which the center points of holes are positioned. The holes are often equally spaced. An example of a bolt circle is the circle of holes on a car's wheel (lug holes) that accepts the lug nuts to securely fasten the wheel to the vehicle.

Goodheart-Willcox Publisher

Bolt circles are often drawn with one of the holes placed on the vertical or horizontal centerline.

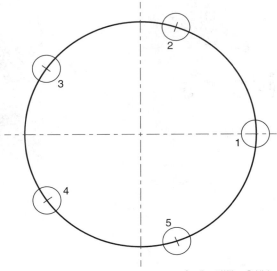

Goodheart-Willcox Publisher

Alternatively, a hole in the bolt circle may be dimensioned angularly to one of the centerlines.

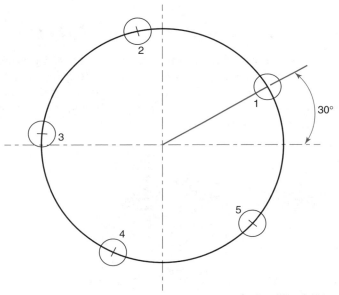

Goodheart-Willcox Publisher

The diameter of the bolt circle is usually given as a dimension, sometimes abbreviated BC, for bolt circle.

When holes are equally spaced on a bolt circle, the angular spacing between holes can be calculated. As a circle contains 360°, divide 360° by the number of holes:

$$\frac{360°}{\text{\# of holes}}$$

For a bolt circle that has five holes equally spaced:

$$\frac{360}{5} = 72°$$

There is a 72° angle between any two holes.

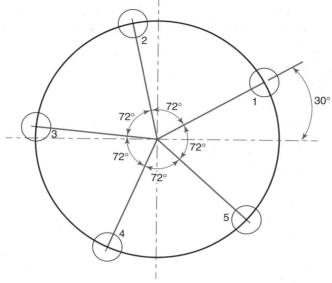

The line drawn from the center of the bolt circle to the center of each hole is a radius. The radius can be calculated by dividing the diameter by two (radius = Ø/2).

Using the radius and the angular position of each hole, a right-angle triangle can be created, and the linear distance of each hole from the center of the circle can be calculated. The radius will form the hypotenuse of each right-angle triangle. In this way, using trigonometry, the X and Y coordinates can then be found for each hole in the circle.

Example 49-1

Calculate the X and Y coordinates for the five holes on the following bolt circle with a diameter of 4″. Round answers to the nearest ten-thousandth of an inch.

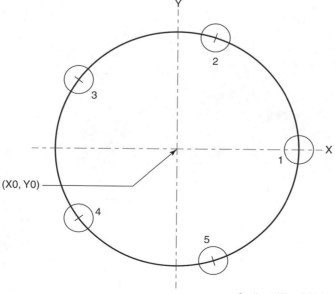

(Continued)

Hole 1

Hole 1 is placed on the horizontal centerline (the X axis). A line drawn from the center of the bolt circle to the center of hole 1 is a radius (which is half the diameter).

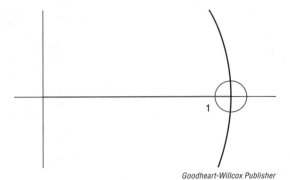

The radius $= \frac{4}{2}$.

The radius $= 2''$.

The coordinate distance from the center of the circle to hole 1 is X = 2.0 and Y = 0.

$$\text{Hole 1: X} = 2.0, \text{Y} = 0$$

Hole 2

A line drawn from the center of the bolt circle to the center of hole 2 is a radius.

The angle between any two holes is $\dfrac{360°}{\text{\# of holes}}$.

The angle between any two holes in this bolt circle is $\dfrac{360°}{5} = 72°$.

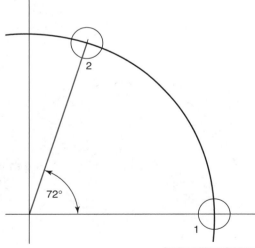

(Continued)

Drawing a vertical line from the center of hole 2 down to the horizontal centerline creates a right-angle triangle. The radius (hypotenuse) is 2″, and the angle is 72°.

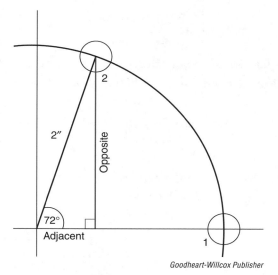

The adjacent side (horizontal axis, or "X") and opposite side (vertical axis, or "Y") can be calculated using trigonometry.

$$\text{adjacent} = \cos(\theta) \times \text{hypotenuse}$$
$$\text{opposite} = \sin(\theta) \times \text{hypotenuse}$$

"X" = adjacent = $\cos(\theta) \times$ hypotenuse.
"X" = adjacent = $\cos(72) \times 2$.

Use a scientific calculator.

The display shows:

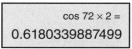

"X" = adjacent = 0.6180 rounded to the nearest ten-thousandth.
"Y" = opposite = $\sin(\theta) \times$ hypotenuse.
"Y" = opposite = $\sin(72) \times 2$.

Use a scientific calculator.

(Continued)

The display shows:

"Y" = opposite = 1.9021 rounded to the nearest ten-thousandth.

The coordinate distance from the center of the circle to hole 2 is X = 0.6180 and Y = 1.9021.

Hole 2: X = 0.6180, Y = 1.9021

Hole 3

As with hole 2, a line drawn from the center of the bolt circle to the center of hole 3 is a radius, and we know the angle between any two holes is 72° (360°/5 = 72°).

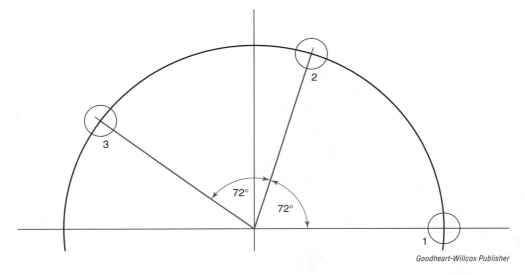

Drawing a vertical line from the center of hole 3 down to the horizontal center-line creates a right-angle triangle. The radius (hypotenuse) is 2", but the angle is unknown.

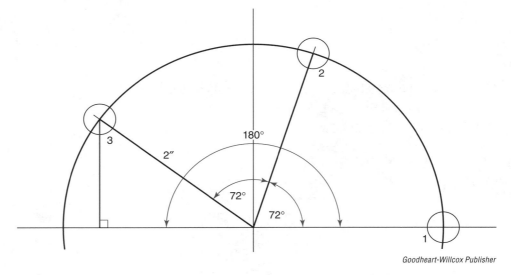

(Continued)

With what we know, we can calculate the angle inside the triangle.

There are 180° in a straight angle.

Hole 3 is 144° from hole 1 (72° + 72°).

The angle in the triangle is 180° − 144° = 36°.

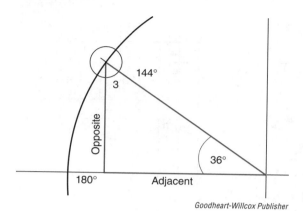

The adjacent side (horizontal axis, or "X") and opposite side (vertical axis, or "Y") can be calculated using trigonometry.

$$\text{adjacent} = \cos(\theta) \times \text{hypotenuse}$$
$$\text{opposite} = \sin(\theta) \times \text{hypotenuse}$$

"X" = adjacent = $\cos(\theta) \times$ hypotenuse.

"X" = adjacent = $\cos(36) \times 2$.

Use a scientific calculator.

The display shows:

"X" = adjacent = 1.6180 rounded to the nearest ten-thousandth.

(Continued)

Hole 3 is in quadrant II using the Cartesian coordinate system, so the "X" value is a negative value!

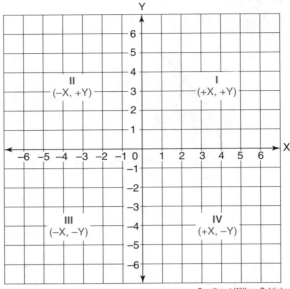

Goodheart-Willcox Publisher

"X" = adjacent = −1.6180

"Y" = opposite = sin(θ) × hypotenuse.

"Y" = opposite = sin(36) × 2.

Use a scientific calculator.

SIN 36° × 2 **ENTER =**

Goodheart-Willcox Publisher

The display shows:

sin 36 × 2 =

1.175570504585

Goodheart-Willcox Publisher

"Y" = opposite = 1.1756 rounded to the nearest ten-thousandth.

The coordinate distance from the center of the circle to hole 3 is X = −1.6180 and Y = 1.1756.

Hole 3: X = −1.6180, Y = 1.1756

(Continued)

Hole 4

As with the other holes, a line drawn from the center of the bolt circle to the center of hole 4 is a radius. We also know the angle between any two holes is 72° (360°/5 = 72°).

Drawing a vertical line from the center of hole 4 to the horizontal centerline creates a right-angle triangle. The radius (hypotenuse) is 2″, but the angle is unknown.

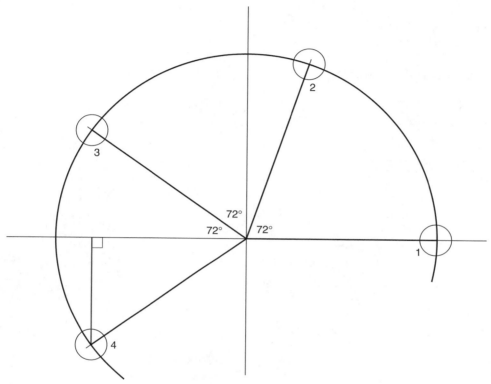

With what we know, we can calculate the angle inside the triangle.

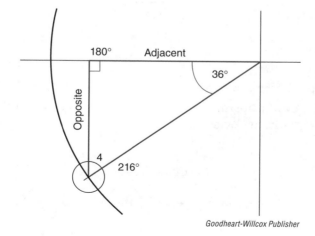

(Continued)

There are 180° in a straight angle.

Hole 4 is 216° from hole 1 (72° + 72° + 72°).

The angle in the triangle is 216° − 180° = 36°.

The adjacent side (horizontal axis, or "X") and opposite side (vertical axis, or "Y") can be calculated using trigonometry.

$$\text{adjacent} = \cos(\theta) \times \text{hypotenuse}$$
$$\text{opposite} = \sin(\theta) \times \text{hypotenuse}$$

"X" = adjacent = $\cos(\theta) \times$ hypotenuse.

"X" = adjacent = $\cos(36) \times 2$.

Use a scientific calculator.

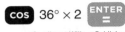

Goodheart-Willcox Publisher

The display shows:

cos 36 × 2 =
1.6180339887499

Goodheart-Willcox Publisher

"X" = adjacent = 1.6180 rounded to the nearest ten-thousandth.

"Y" = opposite = $\sin(\theta) \times$ hypotenuse.

"Y" = opposite = $\sin(36) \times 2$.

Use a scientific calculator.

Goodheart-Willcox Publisher

The display shows:

sin 36 × 2 =
1.175570504585

Goodheart-Willcox Publisher

"Y" = opposite = 1.1756 rounded to the nearest ten-thousandth.

Hole 4 is in quadrant III using the Cartesian coordinate system, so both the "X" value and "Y" value are negative values!

$$\text{"X"} = \text{adjacent} = -1.6180$$
$$\text{"Y"} = \text{opposite} = -1.1756$$

The coordinate distance from the center of the circle to hole 4 is X = −1.6180 and Y = −1.1756.

Hole 4: X = −1.6180, Y = −1.1756

(Continued)

Hole 5

Coordinates for the final hole are determined using the same process as the other holes. A line drawn from the center of the bolt circle to the center of hole 5 is a radius, and the angle between any two holes is 72° (360°/5 = 72°).

Drawing a vertical line from the center of hole 5 to the horizontal centerline creates a right-angle triangle. The radius (hypotenuse) is 2", and the angle is 72°.

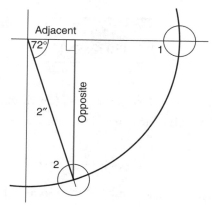

Goodheart-Willcox Publisher

The adjacent side (horizontal axis, or "X") and opposite side (vertical axis, or "Y") can be calculated using trigonometry.

$$\text{adjacent} = \cos(\theta) \times \text{hypotenuse}$$

$$\text{opposite} = \sin(\theta) \times \text{hypotenuse}$$

"X" = adjacent = $\cos(\theta) \times$ hypotenuse.
"X" = adjacent = $\cos(72) \times 2$.

Use a scientific calculator.

Goodheart-Willcox Publisher

The display shows:

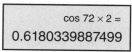

Goodheart-Willcox Publisher

"X" = adjacent = 0.6180 rounded to the nearest ten-thousandth.
"Y" = opposite = $\sin(\theta) \times$ hypotenuse.
"Y" = opposite = $\sin(72) \times 2$.

(Continued)

Use a scientific calculator.

The display shows:

"Y" = opposite = 1.9021 rounded to the nearest ten-thousandth.

Hole 5 is in quadrant IV using the Cartesian coordinate system, so the "Y" value is a negative value!

$$\text{"Y"} = \text{opposite} = -1.9021$$

The coordinate distance from the center of the circle to hole 5 is X = 0.6180 and Y = –1.9021.

$$\text{Hole 5: } X = 0.6180, Y = -1.9021$$

Hole Coordinates for Bolt Circle		
Hole #	**X Coordinate**	**Y Coordinate**
1	2.0000	0.0000
2	0.6180	1.9021
3	–1.6180	1.1756
4	–1.6180	–1.1756
5	0.6180	–1.9021

Example 49-2

Calculate the X and Y coordinates for the six holes on the following bolt circle with a diameter of 6″. Round answers to the nearest ten-thousandth of an inch.

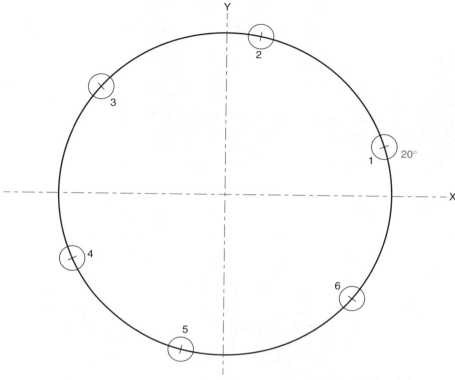

Goodheart-Willcox Publisher

The radius $= \dfrac{6}{2}$.

The radius $= 3″$.

The angle between any two holes in this bolt circle is $\dfrac{360°}{6} = 60°$.

Hole 1

A line drawn from the center of the bolt circle to the center of hole 1 creates the hypotenuse (radius).

Drawing a vertical line from the center of hole 1 to the horizontal centerline creates a right-angle triangle. The radius (hypotenuse) is 3″, and the angle is given as 20°.

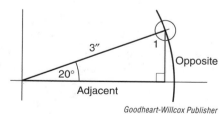

Goodheart-Willcox Publisher

(Continued)

Opposite = sin 20° × 3″ = 1.0261.

Adjacent = cos 20° × 3″ = 2.8191.

The opposite side is a vertical line, so it represents the "Y" axis.

The adjacent side is a horizontal line, so it represents the "X" axis.

Hole 1 is in quadrant I, so the values are both positive.

Hole 1: X = 2.8191, Y = 1.0261

Hole 2

A line drawn from the center of the bolt circle to the center of hole 2 creates the hypotenuse (radius).

Drawing a vertical line from the center of hole 2 to the horizontal centerline creates a right-angle triangle. The radius (hypotenuse) is 3″, and the angle is 80°. We arrive at 80° because we know the angle from the horizontal centerline to hole 1 is 20°, and the angle from hole 1 to hole 2 is 60° (20° + 60° = 80°).

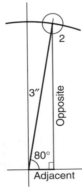

Goodheart-Willcox Publisher

Opposite = sin 80° × 3″ = 2.9544.

Adjacent = cos 80° × 3″ = 0.5209.

The opposite side is a vertical line, so it represents the "Y" axis.

The adjacent side is a horizontal line, so it represents the "X" axis.

Hole 2 is in quadrant I, so the values are both positive.

Hole 2: X = 0.5209, Y = 2.9544

Hole 3

A line drawn from the center of the bolt circle to the center of hole 3 creates the hypotenuse (radius).

Drawing a vertical line from the center of hole 3 to the horizontal centerline creates a right-angle triangle. The radius (hypotenuse) is 3″, and the angle is 40°. There are 180° in a straight angle, and we know hole 3 is 140° from the horizontal centerline (hole 1 is 20° from the centerline, and there are 60° between each hole; 20° + 60° + 60° = 140°).

(Continued)

The angle in the triangle is $180° - 140° = 40°$.

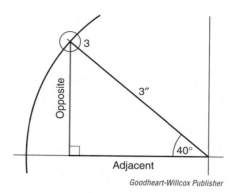

Goodheart-Willcox Publisher

Opposite = $\sin 40° \times 3'' = 1.9284$.

Adjacent = $\cos 40° \times 3'' = 2.2981$.

The opposite side is a vertical line, so it represents the "Y" axis.

The adjacent side is a horizontal line, so it represents the "X" axis.

Hole 3 is in quadrant II, so the X value is negative and the Y value positive.

Hole 3: X = −2.2981, Y = 1.9284

Hole 4

A line drawn from the center of the bolt circle to the center of hole 4 creates the hypotenuse (radius).

Drawing a vertical line from the center of hole 4 to the horizontal centerline creates a right-angle triangle. The radius (hypotenuse) is 3″, and the angle is 20°. There are 180° in a straight angle, and we know hole 4 is 200° from the horizontal centerline from which hole 1 is dimensioned (hole 1 is 20° from the centerline, and there are 60° between each hole; $20° + 60° + 60° + 60° = 200°$).

The angle in the triangle is $200° - 180° = 20°$.

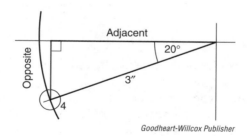

Goodheart-Willcox Publisher

Opposite = $\sin 20° \times 3'' = 1.0261$.

Adjacent = $\cos 20° \times 3'' = 2.8191$.

The opposite side is a vertical line, so it represents the "Y" axis.

The adjacent side is a horizontal line, so it represents the "X" axis.

Hole 4 is in quadrant III, so the values are both negative.

Hole 4: X = −2.8191, Y = −1.0261

(Continued)

Hole 5

A line drawn from the center of the bolt circle to the center of hole 5 creates the hypotenuse (radius).

Drawing a horizontal line from the center of hole 5 to the vertical centerline creates a right-angle triangle. The radius (hypotenuse) is 3″, and the angle is 10°. The vertical line is 270° from the reference line from which hole 1 is dimensioned, and we know hole 5 is 260° from that same reference line, the horizontal centerline (hole 1 is 20° from the centerline, and there are 60° between each hole; 20° + 60° + 60° + 60° + 60° = 260°).

The angle in the triangle is 270° − 260° = 10°.

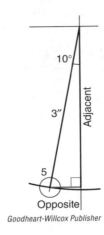

Goodheart-Willcox Publisher

Opposite = sin 10° × 3″ = 0.5209.

Adjacent = cos 10° × 3″ = 2.9544.

The opposite side is a horizontal line, so it represents the "X" axis.

The adjacent side is a vertical line, so it represents the "Y" axis.

Hole 5 is in quadrant III, so the values are both negative.

Hole 5: X = −0.5209, Y = −2.9544

Hole 6

A line drawn from the center of the bolt circle to the center of hole 6 creates the hypotenuse (radius).

Drawing a vertical line from the center of hole 6 to the horizontal centerline creates a right-angle triangle. The radius (hypotenuse) is 3″, and the angle is 40°. The angle between any two holes is 60°, and hole 1 is 20° from the horizontal centerline, so we can find the angle remaining between hole 1 and hole 6.

(Continued)

The angle in the triangle is 60° − 20° = 40°.

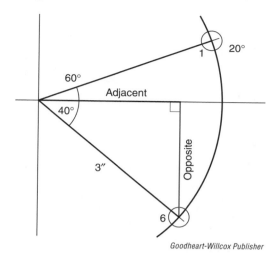

Goodheart-Willcox Publisher

Opposite = sin 40° × 3″ = 1.9284.

Adjacent = cos 40° × 3″ = 2.2981.

The opposite side is a vertical line, so it represents the "Y" axis.

The adjacent side is a horizontal line, so it represents the "X" axis.

Hole 6 is in quadrant IV, so the X value is positive and the Y value is negative.

Hole 6: X = 2.2981, Y = −1.9284

Hole Coordinates for Bolt Circle		
Hole #	X Coordinate	Y Coordinate
1	2.8191	1.0261
2	0.5209	2.9544
3	−2.2981	1.9284
4	−2.8191	−1.0261
5	−0.5209	−2.9544
6	2.2981	−1.9284

Goodheart-Willcox Publisher

Work Space/Notes

Name _____ Date _____ Class _____

Fill in the blanks in the following review questions.

1. Bolt circles are often drawn with one of the holes placed on the vertical or horizontal _____. Alternatively, a hole in the bolt circle may be dimensioned angularly to one of the _____.

2. The _____ diameter of the bolt circle is usually given as a dimension, sometimes abbreviated BC, for bolt circle.

3. When holes are equally spaced on a bolt circle, the angular spacing between holes can be calculated by dividing _____° by the number of holes.

4. A line drawn from the center of the bolt circle to the center of each hole is a _____. The _____ can be calculated by dividing the _____ by two.

5. Using the radius and angular position of each hole, right-angle triangles can be created, in which the radius becomes the _____, and the linear distance of each hole from the _____ of the circle can be calculated.

6. Using _____ with the right-angle triangles, the X and Y coordinates can be found for each hole in the circle.

7. The X and Y coordinates may be positive or negative, depending on the _____ in which the hole is located.

Practice

1. Calculate the X and Y coordinates for the three holes on the following bolt circle that has a diameter of 5″. Round answers to the nearest ten-thousandth of an inch.

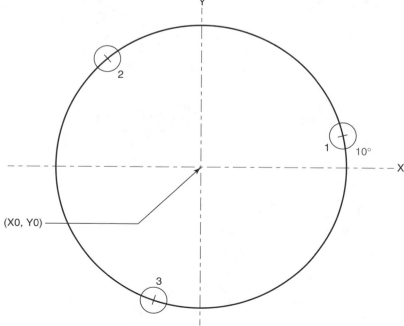

Goodheart-Willcox Publisher

	Hole #	X Coordinate	Y Coordinate
A.	1		
B.	2		
C.	3		

Goodheart-Willcox Publisher

2. Calculate the X and Y coordinates for the seven holes on the following bolt circle that has a diameter of 4.5″. Round answers to the nearest ten-thousandth of an inch.

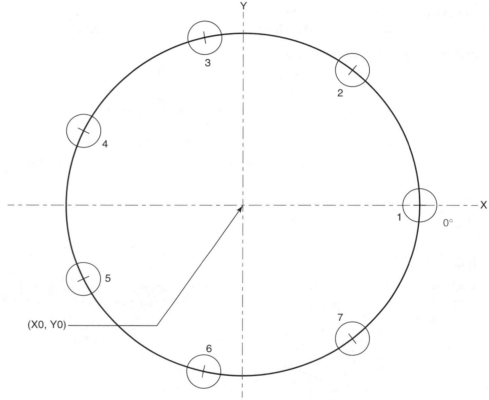

Goodheart-Willcox Publisher

	Hole #	X Coordinate	Y Coordinate
A.	1		
B.	2		
C.	3		
D.	4		
E.	5		
F.	6		
G.	7		

Goodheart-Willcox Publisher

Name _____ **Date** _____ **Class** _____

3. Calculate the X and Y coordinates for the nine holes on the following bolt circle that has a diameter of 7″. Round answers to the nearest ten-thousandth of an inch.

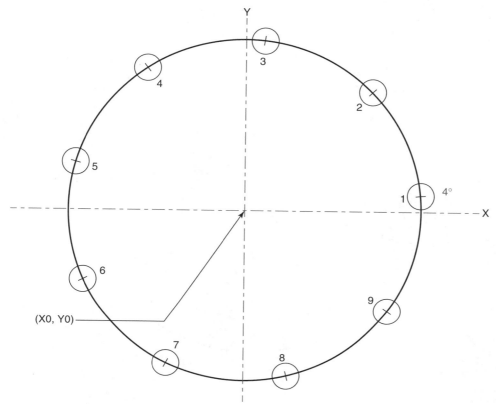

Goodheart-Willcox Publisher

	Hole #	X Coordinate	Y Coordinate
A.	1		
B.	2		
C.	3		
D.	4		
E.	5		
F.	6		
G.	7		
H.	8		
I.	9		

Goodheart-Willcox Publisher

4. Calculate the X and Y coordinates for the 11 holes on the following bolt circle that has a diameter of 8″. Round answers to the nearest ten-thousandth of an inch.

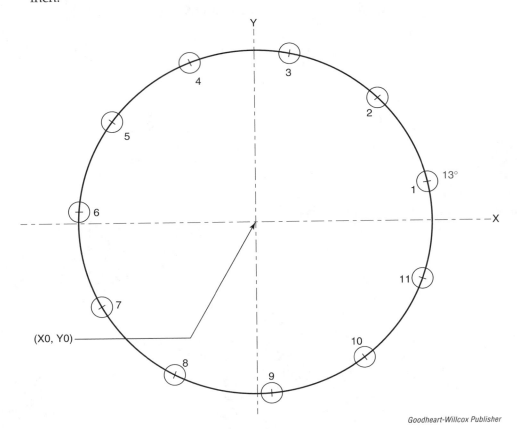

Goodheart-Willcox Publisher

	Hole #	X Coordinate	Y Coordinate
A.	1		
B.	2		
C.	3		
D.	4		
E.	5		
F.	6		
G.	7		
H.	8		
I.	9		
J.	10		
K.	11		

Goodheart-Willcox Publisher

Conversion Tables: US Customary and SI Metric

Converting US Customary to SI Metric

	When You Know:	Multiply By:		To Find:
		Very accurate	**Approximate**	
Length	inches	* 25.4	25.0	millimeters
	inches	* 2.54	2.5	centimeters
	feet	* 0.3048	0.3	meters
	feet	* 30.48	30.5	centimeters
	yards	* 0.9144	0.9	meters
	miles	* 1.609344	1.6	kilometers
Weight	grains	15.43236	15.4	grams
	ounces	* 28.349523125	28.0	grams
	ounces	* 0.028349523125	0.028	kilograms
	pounds	* 0.45359237	0.45	kilograms
	short ton	* 0.90718474	0.9	tonnes
Volume	teaspoons	* 4.92892159	5.0	milliliters
	tablespoons	* 14.7867648	15.0	milliliters
	fluid ounces	* 29.5735296	30.0	milliliters
	cups	* 0.236588236	0.24	liters
	pints	* 0.473176473	0.47	liters
	quarts	* 0.946352946	0.95	liters
	gallons	* 3.785411784	3.8	liters
	cubic inches	* 0.016387064	0.02	liters
	cubic feet	* 0.028316846592	0.03	cubic meters
	cubic yards	* 0.764554857984	0.76	cubic meters
Area	square inches	* 6.4516	6.5	square centimeters
	square feet	* 0.09290304	0.09	square meters
	square yards	* 0.83612736	0.8	square meters
	square miles		2.6	square kilometers
	acres	* 0.40468564224	0.4	hectares
Temperature	Fahrenheit	* $C = \dfrac{(F - 32) \times 5}{9}$	$C = \dfrac{F - 30}{2}$	Celsius

* = Exact

Converting SI Metric to US Customary

When You Know:	Multiply By:		To Find:
	Very accurate	**Approximate**	
Length			
millimeters	0.0393701	0.04	inches
centimeters	0.3937008	0.4	inches
meters	3.280840	3.3	feet
meters	1.093613	1.1	yards
kilometers	0.621371	0.6	miles
Weight			
grains	0.00228571	0.0023	ounces
grams	0.03527396	0.035	ounces
kilograms	2.204623	2.2	pounds
tonnes	1.1023113	1.1	short tons
Volume			
milliliters	0.202884	0.2	teaspoons
milliliters	0.06667	0.067	tablespoons
milliliters	0.03381402	0.03	fluid ounces
liters	61.02374	61.024	cubic inches
liters	2.113376	2.1	pints
liters	1.056688	1.06	quarts
liters	0.26417205	0.26	gallons
liters	0.03531467	0.035	cubic feet
cubic meters	61023.74	61023.7	cubic inches
cubic meters	35.31467	35.0	cubic feet
cubic meters	1.3079506	1.3	cubic yards
cubic meters	264.17205	264.0	gallons
Area			
square centimeters	0.1550003	0.16	square inches
square centimeters	0.00107639	0.001	square feet
square meters	10.76391	10.8	square feet
square meters	1.195990	1.2	square yards
square kilometers	0.3861	0.4	square miles
hectares	2.471054	2.5	acres
Temperature			
Celsius	*F = C × 1.8 + 32	F = C × 2 + 30	Fahrenheit

* = Exact

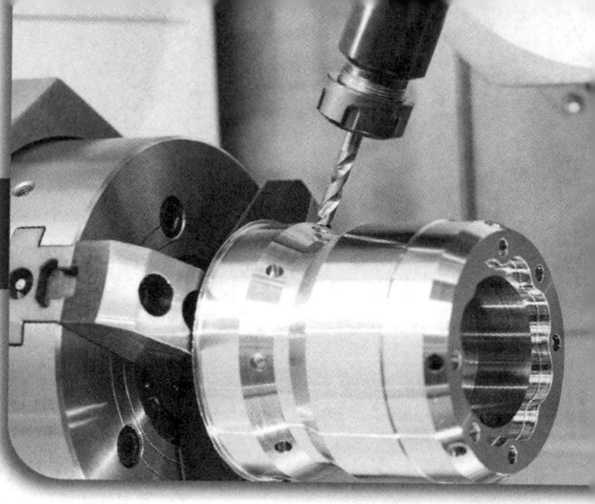

APPENDIX B

Formulas

Percent and Percentage (Unit 20)

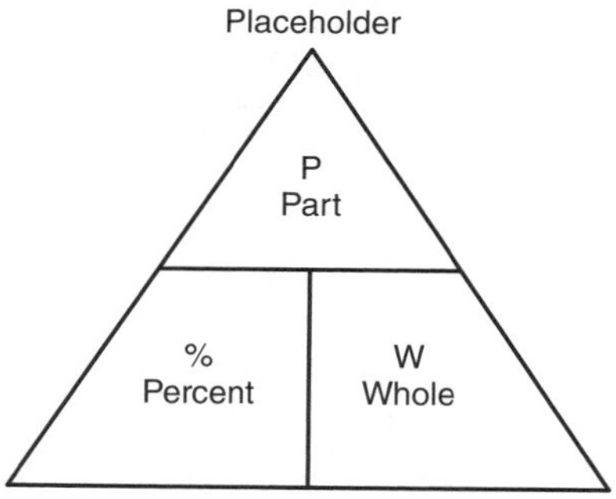
Placeholder
P
Part
%
Percent
W
Whole

Goodheart-Willcox Publisher

Ratio and Proportion (Unit 21)

Unknown mean = Product of extremes ÷ Known mean

Unknown extreme = Product of means ÷ Known extreme

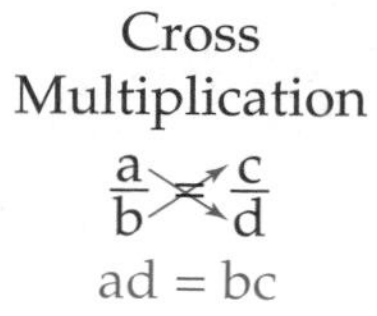
Cross
Multiplication
a
b
c
d
ad = bc

Goodheart-Willcox Publisher

Perimeter and Circumference (Unit 23)

Square	P = 4s
Rectangle or Parallelogram	P = 2l + 2w
Trapezoid	P = sum of all four sides
Circle (circumference)	C = πd

Goodheart-Willcox Publisher

Thread Formulas (Unit 27)

Thread pitch	$p = \frac{1}{tpi}$
Major ∅ of a gage # thread	Major ∅ = Gage # × 0.013 + 0.06
Thread height	H = 0.866 × p
Tap drill size	Tap drill size = Major Ø of thread – pitch

Goodheart-Willcox Publisher

Micrometer Readings (Unit 27)

Mic reading for Unified Thread Standard (UTS) = Pitch diameter $-\dfrac{0.86603}{\text{tpi}} + 3w$

Mic reading for metric threads (mm) = Pitch diameter $- 0.86603 \times \text{pitch} + 3w$

Smallest wire size that can be used for a given thread $= \dfrac{0.56}{\text{tpi}}$

Largest wire size that can be used for a given thread $= \dfrac{0.9}{\text{tpi}}$

Taper Formulas (Unit 28)

Calculating unknown taper dimensions:

Unknown	Known	Formula
tpi	tpf	tpi = tpf/12
tpf	tpi	tpf = tpi × 12
tpf	D, d, l	$\text{tpf} = \dfrac{(D - d) \times 12}{l}$
D	d, l, tpf	$D = d + \left(l\left(\dfrac{\text{tpf}}{12}\right)\right)$
d	D, l, tpf	$d = D - \left(l\left(\dfrac{\text{tpf}}{12}\right)\right)$
l	D, d, tpf	$l = 12\left(\dfrac{D - d}{\text{tpf}}\right)$

Goodheart-Willcox Publisher

Tailstock offset for turning a taper:

$$\text{Offset} = \frac{L \times \text{tpi}}{2}$$

$$\text{Offset} = \frac{L \times (D - d)}{2 \times l}$$

Speeds and Feeds (Unit 29)

Speed of revolving workpiece or tool: $\text{rpm} = \dfrac{\text{fpm} \times 12}{\pi D}$

Simplified formula: $\text{rpm} = \dfrac{\text{fpm} \times 3.82}{D}$

Metric formula: $\text{rpm} = \dfrac{\text{m/min} \times 1000}{\pi D}$

Feed rate for milling cutter: $F = \text{ftr} \times T \times \text{rpm}$

Feed rate for drill: $F = \text{IPR} \times \text{rpm}$

Feed rate for reamer: $F = \text{IPR} \times \text{rpm} \times 2 \quad \text{or} \quad F = \text{IPR} \times \text{rpm} \times 3$

Feed rate for tap: $F = \dfrac{\text{rpm}}{\text{tpi}}$

Where:

rpm = revolutions per minute fpm = feet per minute

m/min = meters per minute F = feed rate measure in inches per minute (IPM)

ftr = feed per tooth per revolution T = teeth or # of flutes

IPR = inches per revolution tpi = threads per inch

Spur Gear Calculations (Unit 30)

Formulas for Spur Gear Calculations

To Find	Rule	Formula
Diametral pitch	Divide π by the circular pitch.	$P = \pi / p$
Circular pitch	Divide π by the diametral pitch.	$p = \pi / P$
Pitch diameter	Divide the number of teeth by the diametral pitch.	$D = N / P$
Outside diameter	Add 2 to the number of teeth and divide the sum by the diametral pitch.	$D_o = \dfrac{N + 2}{P}$
Number of teeth	Multiply the pitch diameter by the diametral pitch.	$N = D \times P$
Tooth thickness	Divide 1.5708 by the diametral pitch.	$t = 1.5708 / P$
Addendum	Divide 1.0 by the diametral pitch.	$a = 1.0 / P$
Minimum dedendum	Divide 1.250 by the diametral pitch.	$b = 1.250 / P$
Working depth	Divide 2 by the diametral pitch.	$h_k = 2 / P$
Minimum whole depth	Divide 2.250 by the diametral pitch.	$h_t = 2.250 / P$
Minimum clearance	Divide 0.250 by the diametral pitch.	$c = 0.250 / P$
Center distance	Add the number of teeth in both gears and divide the sum by two times the diametral pitch.	$C = \dfrac{N_1 + N_2}{2P}$
Length of rack	Multiply the number of teeth in the rack by the circular pitch.	$L = N \times p$

Note: Use 3.1416 for pi when performing calculations.

Goodheart-Willcox Publisher

Right Triangle Trigonometry (Unit 36)

Pythagorean theorem:

$$c^2 = a^2 + b^2 \qquad a^2 = c^2 - b^2 \qquad b^2 = c^2 - a^2$$
$$c = \sqrt{(a^2 + b^2)} \qquad a = \sqrt{(c^2 - b^2)} \qquad b = \sqrt{(c^2 - a^2)}$$

Information Known	Formula to Determine Angle
Opposite and hypotenuse	$\theta = \sin^{-1}\left(\dfrac{\text{opposite}}{\text{hypotenuse}}\right)$
Adjacent and hypotenuse	$\theta = \cos^{-1}\left(\dfrac{\text{adjacent}}{\text{hypotenuse}}\right)$
Opposite and adjacent	$\theta = \tan^{-1}\left(\dfrac{\text{opposite}}{\text{adjacent}}\right)$

Goodheart-Willcox Publisher

Unknown Side	Equation to Use
Hypotenuse	$\text{hypotenuse} = \dfrac{\text{opposite}}{\sin(\theta)}$ or $\text{hypotenuse} = \dfrac{\text{adjacent}}{\cos(\theta)}$
Opposite	$\text{opposite} = \sin(\theta) \times \text{hypotenuse}$ or $\text{opposite} = \tan(\theta) \times \text{adjacent}$
Adjacent	$\text{adjacent} = \cos(\theta) \times \text{hypotenuse}$ or $\text{adjacent} = \dfrac{\text{opposite}}{\tan(\theta)}$

Goodheart-Willcox Publisher

Oblique Triangles (Unit 37)

Law of Sines:

$$\frac{a}{\sin A} = \frac{b}{\sin B} = \frac{c}{\sin C}$$

Law of Cosines:

$$c^2 = a^2 + b^2 - 2ab(\cos C) \qquad b^2 = a^2 + c^2 - 2ac(\cos B) \qquad a^2 = b^2 + c^2 - 2bc(\cos A)$$

Sine Bars and Sine Plates (Unit 38)

To determine angle ($\angle$):

$$\angle = \sin^{-1}\left(\frac{\text{opposite}}{\text{hypotenuse}}\right) \qquad \text{or} \qquad \sin^{-1}\left(\frac{\text{height of gage blocks}}{\text{size of sine bar}}\right)$$

To determine height of gage blocks:

$$\text{opposite} = \sin(\theta) \times \text{hypotenuse} \qquad \text{or} \qquad \sin(\theta) \times \text{size of sine bar}$$

Drill Point Angles (Unit 39)

$$\text{Size of drill point} = \frac{\varnothing \text{ of drill}}{2 \tan\left(\dfrac{\text{drill point angle}}{2}\right)}$$

Size of drill point = Percent (constant) × Tool diameter

Shortcut Formulas for Drill Point Size Using Common Constants	
Drill Point Angle (DPA)	**Shortcut Formula for Size of Drill Point**
60°	0.866 × $\varnothing$ of tool
82°	0.5752 × $\varnothing$ of tool
90°	0.5 × $\varnothing$ of tool
100°	0.4195 × $\varnothing$ of tool
110°	0.3501 × $\varnothing$ of tool
118°	0.3004 × $\varnothing$ of tool
120°	0.2887 × $\varnothing$ of tool
130°	0.2332 × $\varnothing$ of tool
135°	0.2071 × $\varnothing$ of tool
140°	0.1820 × $\varnothing$ of tool
150°	0.134 × $\varnothing$ of tool

Goodheart-Willcox Publisher

To find total depth of a drilled hole:

$$\text{Total depth} = \text{Size of drill point} + \text{Depth of hole}$$

Center-to-Center Distances (Unit 40)

$$radius = \frac{\varnothing \text{ of bolt circle}}{2}$$

For equally spaced holes on a bolt circle:

$$\angle \text{ between holes} = \frac{360°}{\text{\# of holes}}$$

To calculate the opposite side using the hypotenuse and angle:

$$\text{opposite} = \sin(\theta) \times \text{hypotenuse}$$

Center-to-center distance (base) = opposite side × 2

Measurement between pins = Center-to-center distance − Radius of each pin

Measurement over pins = Center-to-center to distance + Radius of each pin

Bolt circle diameter = 2 × radius

To calculate hypotenuse (radius) using opposite and angle:

$$\text{hypotenuse} = \frac{\text{opposite}}{\sin(\theta)}$$

Dovetails (Unit 41)

Calculate width of dovetail, depending on information provided:

$$\text{Dimension X} = \text{width} + (\text{adjacent} \times 2)$$

$$\text{Dimension X} = \text{width} - (\text{adjacent} \times 2)$$

Measurement between gage pins:

$$\text{Dimension X} = \text{width} - (\text{dimension y} \times 2)$$

$$\text{Dimension y} = \text{adjacent} + \text{radius}$$

$$\text{Where: adjacent} = \frac{\text{opposite}}{\tan(\theta)} \qquad \text{radius} = \frac{\varnothing}{2}$$

Measurement over gage pins:

$$\text{Dimension X} = \text{width} + (\text{dimension y} \times 2)$$

$$\text{Dimension y} = \text{adjacent} + \text{radius}$$

$$\text{Where: adjacent} = \frac{\text{opposite}}{\tan(\theta)} \qquad \text{radius} = \frac{\varnothing}{2}$$

Area of Polygons (Unit 43)

Shape	Formula
Square	$A = \pi r^2$
Rectangle	$A = lw$
Parallelogram	$A = bh$
Triangle	$A = 1/2bh$
Trapezoid	$A = 1/2(b_1 + b_2)h$

Goodheart-Willcox Publisher

Area of Circles and Circular Segments (Unit 44)

Shape	Formula
Circle	$A = s^2$
Circular sector	$A = \frac{\theta}{360}(\pi r^2)$
Circular segment	$A = \frac{r^2}{2}\left(\frac{\pi}{180}\right) \times \theta - \sin\theta$

Goodheart-Willcox Publisher

Volume of Cubes, Rectangular Solids, and Prisms (Unit 45)

Shape	Formula
Cube	$V = s^3$
Rectangular solid (cuboid)	$V = l \times w \times d$
Prism	$V = \text{base area} \times \text{length}$

Volume of Cylinders, Cones, and Spheres (Unit 46)

Shape	Formula
Cylinder	$V = \pi r^2 l$
Cone	$V = \frac{1}{3} \pi r^2 l$
Sphere	$V = 4/3\ \pi r^3$

Bolt Circles (Unit 49)

The radius will form the hypotenuse in the right-angle triangle created:

$$\text{radius} = \frac{\varnothing \text{ of bolt circle}}{2}$$

For equally spaced holes on a bolt circle:

$$\angle \text{ between holes} = \frac{360°}{\# \text{ of holes}}$$

To calculate the adjacent side or opposite side using the hypotenuse and angle:

$$\text{adjacent} = \cos(\theta) \times \text{hypotenuse} \qquad \text{opposite} = \sin(\theta) \times \text{hypotenuse}$$

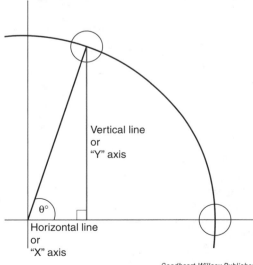

Vertical line
or
"Y" axis

θ°

Horizontal line
or
"X" axis

The vertical line in the created triangle will represent the Y axis.
The horizontal line in the triangle will represent the X axis.

Quadrants:

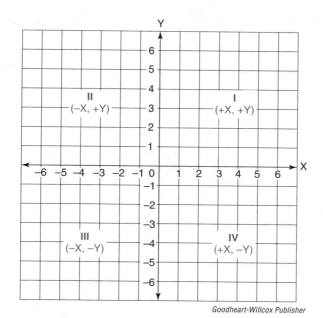

Goodheart-Willcox Publisher

I. Any hole falling in quadrant I has positive values on both X and Y axes.

II. Any hole falling in quadrant II has a negative value on X axis and positive value on Y axis.

III. Any hole falling in quadrant III has negative values on both X and Y axes.

IV. Any hole falling in quadrant IV has a positive value on X axis and negative value on Y axis.

GLOSSARY

A

absolute positioning. A CNC positioning system in which all positions or points are located from the origin, or absolute zero, often referred to as a fixed reference point. (47)

abstract number. A number that has nothing associated with it. (1)

accuracy. The closeness of a measurement to the actual size or dimension. (24)

acute angle. An angle of less than 90°. (33)

acute triangle. A triangle in which all angles are less than 90°. (35)

addendum. The distance a tooth extends above the pitch circle. (30)

adjacent side. The side touching the given angle other than the right angle. (36)

altitude. A line drawn from the vertex of an angle in a triangle to the opposite side to form a right angle. (39)

Arabic number system. A number system using digits 0 through 9. This is the most widely used system today. (1)

B

base of a triangle. Any side of a triangle from which the altitude or height is measured. (35)

bilateral tolerance. A tolerance listed with a separate plus and minus tolerance amount. (24)

bisect. To divide into two congruent, or equal, parts. (41)

C

Cartesian coordinate system. A system made of planes or lines that run horizontally and vertically to one another with perpendicular (90°) intersections. Also called the *rectangular coordinate system.* (47)

chord. A straight line intersecting the circumference of a circle in two places. (44)

circle. A shape enclosed by a continuous curved line on which all points are the same distance from the center. (23)

circular pitch. The distance measured on the pitch circle between the same point on adjacent teeth. (30)

circular segment. A part of a circle that is cut off from the rest of the circle by a chord. (44)

circumference. The perimeter of a circle. (23)

common denominator. A denominator that is divisible by all of the denominators in a problem. This may be greater than the least common denominator. (9)

complementary angles. Two adjacent angles formed by one straight line intersecting another straight line. The sum of these angles is 90°. (33)

concrete number. A number that designates the quantity of something. (1)

cone. A solid shape with a flat circular base and sides tapering to a point or vertex. (46)

constant. A number with a fixed value in a given situation. (39)

cube. A six-sided polyhedron in which all sides are identical squares. (45)

cube of a number. The number to the third power. (31)

cuboid. A polyhedron in which all of the faces are squares or rectangles and all the angles are right angles. (45)

cylinder. A three-dimensional shape with flat ends or bases that are circles and it has length. The flat ends are parallel to one another. (46)

Note: The number in parentheses following each definition indicates the unit in which the term can be found.

D

decimal. A number based on 10. (14)

decimal fraction. A fraction of a whole expressed in decimal numbers. Often referred to as a *decimal*. (14)

decimal number system. A system in which Arabic numbers are divided or multiplied by multiples of 10. (1)

decimal point. A period used to separate decimal fractions from whole numbers. (14)

dedendum. The distance a tooth extends below the pitch circle. (30)

denominate number. A concrete number with designated units of measurement. (1)

denominator. The number written below the fraction bar in a common fraction that indicates the number of parts into which one whole is divided. (7)

diameter. The length of a straight line segment that passes through the center of the circle and whose endpoints lie on the circle. (23, 44)

diametral pitch. The number of teeth per inch on the pitch circle. (30)

difference. The result of subtraction. (3)

dimensional tolerance. Tolerance of dimensions describing size of parts. (24)

dividend. The number being divided in a division operation. (5)

divisor. The number by which another number is divided in a division operation. (5)

dovetail slide. A type of plain linear bearing that relies on direct contact between sliding surfaces to support and move a heavier load. (41)

E

equation. A mathematical statement that shows that two things are equal in value. (26)

equilateral triangle. A triangle in which all three sides and all three angles are equal. (35)

equivalent fractions. Fractions that are expressed with different numerals but that are equal in value. (7)

exponent. The number of powers of a number, written above and to the right of the base number. (31)

extremes. In a proportion, the first and fourth terms. (21)

F

feed. The rate at which the cutting tool advances along the workpiece in one revolution. (29)

formula. An equation involving variables and an unknown value. (26)

fraction bar. The small horizontal line in a fraction. (7)

ftr. Feed rate per tooth per revolution of the tool. (29)

G

gage blocks. A set of steel blocks ground to precise measurements that can be stacked in various combinations of sizes to any desired height. Also called *jo blocks*. (38)

graduations. The segments into which a unit of measure is divided for measuring amounts of less than a full unit. (22)

H

height of a triangle. The perpendicular distance from the base of a triangle to the opposite angle. (35)

hexagon. A six-sided polygon. (43)

hypotenuse. The side opposite the right angle of a right triangle. (35)

I

incremental positioning. A CNC positioning system in which the reference point constantly changes, and the machine tool moves to the next position relative to its current position. Also called *relative positioning*. (47)

improper fraction. A fraction that represents a number greater than 1. The numerator is greater than or equal to the denominator. (8)

index. The number of powers involved. (32)

inverse proportion. A situation in which an increase in one variable causes a decrease in the other variable or vice versa. (21)

invert. To turn upside down. In fractions, it is to exchange the numerator and denominator of a fraction, so the denominator becomes the numerator and the numerator becomes the denominator. (13)

irregular polygon. A shape with straight sides of different lengths or corners of different angles. (43)

isosceles triangle. A triangle that has two equal sides and two equal angles. (35)

L

law of cosines. $c^2 = a^2 + b^2 - 2ab\ cosC$ (37)

law of cosines (angle version). (37)

$$cosC = \frac{a^2 + b^2 - c^2}{2ab}$$

law of sines. (37)

$$\frac{a}{sinA} = \frac{b}{sinB} = \frac{c}{sinC}$$

least common denominator. The smallest denominator that is evenly divisible by all of the denominators in a problem. (9)

linear measurement. Measurement of distance. (22)

lowest terms. The smallest possible denominator and numerator with which a fraction can be expressed. (7)

M

major diameter. Largest diameter of threads. (27)

means. In a proportion, the second and third terms. (21)

metric system. A system of measurement based on powers of 10 and standardized throughout the world. (22)

micrometer. An instrument using a closely controlled thread sleeve and thimble for precise measurements. (25)

minuend. The number from which another amount is subtracted. (34)

minute (angular measure). 1/60 of a degree. (33)

mixed number. A number that includes a whole number plus a fraction. (8)

multiplicand. The number being multiplied. (4)

multiplication table. A table showing the products of multiplication. (4)

multiplier. The number by which another number is multiplied. (4)

N

nearer fractional equivalent. The common fractional equivalent that most closely matches the decimal fraction. (15)

negative. Less than 0. (1)

non-polyhedron. A solid shape that has at least one surface that is not flat. (46)

numerator. The number written above the fraction bar in a common fraction that indicates the number of parts of the whole. (7)

O

obtuse angle. An angle of more than 90° but less than 180°. (33)

obtuse triangle. A triangle in which one angle is greater than 90°. (35)

octagon. An eight-sided polygon. (43)

opposite side. The side across from and not touching the given angle other than the right angle. (36)

P

p. Pitch of thread. $1 \div$ threads per inch $\left(\frac{1}{tpi}\right)$. (27)

parallel lines. Lines that are the same distance apart at all points, regardless of their length. (23)

parallelogram. A four-sided shape in which opposite sides are the same length and are parallel to one another, but the corners are not 90°. (23, 43)

pentagon. A five-sided polygon. (43)

percent. Parts per hundred. Percent is used with a specific number. (20)

percentage. A more general term, used with no specific number. (20)

perimeter. The distance around a two-dimensional shape. (23)

pi or π. A constant number of 3.1415926, which describes the relationship between the diameter of a circle and its circumference. Also used to calculate properties of circles. (23, 44)

pitch circle. An imaginary circle that passes through the approximate mid-height of the gear teeth. (30)

pitch diameter. The diameter of the pitch circle. (30)

plane. A two-dimensional flat surface that theoretically extends indefinitely. (47)

point angle. The angle formed by the cutting edges (lips) at the point of a drill. Also called *included angle*. (39)

point of tangency. The point of contact between a tangent line and a circle. Also called *tangent point*. (41)

polygon. A two-dimensional shape with straight sides. (43)

polyhedron. A three-dimensional shape that has flat faces. (45)

positional tolerance. Tolerance of dimensions describing the location or position of pieces or features. (24)

positive. More than 0. (1)

power of a number. The result of multiplying a number by itself. (31)

precision. In a process, the quality of being repeatable with the same or very similar results. (24)

prism. A solid shape with identically shaped ends and sides that are rectangles. (45)

product. The result of multiplication. (4)

proper fraction. A fraction that represents a number less than 1. The numerator is smaller than the denominator. (8)

proportion. An expression that two ratios are equal. (21)

Pythagorean theorem. A mathematical principle that in right triangles the square of the hypotenuse is equal to the sum of the squares of the other two sides, often stated as $a^2 + b^2 = c^2$. (36)

Q

quotient. The result of division. (5)

R

radical symbol. The $\sqrt{}$ symbol used to indicate the root of a number. (32)

radicand. The number of which the root is to be calculated. (32)

radius. The distance from the center of a circle to its edge. Equal to one-half the circle's diameter. (23, 44)

ratio. A comparison of two values. (21)

rays. The lines forming the sides of an angle. (33)

rectangle. A four-sided shape in which opposite sides are equal and all four corners are 90°. (23)

reducing. Expressing a fraction in its lowest terms. (7)

reflex angle. An angle of more than 180°. (33)

regular polygon. A shape with straight sides of equal length and in which all corners are the same angle. (43)

remainder. An amount left over after all the places in the dividend have been divided. A remainder is always smaller than the divisor. (5)

right angle. A 90° angle. (33)

right triangle. A triangle with a 90° angle. (35)

rpm. Revolutions per minute. (29)

rule of three. A mathematical rule that allows you to solve for the fourth term in a proportion given the other three using the principle that the product of the extremes is equal to the product of the means. (21)

S

scalene triangle. A triangle in which no sides or angles are equal. The angles still equal 180°. (35)

second (angular measure). 1/60 of a minute. (33)

sector. A slice of a circle formed by drawing two straight lines from the center to the circumference. (44)

sine bar. A precisely machined tool consisting of a hardened, precision-ground flat body typically no wider than 1″, with two precision-ground cylinders of equal diameter, one attached at each end. (38)

sine plate. A device similar to a sine bar but much wider, with a top plate that tilts out from the base plate to hold workpieces or fixtures at an exact angle to guide accurate machining. Also called *sine table*. (38)

soh cah toa. Sine equals opposite over hypotenuse. Cosine equals adjacent over hypotenuse. Tangent equals opposite over adjacent. (36)

speed. The speed difference between the cutting tool and the surface of the workpiece. Also called *cutting speed*. (29)

sphere. A three-dimensional figure with all points the same distance from the center. (46)

square. A four-sided shape in which all four sides are equal and all corners are 90°. (23)

square of a number. The number to the second power. (31)

straight angle. Angle of 180°. (33)

subtrahend. The number being subtracted from another number. (34)

sum. The result of addition. (2)

supplementary angles. Two angles that add up to 180°. (33)

Système International d'Unités (SI). Official name of the metric system. (22)

T

tailstock offset. The distance the tailstock of a lathe is moved off the centerline of the lathe to machine a taper. (28)

tangent line. In plane geometry, a line that touches a circle at exactly one point or position and never enters the circle's interior. (41)

tangent to a circle theorem. The tangent of a circle is perpendicular to the radius of that circle at the point of tangency. (41)

taper. A conical surface formed by gradual reduction or increase in diameter from a cylindrical workpiece. (28)

θ (theta). Greek letter used to denote the angle at the center of a circular segment. (44)

tolerance. The allowable limits of difference between the design dimension and actual dimension of a piece or feature. (24, 34)

tpi. Threads per inch. (27)

transversal. A line intersecting two parallel lines, creating corresponding congruent angles. (42)

trapezoid. A four-sided figure in which two sides are parallel and of unequal length and the other two sides are not parallel. (23, 43)

trigonometry. The study of the relationships of sides and angles in a triangle. (36)

two-tangent theorem. If two lines are drawn from the same point outside a circle such that both lines are tangent to the circle, the two lines are congruent, or equal. (41)

U

US Customary System. A system of measurement that includes length measurement based on yards, feet, and inches. (22)

V

vernier caliper. A measuring device that has a sliding jaw and a fixed jaw and uses a vernier scale for precise measurements. (25)

vernier scale. A scale with graduations slightly smaller than the principal graduations on an instrument, so that precision of .0001" is possible. (25)

vertex. The point at which two lines meet to form an angle. (33)

vinculum. The horizontal line that extends from the top of the radical symbol in a square root. (32)

W

whole. The entire unit that is divided to yield a fraction. (7)

Z

zero. The number between all positive numbers and all negative numbers. (1)

INDEX

ANSWERS TO ODD-NUMBERED QUESTIONS

UNIT 1
Number Systems

Practice

1. 3
3. 9
5. −4
7. −9
9. 6
11. 3
13. −3
15. −8
17. −2
19. 3
21. 52
    5: Tens
    2: Ones
23. 863
    8: Hundreds
    6: Tens
    3: Ones
25. 1,284
    1: Thousands
    2: Hundreds
    8: Tens
    4: Ones
27. 10,218
    1: Ten Thousands
    0: Thousands
    2: Hundreds
    1: Tens
    8: Ones
29. 410 528
    4: Hundred Thousands
    1: Ten Thousands
    0: Thousands
    5: Hundreds
    2: Tens
    8: Ones
31. 3,482,951
    3: Millions
    4: Hundred Thousands
    8: Ten Thousands
    2: Thousands
    9: Hundreds

    5: Tens
    1: Ones
33. Abstract
35. Denominate
37. Concrete
39. Abstract
41. Denominate
43. Abstract
45. Abstract
47. Denominate
49. Concrete
51. Denominate
53. Concrete
55. Denominate
57. True
59. True
61. True
63. False
65. False
67. True
69. True

UNIT 2
Adding Whole Numbers

Practice

1. 7
3. 15
5. 13
7. 8
9. 41
11. 79
13. 587
15. 1,366
17. 3,670
19. 12,483
21. 16,216
23. 859
25. 33
27. 81
29. 101
31. 248
33. 598

Applications

1. 29 inches
3. 43 inches
5. 50 mm
7. 13 inches
9. 7 inches
11. 136 mm
13. 28 tools
15. 33 parts
17. 15 mm
19. 19 mm
21. 19 mm
23. 192 mm
25. 52 mm
27. 58 mm
29. 156 mm
31. 110 mm
33. 84 mm

UNIT 3
Subtracting Whole Numbers

Practice

1. 2
3. 5 man-hours
5. 3 gallons
7. 95
9. 31
11. 4 feet
13. 18
15. 53 inches
17. 65
19. 108 meters
21. 179
23. 366 miles
25. 787 yards
27. 1,791
29. 1
31. 14
33. 7
35. 3
37. 24

39. 23
41. 457
43. 3,591
45. 12
47. 51
49. 225 mm
51. 10 pieces

Applications

1. 112 ounces
3. 47 gallons
5. 26 parts
7. 66 mm
9. 5 inches
11. 22 mm
13. 20 mm
15. 3 inches
17. 123 mm
19. 54 mm
21. 18 mm

UNIT 4
Multiplying Whole Numbers

Practice

1. 45 inches
3. 176 threads
5. 5,152 holes
7. 16
9. 33
11. 9
13. 9
15. 35
17. 50
19. 102
21. 28
23. 77
25. 144 inches
27. 215
29. 188 dollars
31. 470
33. 943 man-hours
35. 1,081
37. 4,032
39. 972 feet
41. 10,366
43. 376,376
45. 15,792
47. 53,694
49. 335,936
51. 2,552,954
53. 3,241,758
55. 22,813,083

Applications

1. 45 inches
3. 176 threads
5. 104 inches
7. 1,248 seconds
9. 288 pounds
11. 56 threads total
13. 196 seconds
15. 12 inches
17. 42 pecks per plate
19. 450 inches
21. 1,560 parts per day

UNIT 5
Dividing Whole Numbers

Practice

1. 7
3. 12
5. 7
7. 27
9. 6
11. 8
13. 18
15. 3
17. 5
19. 8
21. 9
23. 13
25. 22
27. 22
29. 21
31. 42 r1
33. 611
35. 180
37. 2,729
39. 36
41. 33
43. 61
45. 51

Applications

1. 12 inches
3. 9 millimeters
5. 12 minutes
7. 50 studs
9. 28 parts
11. 40 parts total
13. $144 per day
15. 26 cents per hex nut
17. 12 parts

15 parts
20 parts
30 parts
40 parts
19. 12 parts per hour
21. 48 hours
23. 96 parts

UNIT 6
Combined Operations

Practice

1. 11
3. 46
5. 8
7. 61
9. 75
11. 512
13. 61
15. 10
17. 26

Applications

1. $490
3. $7,280
5. 13 parts per bar with 3 inches left over
7. 6,624 mm of material
9. 2,664 millimeters
11. 1 millimeter
13. 16 hours total
15. $352 total
17. $17,344
19. 300 castings
21. 6 parts

UNIT 7
Parts of a Fraction

Practice

1. numerator 3, denominator 8
3. numerator 121, denominator 500
5. numerator 21, denominator 96
7. 24
9. 28
11. 18
13. 56
15. 4
17. 6
19. 8
21. 24

23. 8
25. 10
27. 6
29. 3
31. 4
33. 2
35. 3
37. 7
39. 6
41. 3
43. 6
45. 100
47. 30
49. 16
51. 9
53. 3
55. 2
57. 8
59. 6
61. 3
63. 3/10
65. 1/5
67. 1/25
69. 8/9
71. 1/5
73. 1/11
75. 3/173
77. 1/25
79. 16/25
81. 7/15
83. 3/4
85. 13/25
87. 1/8

Applications

1. 1/4
3. 3/8
5. 1/2

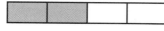

7. 4/10

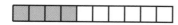

9. 1/4
11. 5/6
13. 1/5

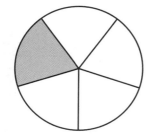

15. 3/8

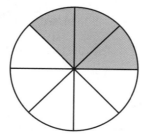

17. 1/4
19. 4/9
21. 1/2

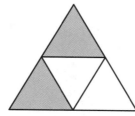

23. 2/9

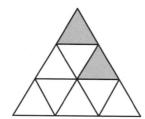

25. 3/4
27. 3/4
29. 5/9

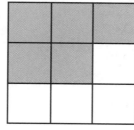

31. 11/16

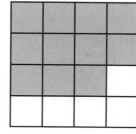

33. 18/55
    37/55
35. 1/4
    7/20
    1/2

33/50
18/25
17/20
37. 3/10
    7/10
39. 3/8
41. 28/36 or 7/9

UNIT 8
Proper Fractions, Improper Fractions, and Mixed Numbers

Practice

1. proper
3. improper
5. proper
7. mixed number
9. improper
11. proper
13. improper
15. mixed number
17. mixed number
19. improper
21. mixed number
23. improper
25. mixed number
27. proper
29. improper
31. proper
33. improper
35. proper
37. mixed number
39. proper
41. 2 1/4
43. 4 11/16
45. 6 12/35
47. 1 1/2
49. 10 1/11
51. 2 1/4
53. 1 3/34
55. 2 1/25
57. 1 9/32
59. 1 13/15
61. 2 15/16
63. 6 5/16
65. 2 2/5
67. 8 2/5
69. 6 1/2
71. 2 10/11
73. 10/3
75. 415/32
77. 555/32

79. 127/100
81. 632/3
83. 13/6
85. 27/5
87. 27/7
89. 31/6
91. 97/16
93. 101/5
95. 441/4
97. 166/3
99. 32/11

unit 9
Least Common Denominator

Practice

1. LCD = 6; 3/6 and 2/6
3. LCD = 80; 48/80 and 5/80
5. LCD = 36; 9/36 and 4/36
7. LCD = 320; 56/320 and 190/320
9. LCD = 20; 25/20 and 28/20
11. LCD = 10; 5/10 and 6/10
13. LCD = 40; 5/40 and 16/40
15. LCD = 60; 20/60 and 27/60
17. LCD = 18; 5/18 and 6/18
19. LCD = 45; 35/45 and 6/45
21. LCD = 44; 16/44 and 11/44
23. LCD = 77; 42/77 and 44/77
25. LCD = 95; 45/95 and 57/95
27. LCD = 10; 26/10 and 45/10
29. LCD = 21; 45/21 and 84/21
31. LCD = 120; 40/120, 48/120, 45/120
33. LCD = 36; 24/36, 18/36, 12/36
35. LCD = 120; 105/120, 264/120, 96/120, 80/120
37. LCD = 36; 20/36, 15/36, 132/36
39. LCD = 1,050; 1,500/1,050, 1,197/1,050, 6,650/1,050
41. LCD = 70; 35/70, 42/70, and 40/70
43. LCD = 105; 70/105, 42/105, and 45/105
45. LCD = 80; 64/80, 72/80, and 65/80
47. LCD = 12; 6/12, 8/12, 3/12, and 4/12
49. LCD = 240; 360/240, 135/240, 336/240, and 40/240
51. LCD = 48; 72/48, 32/48, and 3/48
53. LCD = 80; 170/80, 15/80, and 32/80

55. LCD = 12; 27/12, 16/12, and 50/12
57. LCD = 400; 2,150/400, 1,208/400, and 3,675/400
59. LCD = 32; 6/32, 1/32, and 400/32

unit 10
Adding Fractions

Practice

1. 14/15
3. 7/8
5. 22/105
7. 23/24
9. 19/42
11. 7/12
13. 37/90
15. 1 7/12
17. 9/32
19. 35/48
21. 1 11/24
23. 1 7/16
25. 9 5/12
27. 1 49/60
29. 1 11/40
31. 1 7/32
33. 5 1/16
35. 12 31/32
37. 37 29/40
39. 15 83/120
41. 4 11/24
43. 2 5,111/6,000
45. 51/1,000
47. 15 17/45
49. 15 11/48
51. 17 791/1650
53. 21 1/100
55. 57 5/16
57. 128 25/64
59. 903/1,000

Applications

1. 35 1/8 hours
   33 1/6 hours
   32 11/12 hours
   27 1/3 hours
   128 13/24 hours
3. 1 1/16 inches
5. 9 143/144 inches
   10 127/144 inches
   11 11/16 inches
   11 157/192 inches
   12 61/64 inches
7. 4 1/8 inches

9. 3 1/4 inches
11. 3 1/8 inches
13. 5 15/32 inches
15. 1 1/16 inches
17. 1 9/16 inches
19. 1,351 3/4 pounds
21. 31/64 inch

unit 11
Subtracting Fractions

Practice

1. 1/2
3. 1/10
5. 2 3/5
7. 5/16
9. 3/13
11. 1/2
13. 11/32
15. 3 7/12
17. 4 21/40
19. 3 251/576
21. 21 7/8
23. 4 37/140
25. 2 37/124
27. 24 23/24
29. 2 303/320
31. 22 29/32
33. 5 39/200
35. 77 13/20
37. 61 483/500
39. 4 45/364

Applications

1. Dim A: 2 1/2
   Dim B: 1 1/2
3. Dim A: 4 1/2
   Dim B: 7/8
5. Dim A: 6
   Dim B: 1 1/8
7. Dim A: 9 3/4
   Dim B: 4 3/64
9. 5/16"
11. 7/16"
13. A: 7/16"
    B: 1 5/16"
    C: 2 1/16"
15. 15 19/48"
17. 3/16"
19. 2 11/32"
21. 3/16"
23. $16.75 or 16 3/4 dollars

UNIT 12
Multiplying Fractions

Practice

1. 1/8
3. 21/25
5. 7/10
7. 1/3
9. 2/35
11. 2/7
13. 32/45
15. 14/27
17. 1 7/12
19. 135/512
21. 1/16
23. 39/320
25. 2 3/5
27. 3/28
29. 1/80
31. 7/64
33. 39/80
35. 125/768
37. 2 11/12
39. 75 25/48
41. 202 1/2
43. 6 5/12
45. 17 13/18
47. 41 9/40
49. 269 63/80
51. 28 31/48
53. 72 95/96

Applications

1. 14 5/8 inches
3. 47 11/16 inches
5. 22 1/2 boxes
7. 7 1/32 inches
9. 6 7/8 inches
11. 51/64 inches
13. 500 revolutions per minute
15. 7 7/8 pounds
17. 10 days
19. 1,320 parts

UNIT 13
Dividing Fractions

Practice

1. 2
3. 40/63
5. 7/16
7. 1/2
9. 2/3
11. 5/6
13. 9/14
15. 21/40
17. 1 11/49
19. 5/6
21. 15/28
23. 9/100
25. 5 1/4
27. 2 5/24
29. 2 7/15
31. 2 7/10
33. 8 11/18
35. 5 5/8
37. 1 71/104
39. 3 9/10
41. 2 3/10
43. 3 5/9
45. 5 13/33
47. 3 1/24

Applications

1. 5 53/56 pounds
3. 14
5. 11 pins and 2 15/32 inches remain
7. 431 1/4 revolutions
9. 5 5/9 minutes
11. 39 castings
13. 39 parts
15. A. 7/8"
    B. 1 1/20"

UNIT 14
Decimal System

Practice

1. 2: tenths
   5: hundredths
3. 4: tenths
   7: hundredths
   8: thousandths
   5: ten-thousandths
   6: hundred-thousandths
   3: millionths
5. 3: three hundred thousandths
   9: ninety thousandths
   2: two thousandths
   1: one tenth
7. 5: five inches
   0: zero hundred thousandths
   0: zero ten thousandths
   6: six thousandths

1: one tenth
2: twenty millionths
4: four millionths
9. Five hundred fifty-eight thousandths
11. Five and eighty-seven hundredths
13. Seven and thirty thousandths
15. One and two hundred forty thousandths
17. Four inches, six hundred thousandths
19. Eight hundred thousandths
21. Five inches, ninety thousandths
23. Six hundred thirty thousandths
25. Four hundred sixty-one thousandths
27. One inch, two hundred eighty-four thousandths
29. One hundred eight thousandths and seven tenths
31. Seven inches, two hundred six thousandths, and nine tenths
33. Four inches, two hundred seventy-four thousandths, nine tenths, and twenty millionths
35. Nine inches, one hundred seventy-two thousandths, three tenths, and eighty millionths
37. Three inches, ninety-nine thousandths, one tenth, and eighty-six millionths
39. Two hundred thirteen thousandths, five tenths, and seven millionths
41. 0.77
43. 15.025
45. 0.00025
47. 0.752
49. 0.002
51. 0.011
53. 0.091
55. 3.0008
57. 0.037
59. 3.043
61. 10.0171
63. 0.0986
65. 2.4001
67. 5.136502
69. 20.900009
71. 5.217236
73. 1.081344
75. 0.000008
77. 0.000004

UNIT 15
Converting between Common Fractions and Decimal Fractions

Practice

1. 25/100
3. 875/1,000
5. 4/1,000
7. 8 600/1,000
9. 47/100
11. 1/8
13. 833/2,000
15. 1/100
17. 2,563/5,000
19. 1 16/125
21. 1/16
23. 3 1/10
25. 4 23/125
27. 2 11/20
29. 3 127/500
31. 0.75
33. 0.2
35. 0.125
37. 4.8
39. 0.333333
41. 0.625
43. 0.875
45. 0.285714
47. 0.5625
49. 0.046875
51. 0.3
53. 1.25
55. 5.666666
57. 6.75
59. 3.25
61. 0.5
63. 0.78
65. 0.56
67. 4.3
69. 1.55556
71. 3.14
73. 0.25
75. 0.1875
77. 0.06
79. 3.67
81. 1.12
83. 5.09
85. 5/16
87. 59/64
89. 6 7/16
91. 7/16
93. 1 1/64
95. 4 31/64

UNIT 16
Adding Decimal Fractions

Practice

1. 0.77
3. 10.59
5. 0.59
7. 5.2
9. 6.03
11. 6.0469
13. 17.0906
15. 19.5469
17. 35.8581
19. 3.080
21. 564.275
23. 30.48
25. 48.8063
27. 356.0492
29. 1.226 + 0.133 = 1.359   (one inch, three hundred fifty-nine thou)
31. 4.651 + 3.891 = 8.542   (eight inches, five hundred forty-two thou)
33. 10.3423 + 6.3895 = 16.7318 (sixteen inches, seven hundred thirty-one thou, and eight tenths)
35. 1.155301 + 4.765608 = 5.920909 (five inches, nine hundred twenty thou, nine tenths, and nine millionths)

Applications

1. A. $707.18
   B. $633.66
   C. $629.84
   D. $699.61
   E. $464.92
   F. $566.14
   G. $582.92
   H. $565.19
   I. $491.12
   J. $2670.29
3. 19.527 inches
5. 19.188 inches
7. 0.810"
9. 3.368"
11. $56.38
13. $55.18
15. 2.03"

UNIT 17
Subtracting Decimal Fractions

Practice

1. 0.27
3. 1.185
5. 3.13
7. 0.87
9. 2.04
11. 9.0156
13. 3.4224
15. 6.8125
17. 9.1245
19. 26.6975
21. 59.303
23. 2.625
25. 5.5691
27. 2.0826
29. 130.9156
31. 197.6608
33. 72.4742
35. 12.625 − 3.400 = 9.225 (Nine inches, two hundred twenty-five thou)
37. 6.519 − 1.374 = 5.145 (Five inches, one hundred forty-five thou)
39. 8.2923 − 2.4272 = 5.8651 (Five inches, eight hundred sixty-five thou, and one tenth)

Applications

1. 18.375"
3. 25.125 − 14.250 = 10.875 (Ten inches, eight hundred seventy-five thou)
5. 3.1562 (Three inches, one hundred fifty-six thou, and two tenths)
7. 1.75"
9. 0.1875"
11. 0.1872"
13. 0.065"
15. 9.110"
17. 0.8125 − 0.0469 = 0.7656 (Seven hundred sixty-five thou and six tenths)
19. $200 − $124.95 − $4.95 = $70.10
21. 1.396" − 1.375" = 0.021" (Twenty-one thou)
23. .125"
25. 2.25" − 1.125" − 0.563" = 0.562"

UNIT **18**
Multiplying Decimal Fractions

Practice

1. 0.225
3. 2.625
5. 8.4375
7. 0.076
9. 0.04
11. 0.4125
13. 12
15. 37.275
17. 2.945
19. 1.48
21. 6.048
23. 9.22
25. 1,296.9
27. 1.820
29. 0.0325
31. 0.28875
33. 0.324
35. 0.0384
37. One hundred twenty thou $(0.6 \times 0.2 = 0.12)$
39. Six inches, forty-two thou, and six tenths $(1.62 \times 3.73 = 6.0426)$
41. Four inches, nine hundred twenty-two thou, five tenths, and twenty millionths $(3.876 \times 1.27 = 4.92252)$
43. Four inches, six hundred ninety-three thou, five tenths, and fourteen millionths $(7.133 \times 0.658 = 4.693514)$
45. Thirty-two inches, forty-one thou, seven tenths, and ninety millionths $(20.9 \times 1.5331 = 32.04179)$

Applications

1. 0.012"
3. 0.60"
5. 128.125 square inches
7. 236.25 square inches
9. 4.6875 inches of thread
11. 2.1"
13. 1.7675"
15. $702.15
17. $140.62
19. 63 threads
21. 14.13 pounds
23. 20 pounds
25. 3.37 pounds

UNIT **19**
Dividing Decimal Fractions

Practice

1. 1.833
3. 7.125
5. 0.05875
7. 0.075
9. 5.533
11. 5.433
13. 25.333
15. 4.505
17. 3.427
19. 8.12649
21. 0.0875
23. 1955
25. 39
27. 0.73
29. 0.146718
31. $0.7 \div 0.01 = 70$
33. $1.3 \div 0.9 = 1.44$
35. $8.17 \div 1.4 = 5.8357$
37. $2.08 \div 0.112 = 18.5714$
39. $30.9006 \div 0.3222 = 95.9050$

Applications

1. 2.250"
3. $27.37
5. 2.6"
7. Calculation: $\frac{1}{0.025}$
   TPI: 40
9. Calculation: $\frac{1}{0.05}$
   TPI: 20
11. Calculation: $\frac{1}{0.1}$
    TPI: 10
13. Calculation: $\frac{1}{0.25}$
    TPI: 4
15. Calculation: $\frac{23.8}{0.35}$
    # Threads: 68
17. Calculation: $\frac{15}{0.45}$
    # Threads: 33
19. Calculation: $\frac{30}{0.6}$
    # Threads: 50
21. Calculation: $\frac{35}{0.75}$
    # Threads: 46
23. Calculation: $\frac{25}{1.25}$
    # Threads: 20
25. Calculation: $\frac{100}{1.75}$
    # Threads: 57
27. 2.318"
29. 0.05
31. $5.60
33. $2.09
35. 0.0313

UNIT **20**
Percent and Percentage

Practice

1. 41.1%
3. 83.3
5. 88
7. 52.1
9. 96
11. 41.7%
13. 437.5
15. 4.4%
17. 4/5
19. 27/50
21. 1/4
23. 3/25
25. 33%
27. 6%
29. 5%

Applications

1. $11,900
3. $2,550
5. $8.00
7. 26.9%
9. 16.4%
11. 18.5%
13. 6
15. 3.8
17. 1.7
19. 21.82%
21. A. 17.86%
    B. 82.14%
23. 1.28 horsepower
25. 66.67%
27. 14.4 ounces
29. 8 ounces
31. 29.12
33. 4
35. 446 lb

UNIT **21**
Ratio and Proportion

Practice

1. 3:1
3. 0.42:1
5. 1.67:1
7. 13:1
9. 24:1

11. 20
13. 3
15. 6.89
17. x = 40
19. x = 5
21. x = 40

Applications

1. Ratio: 18:40
   Reduced: 0.45:1
3. Ratio: 56.24
   Reduced: 2.33:1
5. Ratio: 44:56
   Reduced: 0.79:1
7. 1:1
9. 0.009:1
11. 46:9800 = 0.005:1
13. 175.510 grams
15. 238.776 grams
17. 131.633 grams
19. 179.082 grams
21. 0.45:1
23. 4.25 inches
25. 44:56 = 500:x = 636.36
27. 15.6:1 = 22:x = 1.41 gallons

UNIT 22
Reading Rulers

Practice

1. A: 1/8
   B: 1/16
   C: 2/8 = 1/4
   D: 3/16
   E: 3/8
   F: 5/16
   G: 5/8
   H: 7/8
   I: 8/16 = 1/2
   J: 11/16
   K: 1 2/16 = 1 1/8
   L: 1 6/8 = 1 3/4
   M: 1 13/16
   N: 2 6/16 = 2 3/8
   O: 5 13/16
   P: 4 7/8
   Q: 4
   R: 5 2/8 = 5 1/4
   S: 4 15/16
   T: 2 14/16 = 2 7/8
   U: 4 12/16 = 4 3/4
   V: 2 4/8 = 2 1/2
   W: 4 4/8 = 4 1/2

X: 3 9/16
Y: 3 4/16 = 3 1/4
Z: 1 3/8
3. A fractional: 8/50 = 4/25
   A decimal: 0.16
   B fractional: 2 5/10 = 2 1/2
   B decimal: 2.5
   C fractional: 8/10 = 4/5
   C decimal: 0.8
   D fractional: 36/50 = 18/25
   D decimal: 0.72
   E fractional: 1 9/10
   E decimal: 1.9
   F fractional: 1 2/50 = 1 1/25
   F decimal: 1.04
   G fractional: 1/10
   G decimal: 0.1
   H fractional: 2 1/50
   H decimal: 2.02
   I fractional: 3 4/10 = 3 2/5
   I decimal: 3.4
   J fractional: 2 47/50
   J decimal: 2.94
   K fractional: 4 6/50 = 4 3/25
   K decimal: 4.12
   L fractional: 3 6/10 = 3 3/5
   L decimal: 3.6
   M fractional: 5 4/50 = 5 2/25
   M decimal: 5.08
   N fractional: 5 1/10
   N decimal: 5.1
   O fractional: 4 3/10
   O decimal: 4.3
   P fractional: 5 46/50 = 5 23/25
   P decimal: 5.92
   Q fractional: 2 39/50
   Q decimal: 2.78
   R fractional: 5 5/10 = 5 1/2
   R decimal: 5.5
   S fractional: 1 3/10
   S decimal: 1.3
   T fractional: 1 27/50
   T decimal: 1.54
   U fractional: 3
   U decimal: 3
   V fractional: 2 29/50
   V decimal: 2.58
   W fractional: 3 18/50 = 3 9/25
   W decimal: 3.36
   X fractional: 4 7/10
   X decimal: 4.7
   Y fractional: 3/50
   Y decimal: 0.06
   Z fractional: 2 2/10 = 2 1/5
   Z decimal: 2.2

UNIT 23
Perimeter and Circumference

Practice

1. 20 mm
3. 72 mm
5. 4.5 inches
7. 10 inches
9. 6.3 inches
11. 5.25 inches
13. 4.5 inches
15. 10.63 inches
17. 2.513 inches
19. 25.918 inches
21. 75.398 mm
23. 345.575 mm

Applications

1. 112 mm
3. 14.106 inches
5. 5.236 inches
7. 79.796 mm
9. 157.080 mm
11. 131.947 mm
13. 200 mm
15. 15 inches
17. 8.5 inches
19. 13.15 inches
21. 128 mm

UNIT 24
Tolerances

Applications

1. A. Tolerance: ±0.010
      Minimum: 3.17
      Maximum: 3.19
   B. Tolerance: ±0.005
      Minimum: 2.427
      Maximum: 2.437
   C. Tolerance: ±0.005
      Minimum: 0.745
      Maximum: 0.755
   D. Tolerance: +0.010/−0.002
      Minimum: 0.498
      Maximum: 0.510
   E. Tolerance: ±0.005
      Minimum: 0.370
      Maximum: 0.380
   F. Tolerance: ±0.010
      Minimum: 3.17
      Maximum: 3.19

G. Tolerance: ±0.005
Minimum: 0.433
Maximum: 0.443

H. Tolerance: ±0.005
Minimum: 1.183
Maximum: 1.193

I. Tolerance: ±0.005
Minimum: 0.801
Maximum: 0.811

J. Tolerance: ±0.005
Minimum: 0.871
Maximum: 0.881

K. Tolerance: ±0.005
Minimum: 2.301
Maximum: 2.311

3. Nominal size: 3.18
Tolerance: ±0.005
Minimum: 3.175
Maximum: 3.185

5. A. Nominal size: 6.00
Tolerance: ±.03
Minimum: 5.97
Maximum: 6.03

B. Nominal size: 2.50
Tolerance: ±.03
Minimum: 2.47
Maximum: 2.53

C. Nominal size: .50
Tolerance: ±.03
Minimum: .47
Maximum: .53

D. Nominal size: .25
Tolerance: ±.03
Minimum: .22
Maximum: .28

E. Nominal size: .5520
Tolerance: ±.001
Minimum: .5510
Maximum: .5530

F. Nominal size: .125
Tolerance: ±.005
Minimum: .120
Maximum: .130

G. Nominal size: .1250
Tolerance: ±.001
Minimum: .1240
Maximum: .1260

H. Nominal size: .220
Tolerance: ±.005
Minimum: .215
Maximum: .225

I. Nominal size: 2.2500
Tolerance: ±.001
Minimum: 2.2490
Maximum: 2.2510

J. Nominal size: .50
Tolerance: ±.03
Minimum: .47
Maximum: .53

K. Nominal size: .5000
Tolerance: ±.001
Minimum: .4990
Maximum: .5010

L. Nominal size: 3.203
Tolerance: ±.005
Minimum: 3.198
Maximum: 3.208

M. Nominal size: .7500
Tolerance: ±.001
Minimum: .7490
Maximum: .7510

N. Nominal size: .250
Tolerance: ±.005
Minimum: .245
Maximum: .255

UNIT 25
Measuring Instruments

Applications

1. 0.1348″
3. 0.3787″
5. 0.4476″
7. 6.746 mm
9. 4.942 mm
11. 10.041 mm
13. 0.7622″
15. 1.3405″
17. 2.8966″
19. 1.2267″
21. 0.5668″
23. 4.961 mm
25. 47.725 mm
27. 87.922 mm
29. 120.523 mm
31. 25.787 mm
33. 3.262″
35. 2.391″
37. 5.758″
39. 83.46 mm

UNIT 26
Equations and Formulas

Practice

1. $x = 21$
3. $x = 42$

5. $x = 31 - a$

7. $\frac{1}{3}x = a + 3$

Step 1: Divide both sides by 1/3.

$$\frac{\frac{1}{3}x}{\frac{1}{3}} = \frac{a+3}{\frac{1}{3}}$$

$x = 3a + 9$

9. $yx + 5 = 4a$

Step 1: Add −5 to both sides.

$yx + 5 + -5 = 4a + -5$

$yx = 4a - 5$

Step 2: Divide both sides by y.

$$\frac{xy}{y} = \frac{4a-5}{y}$$

$x = \frac{4a-5}{y}$

11. $a = 3.34$

13. What is x if $\frac{x}{4} + c = 6$ and $c = 4$?

$\frac{x}{4} + 4 = 6$

Step 1: Subtract 4 from both sides.

$\frac{x}{4} + 4 - 4 = 6 - 4$

$\frac{x}{4} = 2$ or $\frac{1}{4}x = 2$

Step 3: Multiply both sides by 4.

$4x\left(\frac{1}{4}x\right) = 2 \times 4$

$x = 8$

15. What is y if $y - \frac{25}{a} = 4$ and $a = 5$?

$y - \frac{25}{5} = 4$

Step 1: Simplify both sides of the equation.

$y - 5 = 4$

Step 2: Add 5 to both sides.

$y - 5 + 5 = 4 + 5$

$y = 9$

17. $b = 0.28$
19. $x = 8.33$
21. $a = 4.13$

UNIT 27
Thread Formulas

Applications

1. major diameter
3. series (Unified National Coarse)
5. length
7. nominal diameter
9. length
11. 0.062″
13. 0.06
15. 0.086
17. 0.112

19. 0.138
21. 0.19
23. Pitch: 0.25
    Min: 0.14
    Max: 0.225
25. Pitch: 0.1666
    Min: 0.0933
    Max: 0.15
27. Pitch: 0.125
    Min: 0.07
    Max: 0.1125
29. Pitch: 0.0909
    Min: 0.0509
    Max: 0.0818
31. Pitch: 0.0769
    Min: 0.0431
    Max: 0.0692
33. Pitch: 0.0625
    Min: 0.035
    Max: 0.0563
35. Pitch: 0.05
    Min: 0.028
    Max: 0.045
37. Pitch: 0.0417
    Min: 0.0233
    Max: 0.0375
39. Pitch: 0.037
    Min: 0.0207
    Max: 0.0333
41. Pitch: 0.0333
    Min: 0.0187
    Max: 0.03
43. Pitch: 0.0278
    Min: 0.0156
    Max: 0.025
45. Pitch: 0.0227
    Min: 0.0127
    Max: 0.0205
47. Pitch: 0.02
    Min: 0.0112
    Max: 0.018
49. Pitch: 0.0156
    Min: 0.0088
    Max: 0.0141
51. Pitch: 0.0125
    Min: 0.007
    Max: 0.0113
53. 1.6 mm
55. 2.5 mm
57. 3.3 mm
59. 5 mm
61. 7 mm
63. 8.75 mm
65. 10.75 mm
67. 12.5 mm
69. 14.5 mm

71. 16.5 mm
73. 18.5 mm
75. 20.5 mm
77. 22 mm
79. 25 mm
81. 28 mm
83. 31 mm
85. 33 mm
87. Minimum micrometer
    reading = Minimum pitch
    diameter $- \frac{0.86603}{\text{tpi}} + 3w$
    Minimum pitch diameter is
    0.4435"; tpi = 13; w = 0.045"
    Minimum micrometer reading =
    $0.4435 - \frac{0.86603}{13} + 3 \times 0.045$
    Minimum micrometer reading =
    0.5119"
    Maximum micrometer reading =
    Maximum pitch diameter
    $- \frac{0.86603}{\text{tpi}} + 3w$
    Maximum pitch diameter is
    0.4485"; tpi = 13; w = 0.045"
    Maximum micrometer reading =
    $0.4485 - \frac{0.86603}{13} + 3 \times 0.045$
    Maximum micrometer
    reading = 0.5169"

UNIT 28
Taper Formulas

Applications

1. tpf: 1.2"
   d: 0.375"
3. tpi: 0.042"
   d: 0.469"
5. l: 2.083"
   tpi: 0.021"
7. tpf: 6.141"
   tpi: 0.512"
9. tpf: 2.571"
   tpi: 0.214"
11. tpf: 0.504"
    D: 1.142"
13. tpf: 2.4"
    tpi: 0.2"
15. tpf: 0.6"
    D: 1.125"
17. 12.93 mm
19. .188"
21. d = 1.083"
23. tpf = 1.412"
25. d = 1.948"
27. tpf = 2.87"

UNIT 29
Speeds and Feeds

Applications

1. 150 rpm
3. 60 rpm
5. 510 rpm
7. $\text{rpm} = \frac{60 \times 1{,}000}{\pi \times 40}$
   rpm = 477
9. $\text{rpm} = \frac{120 \times 1{,}000}{\pi \times 75}$
   rpm = 509
11. 7 inches per minute
13. 92 inches per minute
15. $\text{rpm} = \frac{\text{fpm} \times 12}{\pi D}$
    $\text{rpm} = \frac{1{,}500 \times 12}{\pi \times 2}$
    rpm = 2,865
    Feed (F) = ftr $\times$ T $\times$ rpm
    $F = 0.011 \times 3 \times 2{,}865$
    F = 95 IPM
17. $\text{rpm} = \frac{\text{fpm} \times 12}{\pi D}$
    $\text{rpm} = \frac{1{,}200 \times 12}{\pi \times 3}$
    rpm = 1,528
    Feed (F) = ftr $\times$ T $\times$ rpm
    $F = 0.028 \times 5 \times 1{,}528$
    F = 214 IPM
19. $\text{rpm} = \frac{\text{fpm} \times 12}{\pi D}$
    $\text{rpm} = \frac{100 \times 12}{\pi \times 2.5}$
    rpm = 153
    Feed (F) = ftr $\times$ T $\times$ rpm
    $F = 0.004 \times 30 \times 153$
    F = 18 IPM
21. Decimal: 0.1875
    RPM: 1426
    Feed: 2.8520
23. Decimal: 0.4375
    RPM: 1310
    Feed: 65.5
25. Decimal: 0.65625
    RPM: 175
    Feed: 3.5/5.25
27. Decimal: 0.413
    RPM: 2081
    Feed: 12.486
29. Decimal: 0.25
    RPM: 2292
    Feed: 81.8571

UNIT 30
Spur Gear Calculations

Applications

1. 7.077"
3. 0.141"
5. 0.305"
7. 0.222"
9. 0.283"
11. 7.5"
13. 0.145"
15. 0.25"
17. 0.393"
19. 4.2"
21. 0.1"
23. 0.216"
25. 0.157"
27. 0.016"

UNIT 31
Powers

Practice

1. 25
3. 7,776
5. 4/5
7. 7/256
9. 8/343
11. 2,097,152
13. 1/9
15. 1/8
17. 1/25
19. $6^8 = 1,679,616$
21. $3^9 = 19,683$
23. $5^5 = 3,125$
25. $4^2 = 16$
27. $12^3 = 1,728$
29. $8^4 = 4,096$
31. $15^2 = 225$
33. $\left(\frac{2}{3}\right)^2 = \frac{4}{9}$
35. $\left(\frac{3}{4}\right)^4 = \frac{81}{256}$
37. $\left(\frac{1}{5}\right)^5 = \frac{1}{3,125}$

Applications

1. 262.4 mm²
3. 4 in²
5. 625 mm²
7. $5\frac{4}{9}$ in²
9. 9.766 in²
11. $\frac{4}{9}$ in²

13. 405.22 in³
15. 42,875 mm³
17. 125 in³
19. $190\frac{7}{64}$ in³
21. 14,706.125 mm³
23. $\frac{64}{125}$ in³

UNIT 32
Roots of Numbers

Practice

1. 6
3. 3
5. 7.746
7. 7
9. 8
11. 4
13. 9.644
15. 2
17. 11
19. 4.642
21. 2/3 or 0.667
23. 2
25. 6.910
27. 1
29. 9
31. 7/8 or 0.875
33. 3/4 or 0.75
35. 1 4/5 or 1.8

Applications

1. 14.68"
3. 75 mm
5. 7.071 inches
7. 15 mm
9. 12.7 mm
11. 8.5 mm
13. 3.250 inches
15. 2.250 mm
17. 18 mm
19. 5.125 inches
21. 1.5 inches

UNIT 33
Units of Angular Measure

Practice

1. Acute angle
3. Obtuse angle
5. Obtuse angle
7. Acute angle

9. Reflex angle
11. Thirty-four degrees, thirty minutes, forty-five seconds
13. One hundred eighteen degrees, thirteen minutes, fifty-one seconds
15. Sixty-one degrees, forty-three minutes, twenty-two seconds
17. One hundred ten degrees, two minutes, fifty-eight seconds
19. Two hundred thirty-three degrees, thirty-one minutes, forty-five seconds
21. 212° 58′ 34″
23. 61° 34′ 42″
25. 123° 43′ 12″
27. 307° 1′ 8″
29. B
31. D
33. D
35. A
37. A
39. A
41. B
43. 39°
45. 42° 30′
47. 137° 30′
49. 48° 14′ 23″
51. 70° 51′ 48″

Applications

1. complementary
3. 66°

UNIT 34
Basic Math Operations with Angles

Practice

1. 28° 31′ 47″
3. 126° 57′ 57″
5. 213° 17′ 2″
7. 21° 7′ 3″
9. 318° 15′ 21″
11. 97° 5′ 44″
13. 31° 19′ 14″
15. 15° 7′ 2″
17. 104° 28′ 49″
19. 96° 23′ 44″
21. 93° 30′ 58″
23. 1° 7′ 47″
25. 60° 45′ 54″
27. 90°
29. 166° 30′ 12″

31. 251° 22′ 0″
33. 171° 23′ 10″
35. 168° 34′ 21″
37. 5° 20′
39. 12° 7′ 40″
41. 30° 25′ 28″
43. 13° 33′ 16″
45. 39° 50′ 16″
47. 35° 50′ 34″
49. Maximum angle: 47°
    Minimum angle: 43°
51. Maximum angle: 61°
    Minimum angle: 60°
53. Maximum angle: 35° 7′ 35″
    Minimum angle: 34° 17′ 35″
55. Maximum angle: 35° 15′ 30″
    Minimum angle: 35° 14′ 30″
57. Maximum angle: 90°
    Minimum angle: 89° 30′
59. Maximum angle: 75° 40′ 30″
    Minimum angle: 75° 39′ 30″
61. Maximum angle: 115° 30′ 45″
    Minimum angle: 115° 30′ 15″
63. Maximum angle: 16° 30′
    Minimum angle: 15°
65. Maximum angle: 30° 30′
    Minimum angle: 30° 29′ 20″
67. 96.733°
69. 102.028°
71. 37.66°
73. 9.513°
75. 338.96°
77. 58° 46′ 48″
79. 27° 57′ 54″
81. 334° 46′ 30″
83. 93° 19′ 30″
85. 9° 17′ 46″

Applications

1. 120°
3. 72°
5. 51° 25′43″
7. 40°
9. 70°
11. 1° 45′
    34° 55′
    31° 25′

UNIT 35
Triangles
Practice

1. Isosceles
3. Equilateral
5. Isosceles
7. Isosceles
9. Equilateral
11. Right
13. Obtuse
15. Obtuse
17. Obtuse
19. Acute
21. 1,050 mm²
23. 1.3 sq. in
25. 308 mm²
27. 11.3 sq. in
29. 4.5 in²
31. 1,925 mm²
33. 65°
    Acute
35. 74°
    Acute
37. 50°
    Acute
39. 63°
    Obtuse
41. 124°
    Obtuse
43. 45°
    Right

UNIT 36
Right Triangle Trigonometry
Practice

1. 5″
3. 11.6 mm
5. 4.9″
7. 1.2″
9. 3.8″
11. Side a = adjacent
    Side b = opposite
    Side c = hypotenuse
13. Side a = adjacent
    Side b = hypotenuse
    Side c = opposite
15. Side a = hypotenuse
    Side b = opposite
    Side c = adjacent
17. Side a = hypotenuse
    Side b = adjacent
    Side c = opposite
19. Side a = adjacent
    Side b = opposite
    Side c = hypotenuse
21. Side a = opposite
    Side b = hypotenuse
    Side c = adjacent

23. Side a = opposite
    Side b = hypotenuse
    Side c = adjacent
25. Name of side a: opposite
    Name of side that measures 13: adjacent
    Name of side that measures 17: hypotenuse
    Equation to solve:
    $\theta = \cos^{-1}\left(\dfrac{\text{adjacent}}{\text{hypotenuse}}\right)$
    ∠A (θ) = 40° 7′ 9″
27. Name of side a: opposite
    Name of side that measures 5: adjacent
    Name of side that measures 11: hypotenuse
    Equation to solve:
    $\theta = \cos^{-1}\left(\dfrac{\text{adjacent}}{\text{hypotenuse}}\right)$
    ∠A (θ) = 62° 57′ 52″
29. Name of side a: hypotenuse
    Name of side that measures 0.7: opposite
    Name of side that measures 1.44: adjacent
    Equation to solve:
    $\theta = \tan^{-1}\left(\dfrac{\text{opposite}}{\text{adjacent}}\right)$
    ∠B (θ) = 25° 55′ 30″
31. Name of side a: adjacent
    Name of side that measures 0.875: opposite
    Name of side that measures 1.592: hypotenuse
    Equation to solve:
    $\theta = \sin^{-1}\left(\dfrac{\text{opposite}}{\text{hypotenuse}}\right)$
    ∠A (θ) = 33° 20′ 28″
33. Name of side b: opposite
    Name of side x: hypotenuse
    Name of side that measures 8.5: adjacent
    Equation to solve:
    opposite = tan(θ) × adjacent
    Length of side x = 7.1323
35. Name of side b: hypotenuse
    Name of side x: opposite
    Name of side that measures 6.25: adjacent
    Equation to solve:
    opposite = sin(θ) × hypotenuse
    Length of side x = 4.6810
37. Name of side b: hypotenuse
    Name of side x: adjacent
    Name of side that measures 3.778: opposite

Equation to solve:

$$adjacent = \frac{opposite}{\tan(\theta)}$$

Length of side x = 4.7326

39. Name of side a: adjacent
Name of side x: opposite
Name of side that measures
7.125: hypotenuse
Equation to solve:
opposite = sin(θ) × hypotenuse
Length of side x = 4.5957

41. Unknown angle (A) = 30° 40′
Name of side a: adjacent
Name of side c: hypotenuse
Name of side that measures
18.675: opposite
Equation to solve for side a:

$$adjacent = \frac{opposite}{\tan(\theta)}$$

Length of side a = 11.0737
Equation to solve for side c:

$$hypotenuse = \frac{opposite}{\sin(\theta)}$$

Length of side c = 21.7114

43. Unknown angle (A) = 50° 29′ 30″
Name of side a: adjacent
Name of side c: hypotenuse
Name of side that measures
18.192: opposite
Equation to solve for side a:

$$adjacent = \frac{opposite}{\tan(\theta)}$$

Length of side a = 22.0621
Equation to solve for side c:

$$hypotenuse = \frac{opposite}{\sin(\theta)}$$

Length of side c = 28.5952

45. Unknown angle (B) = 63° 7′ 42″
Name of side a: opposite
Name of side c: hypotenuse
Name of side that measures
0.625: adjacent
Equation to solve for side a:
opposite = tan(θ) × adjacent
Length of side a = 0.3167
Equation to solve for side c:

$$hypotenuse = \frac{adjacent}{\cos(\theta)}$$

Length of side c = 0.7007

47. Unknown angle (B) = 48°
Name of side a: opposite
Name of side b: adjacent
Name of side that measures 6.05:
hypotenuse
Equation to solve for side a:
opposite = sin(θ) × hypotenuse

Length of side a = 4.0482
Equation to solve for side b:
adjacent = cos(θ) × hypotenuse
Length of side b = 4.4960

49. Angle A = 36° 52′ 12″
Angle B = 53° 7′ 48″
Side b = 4″

51. Angle A = 28° 48′ 39″
Angle B = 61° 11′ 21″
Hypotenuse = 45.7 mm

53. Angle B = 49°
Side a = 4.7″
Hypotenuse = 7.2″

55. Angle B = 56° 34′ 50″
Side b = 4.8″
Hypotenuse = 5.8″

57. Angle A = 48° 47′ 42″
Side a = 4.3″
Hypotenuse = 5.7″

Applications

1. 0.71″
3. 3.712″

UNIT 37
Oblique Triangles

Practice

1. 64°
3. 3.72″
5. 13.81 mm
7. 73°
9. 61.36 mm
11. 0.80″
13. 60.2°
15. 63.9°
17. 26.4°
19. 1.10″

Applications

1. 2.464″
3. 40°
5. 60°

UNIT 38
Sine Bars and Sine Plates

Practice

1. 7° 48′ 43″
3. 11° 49′ 23″
5. 15.441°

7. 6.5946″
9. 48.984 mm

Applications

1. 20° 14′ 14″
3. 3.1694″
5. 24.740°
7. A. 5.7358″
   B. 2.2778″
   C. 5.9949″
   D. 1.6906″
   E. 4.7767″

UNIT 39
Drill Point Angles

Practice

1. Formula: 0.3004 × ⌀ of tool
   Drill point size: 0.0939
   Total depth: 0.5939
3. Formula: 0.2071 × ⌀ of tool
   Drill point size: 0.0259
   Total depth: 0.3009
5. Formula: 0.5 × ⌀ of tool
   Drill point size: 0.1285
   Total depth: 1.6285
7. Formula: 0.866 × ⌀ of tool
   Drill point size: 0.1741
   Total depth: 0.9241
9. Formula: 0.134 × ⌀ of tool
   Drill point size: 0.0285
   Total depth: 0.7285
11. Formula: 0.2071 × ⌀ of tool
    Drill point size: 0.0646
    Total depth: 1.0646
13. Formula: 0.3004 × ⌀ of tool
    Drill point size: 2.553
    Total depth: 21.553
15. Formula: 0.2071 × ⌀ of tool
    Drill point size: 2.485
    Total depth: 30.985
17. Formula: 0.5 × ⌀ of tool
    Drill point size: 3.000
    Total depth: 19.000

UNIT 40
Center-to-Center Distances

Practice

1. SIN 60° × 1.5″ = 1.299038106″
   (C-to-C) = 2 × 1.299038106 =
   2.598076211 = 2.5981″

3. Hole 1 to 2: SIN 36° × 50 mm =
   29.38926261 mm

   (C-to-C) = 2 × 29.38926261 =
   58.77852523 mm = 58.779 mm

   Hole 1 to 3: SIN 72° × 50 mm =
   47.55282581 mm

   (C-to-C) = 2 × 47.55282581 mm =
   95.10565163 mm = 95.106 mm

5. Hole 1 to 2: SIN 30° × 75 mm =
   37.5 mm

   (C-to-C) = 2 × 37.5 = 75 mm

   Hole 1 to 3: SIN 60° × 75 mm =
   64.95190528 mm

   (C-to-C) = 2 × 64.95190528 =
   129.9038106 mm = 129.904 mm

   Hole 1 to 4: The angle between
   hole 1 and hole 4 is 180°. The
   linear distance between hole 1
   and hole 4 is the diameter of the
   bolt circle: 150 mm.

7. Hole 1 to 2: SIN 18° × 112.5 mm =
   34.76441187 mm

   (C-to-C) = 2 × 34.76441187 mm =
   69.52882373 mm = 69.529 mm

   Hole 1 to 3: SIN 36° × 112.5 mm =
   66.12584088 mm

   (C-to-C) = 2 × 66.12584088 mm =
   132.2516818 mm = 132.252 mm

   Hole 1 to 4: SIN 54° × 112.5 mm =
   91.01441187 mm

   (C-to-C) = 2 × 91.01441187 mm =
   182.0288237 mm = 182.029 mm

   Hole 1 to 5: SIN 72° × 112.5 mm =
   106.9938581 mm

   (C-to-C) = 2 × 106.9938581 mm =
   213.9877162 mm = 213.988 mm

   Hole 1 to 6: The angle between
   hole 1 and hole 6 is 180°. The
   linear distance between hole 1
   and hole 6 is the diameter of the
   bolt circle: 225 mm.

9. Radius: 2.74005″ ÷ SIN 45° =
   3.875015872″ = R3.8750″

   Diameter: Radius × 2 =
   3.875015872″ × 2 = 7.750031743″ =
   Ø7.7500″

Applications

1. Angle between adj holes: 120°
   Angle in right triangle: 60°
   Chord length: 0.866025403

3. Angle between adj holes: 72°
   Angle in right triangle: 36°
   Chord length: 0.587785252

5. Angle between adj holes:
   51.42857143°
   Angle in right triangle:
   25.71428571°
   Chord length: 0.433883739

7. Angle between adj holes: 40°

Angle in right triangle: 20°
Chord length: 0.342020143

9. Angle between adj holes:
   32.72727273°
   Angle in right triangle:
   16.36363636°
   Chord length: 0.281732556

11. Angle between adj holes:
    27.69230769°
    Angle in right triangle:
    13.84615385°
    Chord length: 0.239315664

13. Angle between adj holes: 24°
    Angle in right triangle: 12°
    Chord length: 0.20791169

15. Angle between adj holes:
    21.17647059°
    Angle in right triangle:
    10.58823529°
    Chord length: 0.183749517

17. Angle between adj holes:
    18.94736842°
    Angle in right triangle:
    9.473684211°
    Chord length: 0.16459459

19. Angle between adj holes:
    17.14285714°
    Angle in right triangle:
    8.571428571°
    Chord length: 0.149042266

21. Angle between adj holes:
    15.65217391°
    Angle in right triangle:
    7.826086957°
    Chord length: 0.136166649

23. Angle between adj holes: 14.4°
    Angle in right triangle: 7.2°
    Chord length: 0.125333233

25. 1.414213562
27. 1.0
29. 0.765366864
31. 0.618033988
33. 0.51763809
35. 0.445041868
37. 0.390180644
39. 0.347296354
41. 0.31286893
43. 0.284629676
45. 0.261052384

UNIT 41
Dovetails

Applications

1. A. 3″
   B. 3.281575645″; 3.2816″ rounded

C. 3.524636808″; 3.5246″ rounded
D. 3.739637029″; 3.7396″ rounded

3. 1.14362915″, or 1.1436″ rounded
5. 15.2987673″, or 15.2988″ rounded
7. W = 3.491025404″, or 3.4910″
   rounded
   X = 3.649519053″, or 3.6495″
   rounded

UNIT 42
Tapers

Practice

1. 4.0156″
3. 38.4013″
5. 1.0000″
7. 11.5197″

Applications

1. 1.5420″
3. 46° 23′ 50″
5. 0.9663″
7. 2.3906″

UNIT 43
Area of Polygons

Practice

1. A. 1.56 in²
   B. 153 mm²
3. A. 451.4 mm²
   B. 4 in²
5. A. 351 mm²
   B. 2.98 in²
7. A. 200.72 in²
   B. 1,406.25 mm²
9. A. 243.64 mm²
   B. 10.16 in²

Applications

1. 245 in² (196 + 49 = 245)
3. 20 in² (25 − 5 = 20)
5. 4.24 in² (4.5 − 0.2625 = 4.24)

UNIT 44
Area of Circles and Circular Segments

Practice

1. Circle
3. Chord
5. Circumference

7. Radius
9. 7.069 in^2
11. 254.469 mm^2
13. .468 in^2
15. 1.77 in^2
17. 0.487 in^2
19. .244 in^2
21. 14.296 mm

Applications

1. 44.1786 − 4.909 = 39.270 in^2
3. 113.097 − 28.274 = 82.823 in^2
5. 12.566 − 4 = 8.566 in^2
7. 96,211.194 mm^2
9. 2,375.829 mm^2
11. 81,863.045 mm^2
13. 4 911 783 mm^2

UNIT 45
Volume of Cubes, Rectangular Solids, and Prisms

Practice

1. 15.625 in^3
3. 3.15 in^3
5. 30 000 mm^3
7. .495 in^3
9. 55.801 in^3
11. 23 176.450 mm^3

Applications

1. 240 in^3
3. 260 in^3
5. 12,672 in^3
7. Medium: 7 1/4″ × 7 1/4″ × 7 1/4″
9. .339 in^2
11. 3.783 lb

UNIT 46
Volume of Cylinders, Cones, and Spheres

Practice

1. 2,624.077 in^3
3. 2.142 in^3
5. 53.996 in^3
7. 6,844.74 mm^3
9. 33.510 in^3
11. .148 in^3

Applications

1. 12,924.512 in^3
3. 1.43581 + 0.15033 = 1.586 in^3

5. 5.59596 in^3
7. 27 lb 1 oz
9. 58.90 in^3

UNIT 47
CNC Milling

Practice

1.

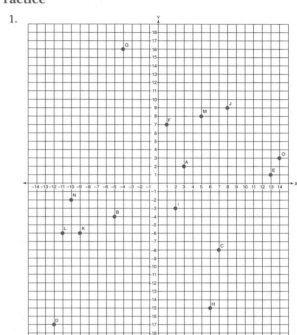

3. A. B = X−18.0, Y1.0
   B. C = X1.0, Y−7.0
   C. D = X9.0, Y−8.0
   D. E = X−11.0, Y−9.0
   E. F = X7.0, Y10.0
   F. G = X13.0, Y3.0
   G. H = X−13.0, Y6.0
   H. I = X8.0, Y−12.0
   I. J = X−13.0, Y1.0
   J. K = X3.0, Y−3.0
   K. L = X8.0, Y−3.0
   L. M = X10.0, Y13.0
   M. N = X−3.0, Y−13.0
   N. O = X−3.0, Y16.0
   O. P = X−19.0, Y−2.0
   P. Q = X25.0, Y−6.0
   Q. R = X1.0, Y−9.0
   R. S = X−11.0, Y16.0
   S. T = X−2.0, Y8.0
   T. A = X8.0, Y0.0

Applications

1. A = X−3.0, Y4.0
   B = X9.0, Y2.0
   C = X5.0, Y−4.0

D = X–5.0, Y–1.0
E = X3.0, Y2.0
F = X7.0, Y–2.0
G = X–6.0, Y–4.0
H = X0.0, Y4.0
I = X9.0, Y–5.0
J = X–6.0, Y1.0
K = X11.0, Y4.0
L = X–1.0, Y–5.0
M = X3.0, Y5.0
N = X4.0, Y0.0
O = X10.0, Y–1.0
P = X–5.0, Y3.0
Q = –7.0, Y5.0
R = X2.0, Y–3.0
S = X6.0, Y3.0
T = X–3.0, Y–3.0
3. A = X0.0, Y2.0
B = X0.0, Y–2.0
C = X–2.5, Y–2.0
D = X–2.5, Y–1.5
E = X–3.5, Y–1.5
F = X–3.5, Y–2.0
G = X–5.5, Y–2.0
H = X–6.0, Y–1.5
I = X–6.0, Y1.5
J = X–5.5, Y2.0
K = X–1.5, Y1.25
L = X–2.25, Y1.25
M = X–3.0, Y1.25
N = X–3.75, Y1.25
O = X–4.5, Y1.25
P = X–1.75, Y0.5
Q = X–4.0, Y0.5
R = X–4.25, Y0.25
S = X–4.25, Y–0.25
T = X–4.0, Y–0.5
U = X–1.75 ,Y–0.5
V = X–1.5, Y–0.25
W = X–1.5, Y0.25

UNIT 48
CNC Turning
Practice

1. A = X7.0, Z–10.0
B = X4.0, Z–16.0
C = X9.0, Z–14.0
D = X2.0, Z–12.0
E = X2.0, Z–3.0
F = X10.0, Z–4.0
G = X8.0, Z–7.0
H = X12.0, Z–10.0
I = X13.0, Z–18.0
J = X3.0, Z–8.0
3. Pos 1 = U0.0, W0.0
Pos 2 = U0.525, W0.0
Pos 3 = U0.1, W–0.05
Pos 4 = U0.0, W–0.775
Pos 5 = U0.1, W–0.05
Pos 6 = U0.175, W0.0
Pos 7 = U0.1, W–0.05
Pos 8 = U0.0, W–0.7
Pos 9 = U0.1, W–0.05
Pos 10 = U0.25, W0.0
Pos 11 = U0.1, W–0.05
Pos 12 = U0.0, W–0.775

Applications

1. A = X0.0, Z0.0
B = X0.5, Z0.0
C = X0.5, Z–0.75
D = X0.75, Z–0.75
E = X0.75, Z–1.5
F = X1.0, Z–1.5
G = X1.0, Z–1.88
3. A = X0.0, Z0.0
B = X0.4, Z0.0
C = X0.5, Z–0.1
D = X0.5, Z–0.75

E = X0.65, Z–0.75
F = X0.75, Z–0.85
G = X0.75, Z–1.5
H = X0.8, Z–1.5
I = X1.0, Z–1.6
J = X1.0, Z–1.88
5. A = X0.0, Z0.0
B = X0.225, Z0.0
C = X0.375, Z–0.075
D = X0.375, Z–0.35
E = X0.475, Z–0.4
F = X0.57, Z–0.4
G = X0.57, Z–0.75
H = X1.0, Z–1.5
I = X1.0, Z–1.95
J = X1.1, Z–2.0
K = X1.4, Z–2.0
L = X1.5, Z–2.05
M = X1.5, Z–2.5

UNIT 49
Bolt Circles
Practice

1. Hole 1: X2.4620, Y0.4341
Hole 2: X–1.607, Y1.9151
Hole 3: X–0.8551, Y–2.3492
3. Hole 1: X3.4915, Y0.2441
Hole 2: X2.5177, Y2.4313
Hole 3: X0.3658, Y3.4808
Hole 4: X–1.9572, Y2.9016
Hole 5: X–3.3644, Y0.9647
Hole 6: X–3.1974, Y–1.4236
Hole 7: X–1.5343, Y–3.1458
Hole 8: X0.8467, Y–3.3960
Hole 9: X2.8316, Y–2.0572